PTOLEMY'S PHILOSOPHY

Ptolemy's Philosophy

MATHEMATICS AS A WAY OF LIFE

JACQUELINE FEKE

PRINCETON UNIVERSITY PRESS

PRINCETON & OXFORD

Copyright © 2018 by Princeton University Press

Published by Princeton University Press
41 William Street, Princeton, New Jersey 08540
6 Oxford Street, Woodstock, Oxfordshire OX20 1TR

press.princeton.edu

Library of Congress Control Number: 2017962530
First paperback printing, 2020
Paperback ISBN 978-0-691-21039-1
Cloth ISBN 978-0-691-17958-2

British Library Cataloging-in-Publication Data is available

Editorial: Al Bertrand and Kristin Zodrow
Production Editorial: Jenny Wolkowicki
Cover image: *Ptolemy's triquetrum (or "parallactic instrument")*
according to William Cunningham's The Cosmographical Glasse,
conteinying the Pleasant Principles of Cosmographie, Geographie,
Hydrographie, or Navigation (London: John Day, 1559)
Production: Erin Suydam
Publicity: Alyssa Sanford
Copyeditor: Maia Vaswani

This book has been composed in Arno Pro

Printed in the United States of America

For Carol and Gilbert T. Feke

CONTENTS

ILLUSTRATIONS

ACKNOWLEDGMENTS

IN 2002 as an undergraduate student at Brown University, I was encouraged by Debby Boedeker, my then honors thesis supervisor, to meet David Pingree. I asked him where I could go to get a PhD studying the history of ancient science, and he gave me a simple answer. He told me that there was only one place to go in North America and that was the University of Toronto to study under Brown alumnus Alexander Jones. To this day, I don't know if Pingree was exaggerating or if he really believed that U of T was the only place to be, but I followed his wise counsel. Alex Jones taught me how to be a historian. Brad Inwood guided my research in ancient philosophy and encouraged me to make my mark by focusing my research on Ptolemy. Craig Fraser mentored my research in the history of mathematics more broadly. The result, so many years later, is this book.

I am grateful to more people than I can mention for their instruction, mentorship, and friendship. In addition to Debby, David, Alex, Brad, and Craig, without whom this book would not be possible, I would like to thank Vincenzo De Risi for funding my research at the Max Planck Institute for the History of Science; Orna Harari for her professional and scholarly support; Stephan Heilen for giving feedback, especially on my analysis of Ptolemy's astrology; Alain Bernard for motivating me to ask new questions about Ptolemy's philosophy; Nathan Sidoli for showing me that success as an emerging scholar of ancient mathematics is possible; Vicky Albritton, Fredrik Albritton Jonsson, Daryn Lehoux, Patricia Marino, and Paul Thagard for coaching me through the early stages of book publishing; and all of the colleagues and friends from whom I've learned and on whom I've leaned from the University of Toronto, Stanford University, the Max Planck Institute for the History of Science, the University of Chicago, and now the University of Waterloo. I am grateful to the many scholars who have asked questions and made comments on my work at numerous conferences and colloquia over the years, and I also would like to thank the anonymous referees for their suggestions. Above all, I'd like to thank my parents, Carol and Gilbert T. Feke. It is to them that I dedicate this book.

PTOLEMY'S PHILOSOPHY

1

Introduction

CLAUDIUS PTOLEMY is one of the most significant figures in the history of science. Living in or around Alexandria in the second century CE, he is remembered most of all for his contributions in astronomy. His *Almagest*, a thirteen-book astronomical treatise,[1] was authoritative until natural philosophers in the sixteenth and seventeenth centuries repudiated the geocentric hypothesis and appropriated Nicolaus Copernicus's heliostatic system of *De revolutionibus*. Ptolemy also composed texts on harmonics, geography, optics, and astrology that influenced the study of these sciences through the Renaissance.

Ptolemy's contributions in philosophy, on the other hand, have been all but forgotten. His philosophical claims lie scattered across his corpus and intermixed with technical studies in the exact sciences. The late nineteenth and early twentieth centuries' development of discrete academic disciplines let the study of Ptolemy's philosophy fall through the cracks. When scholars do make reference to it, they tend to portray Ptolemy as either a practical scientist—mostly unconcerned with philosophical matters, as if he were a forerunner to the modern-day scientist—or a scholastic thinker who simply adopted

---

1. "Almagest" is not the text's original title, but rather "Mathematical Composition" (μαθηματικὴ σύνταξις), to which Ptolemy makes reference in Book 1 of the *Planetary Hypotheses* as well as *Geography* 8.2.3. Cf. *Tetrabiblos* 1.1.1, H3. The name "Almagest" comes from the Arabic *al-Majistî*, which derives from the Greek μεγίστη ("the biggest"). The designation "biggest" does not occur in the Greek tradition but instead in the Arabic, although "big composition" (μεγάλη σύνταξις) does appear in the Greek. See Tihon, "Alexandrian Astronomy," 74. For the *Almagest* and Ptolemy's other texts, I will use the name in common usage today rather than, in some cases, the likely original. Notably, "Tetrabiblos" (Treatise in four books) is probably not the original title of Ptolemy's astrological text—it is likely *Apotelesmatika* ([Books on] effects)—but again I will use the more common title. On the original title of the *Tetrabiblos*, see Hübner, *Apotelesmatika*, XXXVI–XXXIX.

the philosophical ideas of authoritative philosophers, especially Aristotle.[2] This latter portrayal no doubt evolved in part because Ptolemy cites Aristotle in the first chapter of the *Almagest*. Liba Taub proved that the philosophical claims in *Almagest* 1.1, as well as in Ptolemy's cosmological text, the *Planetary Hypotheses*, are not Aristotle's, and with this debunking of the assumed view Taub opened the door for my own analysis of Ptolemy's philosophy, including how it manifests throughout his corpus and how it relates to several ancient philosophical traditions.[3] This monograph is the first ever reconstruction and intellectual history of Ptolemy's general philosophical system.

Concerning Ptolemy's life we know nothing beyond approximately when and where he lived. In the *Almagest*, he includes thirty-six astronomical observations that he reports he made in Alexandria from 127 to 141 CE. Another unaccredited observation from 125 CE may be his as well.[4] The *Canobic Inscription*, a list of astronomical parameters that Ptolemy erected at Canopus, Egypt, provides a slightly later date: 146/147 CE. Because the *Canobic Inscription* contains numerical values that Ptolemy corrects in the *Almagest*, it must predate the *Almagest*.[5] Therefore, Ptolemy completed the *Almagest* sometime after 146/147 CE. In addition, Ptolemy makes reference to the *Almagest* in several of his later texts. The life span that this chronology requires is consistent with a scholion attached to the *Tetrabiblos*, Ptolemy's astrological text, indicating that he flourished during Hadrian's reign and lived until the reign of Marcus Aurelius, who became Roman emperor in 161 CE but ruled jointly with Lucius Verus until 169 CE. Thus, we can estimate that Ptolemy lived from approximately 100 to 170 CE.

Concerning any philosophical allegiance, Ptolemy says nothing. In his texts, he does not align himself with a philosophical school. He does not state who his teacher was. He does not indicate in what his education consisted or even what philosophical books he read. In order to discern where his philosophical ideas came from, one must mine his corpus, extract the philosophical content, and, with philological attention, relate his ideas to concepts presented

---

2. A. A. Long emphasizes Ptolemy's practicality when examining his *On the Kritêrion and Hêgemonikon*: "His little essay should be read, I suggest, as a practising scientist's statement of where he stands on the epistemological issues that arise in his day-to-day work." See Long, "Ptolemy on the Criterion," 163.

3. Taub, *Ptolemy's Universe*. The most comprehensive analysis of Ptolemy's philosophy is the philological study of Franz Boll, "Studien über Claudius Ptolemäus."

4. For a chronological list of dated observations in the *Almagest*, see Pedersen, *Survey of the Almagest*, 408–22.

5. See Hamilton, Swerdlow, and Toomer, "Canobic Inscription." See also A. Jones, "Ptolemy's *Canobic Inscription*."

in texts that are contemporary with his own or that were authoritative in the second century. Unfortunately, what survives of the ancient Greek corpus is but a fraction of what was written and we have very little from Ptolemy's time. It is impossible to determine what exactly he read or even where he read it, as it is dubious that the great Alexandrian library was still in existence. At best we can place Ptolemy's thought in relation to prevailing ancient philosophical traditions.

The first century BCE to the second century CE is distinguished by the eclectic practice of philosophy. The Greek verb *eklegein* means to pick or choose, and the philosophers of this period selected and combined concepts that traditionally were the intellectual property of distinct schools of thought. Mostly, these philosophers blended the Platonic and Aristotelian traditions, but they also appropriated ideas from the Stoics and Epicureans. The label "eclecticism" has long held a pejorative connotation in philosophy, as if eclectic philosophers were not sufficiently innovative to contribute their own ideas, and the philosophy of the periods before and after this seemingly intermediate chapter in ancient philosophy were comparatively inventive, with the development of the Hellenistic movements, including the Stoic, Epicurean, and Skeptic, and the rise of Neoplatonism, respectively. Nevertheless, John Dillon and A. A. Long revitalized the study of eclectic philosophy.[6] So-called middle Platonism and the early Aristotelian commentary tradition have received more attention in recent years, and their study has demonstrated that the manners in which these philosophers integrated authoritative ideas are themselves noteworthy.

I aim to prove that Ptolemy was very much a man of his time in that his philosophy is most similar to middle Platonism, the period in Platonic philosophy that extended from the first century BCE—with Antiochus of Ascalon, who was born near the end of the second century BCE and moved from Ascalon, in present-day Israel, to Athens to join the Academy—to the beginning of the third century CE, with Ammonius Saccas, the Alexandrian philosopher and teacher of Plotinus, the founder of Neoplatonism. Both Antiochus and Ammonius Saccas are known for their syncretic tendencies. In response to Academic skepticism, Antiochus argued not only that knowledge is possible but also that the old, pre-skeptical Academy was in broad agreement with the Aristotelian and Stoic schools. Centuries later, Ammonius Saccas argued that

---

6. Dillon and Long, *Question of "Eclecticism."* In their studies of *On the Kritērion* and the *Optics*, respectively, A. A. Long and A. Mark Smith describe Ptolemy's philosophy as eclectic. See Long, "Ptolemy on the Criterion," 152; Smith, *Ptolemy's Theory*, 18. For other interpretations of Ptolemy's philosophy, see Lammert, "Philosophie der mittleren Stoa"; de Pace, "Elementi Aristotelici nell'*Ottica*."

Plato's and Aristotle's philosophies were in fundamental agreement. Middle Platonism manifested in a variety of literary forms, styles of argument, and attitudes toward authoritative figures, but a significant trend emerged in this period where philosophers asserted the harmony of previously distinct schools of thought. They drew concepts, theories, and arguments from philosophers attached to once competing schools. To be a Platonist at this time entailed not only clarifying the meaning of Plato's texts but also appropriating ideas from the Aristotelian and Stoic traditions in the course of developing Platonic philosophy. Epicurean philosophy had less of an impact, but several of its terms had by this time become common intellectual property. It is this harmonizing tendency of middle Platonism, coupled with its emphasis on certain key themes in Platonic philosophy, that fundamentally influenced Ptolemy's own contributions in philosophy.

Ptolemy's seamless blending of concepts from the Platonic and Aristotelian traditions and, to a lesser extent, the Stoic and Epicurean, is itself impressive, but its greater significance lies in its radical and even subversive character. Ptolemy adopted ideas from these many traditions but his integration of them yielded a philosophical system that upended the entire edifice of ancient philosophy. In *Almagest* 1.1, Ptolemy denounces attempts by philosophers to answer some of the most central questions of philosophy, and he argues that the fields of inquiry that philosophers study are merely conjectural. Against the vast current of ancient Greek philosophy, Ptolemy maintains that theology and physics are essentially guesswork and that mathematics alone generates sure and incontrovertible knowledge. This epistemological position—that mathematics alone, and neither physics nor theology, yields knowledge—is unprecedented in the history of philosophy and would have been extraordinarily controversial. Moreover, Ptolemy's appropriation of ancient virtue ethics is equally subversive. He maintains that the best life is one where the human soul is in a virtuous, or excellent, condition, and in his adaptation of Platonic ethics he affirms that the highest goal of human life is to resemble the divine—to be, as much as humanly possible, like the gods—but, according to Ptolemy, the one and only path to the good life is through mathematics.

Ptolemy deems mathematics epistemologically and ethically superior to every other field of inquiry, but that is not to say that he eschewed philosophy. For Ptolemy, mathematics is philosophy or, rather, a part of philosophy. It is one of the three parts of theoretical philosophy, alongside physics and theology. In addition to these three theoretical sciences—where, in ancient Greek philosophy, a science is simply a branch of knowledge—there are the three practical parts of philosophy: ethics, domestics, and politics. Ptolemy argues in *Almagest* 1.1 that the theoretical part of philosophy is more valuable than the practical, and that, of the three theoretical sciences, mathematics is the best in its abilities to render knowledge and transform the human soul into its most perfect condition. Mathematics reveals the objective of human life, to be like the heavenly divine, and it provides the means to achieve it. Ptolemy does not claim, however, that one should study only mathematics. He argues that mathematics contributes to physics and theology, and, furthermore, that it guides practical philosophy and even the ordinary affairs of life. Positioning mathematics at the foundation of every one of life's activities, Ptolemy advances the mathematical way of life.

Consistent with Plato's account of the philosopher's education in Book VII of the *Republic*, Platonists upheld mathematics as a useful means of training the soul, where mathematics is propaedeutic, preparing the way for other, higher, more valuable studies, such as dialectic or metaphysics. Yet, for Ptolemy, mathematics is not simply useful; it is not merely a path to another science. For Ptolemy, it is the highest science. Only mathematics yields knowledge. Through its study alone human beings achieve their highest objective, to become like the divine. Human beings come to comprehend, love, and resemble divinities through the study of astronomy and harmonics, which, according to Ptolemy, are both mathematical sciences. Astronomy is the study of the movements and configurations of the stars; harmonics is the study of the ratios that characterize the relations among musical pitches. Astronomical objects serve as ethical exemplars for human souls, and both astronomy and harmonics give rise to souls' virtuous transformation.

Ptolemy's texts testify to his additional interest in mathematics' application to theology and physics, especially. In the *Almagest*, Ptolemy's astronomy informs his theology, and his natural philosophical investigations are extensive. Just as he argues in *Almagest* 1.1 that mathematics contributes significantly to physics, time and again Ptolemy studies bodies mathematically before investigating their physical properties. Mathematical study informs the analysis of bodies' physical qualities, and, though physics is conjectural, the application of mathematics affords the best guesses possible of bodies' physical natures. In the chapters that follow, I examine Ptolemy's applications of geometry to element theory, harmonics to psychology, and astronomy to astrology and cosmology.

The only one of Ptolemy's texts devoid of mathematics is *On the Kritêrion and Hêgemonikon*, an epistemological study that examines the criterion of truth, the method by which a human being generates knowledge, as well as the physical nature and structure of the human soul, including the *hêgemonikon*, its chief part. More than any other text of Ptolemy, *On the Kritêrion* has provoked controversy concerning its authorship, no doubt in part because it contains no mathematics. Nevertheless, thematic, stylistic, and linguistic arguments support Ptolemy's authorship, and I argue that it is one of the earliest, if not

the earliest, of Ptolemy's extant texts.[7] In *On the Kritêrion*, Ptolemy proposes
a dually rational and empirical criterion of truth, where the faculties of sense
perception and thought cooperate in the production of knowledge. Ptolemy
adheres to this criterion in the rest of his corpus, but when he wrote *On the
Kritêrion* he had not yet mandated the application of mathematics to physics.
After he composed it, he devised his mathematical-scientific method, which
he employed in every one of his subsequent studies. Every other of Ptolemy's
texts constitutes an inquiry into or an implementation of mathematics.

In addition to *On the Kritêrion*, the texts I analyze are those of Ptolemy
that contain manifestly philosophical content.[8] Again, the *Almagest* is Ptolemy's most famous astronomical text. It comprises thirteen books—likely
in homage to the thirteen books of Euclid's *Elements*—and it consists in the
deduction of geometric models that, according to Ptolemy, truly describe the
mathematical objects in the heavens, the combinations of rotating spheres
that give rise to the movements of celestial bodies, the fixed and wandering
stars. In the first book, Ptolemy situates astronomy in relation to the other
parts of philosophy, he describes the structure of the ensuing text, and he
establishes the fundamental hypotheses of his astronomical system, such as
the heavens' sphericity and the earth's location at the center of the cosmos. In
the latter part of Book 1 through Book 2, he presents the mathematics necessary for the mathematical deduction, including the "Table of Chords," used
in the trigonometric calculations that follow. The remainder of the *Almagest*,
Books 3 through 13, contains the deduction itself of the astronomical models,
accounting for the movements of the sun, moon, fixed stars, and five planets.
These models are both demonstrative and predictive, since by using the tables
an astrologer would have been able to approximate the perceptible location of
any celestial body on any given date.

The *Planetary Hypotheses* is Ptolemy's cosmological text. In the first of
the two books, he presents astronomical models, mostly consistent with the
Almagest's models; he specifies the order and absolute distances of the celestial
systems; and he determines the diameters of the celestial bodies. In Book 2,
he presents his aethereal physics, describing the heavenly bodies in physical
terms, and he discusses celestial souls, which, in Ptolemy's cosmology, control the aethereal bodies' movements. Only a portion of the first book of the

7. See A. Jones, "Ptolemy," 174; Feke and Jones, "Ptolemy," 199. Boll argues in favor of *On the
Kritêrion*'s authenticity: Boll, "Studien," 78. Against its authenticity, see Toomer, "Ptolemy," 201;
Swerdlow, "Ptolemy's *Harmonics*," 179–80. Taub merely states that the work's attribution to Ptol-
emy has been questioned: Taub, *Ptolemy's Universe*, 9.

8. For a complete list of Ptolemy's texts and their editions, see Feke, "Ptolémée d'Alexandrie
(Claude)."

Planetary Hypotheses exists in the original Greek. The second of the two books and the remainder of the first book exist only in a ninth-century Arabic translation as well as a Hebrew translation from the Arabic.

The *Tetrabiblos* delineates Ptolemy's astrological theory. In the introductory chapters, he defines astrology and defends this physical science's possibility and utility. Thereafter, he summarizes its principles, including the powers of celestial bodies, the rays by which stars transmit their powers, and the effects these powers have on sublunary bodies and souls. Book 2 examines the celestial powers' large-scale effects on geographic regions and meteorological phenomena, and Books 3 and 4 address celestial influences on human beings and their individual lives.

In the *Harmonics*, Ptolemy elaborates on his criterion of truth and employs it in the analysis of the mathematical relations among musical pitches. The text contains three books, and, after completing his study of music theory in *Harmonics* 3.2, he examines the harmonic ratios that exist among psychological, astrological, and astronomical phenomena. Unfortunately, the last three chapters, 3.14–3.16, are no longer extant; only their titles remain. In the chapters that follow, I also make reference to Ptolemy's *Geography*, *Optics*, and two works—*On the Elements* and *On Weights*—that are entirely lost to us but which Simplicius, the sixth-century philosopher, attests to in his commentary on Aristotle's *De caelo*.[9]

Ptolemy's texts offer few clues to their chronology. In the *Tetrabiblos* and *Planetary Hypotheses*, as well as in the *Geography*, he refers to his "syntaxis" or "mathematical composition" (μαθηματικὴ σύνταξις), manifestly the *Almagest*.[10] Consequently, Ptolemy must have completed these texts after the *Almagest*. Noel Swerdlow has argued that the *Harmonics* predates the *Almagest* because the titles of the three lost chapters indicate that they examined the relations between musical pitches and celestial bodies tabulated in the *Canobic Inscription*.[11] Considering that Ptolemy must have written the *Canobic Inscription* before the *Almagest*, the *Harmonics* probably predates the *Almagest* as well, and I argue that Ptolemy completed *On the Kritêrion* before the *Harmonics*. Thus, one reasonably can conclude that Ptolemy composed the texts most relevant to this study in the following order: (1) *On the Kritêrion and*

---

9. For an analysis of the philosophical claims in Ptolemy's *Geography*, see Feke, "Ptolemy's Philosophy of Geography." Harald Siebert has put Ptolemy's authorship of the *Optics* into question in *Die ptolemäische "Optik."* For Simplicius's discussion of Ptolemy's *On the Elements* and *On Weights*, see Simplicius, *In de caelo* 1.2.20.10–25; 4.4.710.14–711.9.

10. Ptolemy, *Tetrabiblos* 1.1.1, H3; *Planetary Hypotheses* 1.1, H70; *Geography* 8.2.3.

11. Swerdlow, "Ptolemy's *Harmonics*," 175.

Hêgemonikon; (2) *Harmonics*; (3) *Almagest*; and (4) *Tetrabiblos* and *Planetary Hypotheses*, in an indeterminate order.

I take *Almagest* 1.1 as the starting point of this study, as it functions as an epitome of Ptolemy's general philosophical system. My chapters 2 through 4 are analyses and intellectual histories of the metaphysical, epistemological, and ethical statements of *Almagest* 1.1. In chapter 2, I argue that the metaphysics Ptolemy presents when differentiating the three theoretical sciences—physics, mathematics, and theology—is Aristotelian, though not Aristotle's, and that Ptolemy underlays his ontology with epistemology. In chapter 3, I show how Ptolemy blends an Aristotelian form of empiricism with a Platonic concern for distinguishing knowledge and opinion, and he thereby produces a new and subversive epistemology where mathematics is the only science that generates knowledge rather than conjecture. Moreover, I analyze Ptolemy's argument for the contribution of mathematics to physics and theology, and I examine the case studies of how astronomy informs his theology and geometry drives his element theory. In chapter 4, I demonstrate how Ptolemy's distinctly mathematical ethics emerges from his response to a contemporary debate over the relationship between theoretical and practical philosophy. Ptolemy argues that practical philosophy is dependent on theoretical philosophy and that mathematics, in particular, reveals the ultimate goal of all philosophy and even directs the ordinary affairs of life.

Thereafter, I address the philosophical statements Ptolemy propounds in the rest of his corpus. In chapter 5, I argue that Ptolemy's concept of *harmonia*, which he examines in the *Harmonics*, is crucial to his ethical system. *Harmonia* is a technical term whose meaning differs from our notion of harmony. I dissect the concept in detail and argue that it is because of *harmonia* that the human soul is able to resemble astronomical objects. In chapter 6, I analyze the relationship between harmonics and astrology, which Ptolemy portrays as complementary mathematical sciences, and I determine whether, when examining these sciences in the *Harmonics* and the remainder of the *Almagest*, Ptolemy maintains his position in *Almagest* 1.1 that mathematics yields sure and incontrovertible knowledge. In chapters 7 and 8, I turn to Ptolemy's application of mathematics to the physics of composite bodies. In the former, I argue for the development of his psychology from *On the Kritêrion* to the *Harmonics*, where he strives to improve his account of the human soul by mathematizing it. The development in his psychological theory, I contend, marks the maturation of his scientific method. In the latter chapter, I argue that Ptolemy maintains the epistemology and scientific method that he articulates in *Almagest* 1.1 and applies in the *Harmonics* in his studies of astrology and cosmology in the *Tetrabiblos* and *Planetary Hypotheses*. Overall, Ptolemy's philosophy remains remarkably consistent across his corpus.

At the foundation of Ptolemy's complex philosophical system is his ethics. The explicit motivation for his study of the theoretical sciences is his objective to transform his soul into a condition that resembles the divine, mathematical objects of the heavens, the movements and configurations of the stars. That Ptolemy required such a motivation for his prodigious and influential scientific investigations may be surprising, but we must remember that in antiquity mathematicians were rare. In any one generation in the ancient Mediterranean, no more than a few dozen individuals studied high-level mathematics.[12] Given the scarcity of advanced mathematical study, an individual who concentrated on it would have made a deliberate choice to disavow more dominant intellectual practices, including the conventions of philosophers, and assume an unconventional way of life. Mathematicians play a special role in the ancient philosophical landscape in that they studied philosophy to varying degrees but they were not philosophers. In Ptolemy's case, he was well versed in the philosophy of his time. He appropriated ideas from authoritative and contemporary philosophical traditions for his own philosophical system. What led him to set aside the nonmathematical study of philosophy and focus on mathematics? We know so little of Ptolemy's life that it is impossible to say for certain. It would be easiest to suppose that he simply found mathematics to be captivatingly interesting. Nevertheless, I aim to present a more complex portrait, where the clues lie in the philosophical claims scattered across his corpus, and I propose that it was Ptolemy's appropriation of Platonic ethics and the formulation of a radical philosophy—the mathematical way of life—that motivated him to devote his life to mathematics.

---

12. Netz, *Shaping of Deduction*, 291.

2

Defining the Sciences

THE FIRST chapter of the *Almagest* epitomizes Ptolemy's philosophical system, and in it he provides one of his few citations of a philosopher—namely, Aristotle. In fact, Ptolemy cites only two philosophers in his corpus: Aristotle in *Almagest* 1.1, as well as in the *Planetary Hypotheses*, and Plato in the *Planetary Hypotheses*. When he cites them in the *Planetary Hypotheses*, he responds to aspects of their cosmological models.[1] Ptolemy's citation of Aristotle in the *Almagest* is his only citation of a philosopher in an unequivocally philosophical context. He proclaims, "For Aristotle divides the theoretical [part of philosophy], too, very fittingly, into three primary genera, the physical, mathematical, and theological."[2] Aristotle divides theoretical understanding into these three sciences in the *Metaphysics*, but Ptolemy's definitions of the sciences are not Aristotle's, and they provide insufficient evidence for supposing that Ptolemy even read Aristotle's *Metaphysics*. The terms Ptolemy employs when defining the sciences indicate that his definitions are Aristotelian but not Aristotle's. They do not derive directly from Aristotle's *Metaphysics* but from his greater corpus. Ptolemy appropriates aspects of Aristotle's metaphysics and epistemology to construct his own, individual definitions of physics, mathematics, and theology.

Aristotle's Accounts

In order to address how Ptolemy's definitions of the sciences diverge from Aristotle's, it is necessary first to examine Aristotle's accounts. In general,

---

1. Ptolemy, *Planetary Hypotheses* 2.4–5, H113–14.

2. Ptolemy, *Almagest* 1.1, H5: καὶ γὰρ αὖ καὶ τὸ θεωρητικὸν ὁ Ἀριστοτέλης πάνυ ἐμμελῶς εἰς τρία τὰ πρῶτα γένη διαιρεῖ τό τε φυσικὸν καὶ τὸ μαθηματικὸν καὶ τὸ θεολογικόν. All translations, including this one, are my own unless otherwise indicated. Note that my translations of Ptolemy's *Harmonics* are modifications of Andrew Barker's translation in *Greek Musical Writings*, 270–391.

Aristotle defines fields of inquiry according to ontological criteria. Each field examines a distinct set of objects, and the properties of an object determine which field studies it. Aristotle delineates three types of intelligence, or thought—the practical, productive, and theoretical—and he distinguishes them according to their objects' principles of motion. In *Metaphysics* K7, he explains as follows:

> For in productive [science] the principle of motion is in the producer and not in the product and is either an art or some other capacity; and similarly in practical [science] the motion is not in the thing acted upon but rather in the agent. But the [science] of the natural philosopher deals with the things that have in themselves a principle of motion. From this it is clear, then, that natural science must be neither practical nor productive but theoretical (since it must fall into one of these genera)[3]

Both productive and practical thought examine objects that have an external source of motion, and theoretical thought concerns objects that possess their own principle of motion, where the principle of motion inheres in the object itself.

Aristotle examines theoretical knowledge, in particular, in *Metaphysics* E1 and K7. Just as thought encompasses three domains, theoretical knowledge consists of three sciences: physics, mathematics, and theology. Each of these sciences studies a set of objects distinguished by the objects' share in two dichotomies: (1) whether the objects are separable (χωριστόν) or inseparable (οὐ χωριστόν), and (2) whether the objects are movable (κινητόν) or immovable (ἀκίνητον). Separable objects are able to exist independently, but inseparable objects do not exist independently of another type of object. Movable objects experience change; immovable objects do not experience any change whatsoever. Aristotle lists the various types of change an object may undergo in *On Generation and Corruption*. They include, as the title suggests, generation and corruption, or coming into and out of being, as well as alteration, growth, diminution, mixture, and motion from place to place. The pairing in an object of the characteristics separable or inseparable with movable or immovable determines which of the three theoretical sciences studies it.

---

3. Aristotle, *Metaphysics* XI, 1064a11–19: ποιητικῆς μὲν γὰρ ἐν τῷ ποιοῦντι καὶ οὐ τῷ ποιουμένῳ τῆς κινήσεως ἡ ἀρχή, καὶ τοῦτ᾽ ἔστιν εἴτε τέχνη τις εἴτ᾽ ἄλλη τις δύναμις· ὁμοίως δὲ καὶ τῆς πρακτικῆς οὐκ ἐν τῷ πρακτῷ μᾶλλον δ᾽ ἐν τοῖς πράττουσιν ἡ κίνησις. ἡ δὲ τοῦ φυσικοῦ περὶ τὰ ἔχοντ᾽ ἐν ἑαυτοῖς κινήσεως ἀρχήν ἐστιν. ὅτι μὲν τοίνυν οὔτε πρακτικὴν οὔτε ποιητικὴν ἀλλὰ θεωρητικὴν ἀναγκαῖον εἶναι τὴν φυσικὴν ἐπιστήμην, δῆλον ἐκ τούτων (εἰς ἕν γὰρ τι τούτων τῶν γενῶν ἀνάγκη πίπτειν)

As for which pairings of characteristics inhere in the objects studied by the sciences, Aristotle unequivocally states that physical objects are separable and movable and that theological objects are separable and immovable. He equivocates on the characteristics of mathematical objects. In *Metaphysics* E1, he reveals his uncertainty, stating, "mathematics is also theoretical; but it is not clear at present whether [its objects] are immovable and separable; it is clear, however, that some branches of mathematics consider [their objects] *qua* immovable and *qua* separable."[4] Whether or not mathematical objects are separable and immovable, mathematics studies these objects as if they are separable and immovable. After confessing this uncertainty, Aristotle proceeds to categorize at least some mathematical objects as inseparable and immovable: "For physics deals with objects that are separable but are not immovable, and some branches of mathematics deal with objects that are immovable but presumably not separable but, rather, present in matter; but the primary [science, viz. theology] deals with objects that are both separable and immovable."[5] Although Aristotle's classification of mathematical objects remains tentative, he elects this categorization of their inseparability and immovability, presumably in order to keep the three domains of theoretical knowledge and the objects they study distinct. If mathematical objects are not only immovable but also separable, then they are indistinguishable from theological objects, but—rejecting the Platonic theory of transcendent mathematical objects— Aristotle wishes to keep the mathematical and theological domains distinct.

For Aristotle, theology is "the primary science." It deals with objects that are divine and, therefore, prior. He justifies the privileged position of theology as follows:

> Now all causes must be eternal, but these especially; for they are the causes of what is visible of divine things. Therefore there must be three theoretical philosophies—the mathematical, physical, and theological—since it is obvious that if the divine is present anywhere, it is present in this kind of nature; and the most honorable [science] must deal with the most honorable genus.[6]

---

4. Ibid. VI, 1026a7–10: ἀλλ' ἔστι καὶ ἡ μαθηματικὴ θεωρητική· ἀλλ' εἰ ἀκινήτων καὶ χωριστῶν ἐστί, νῦν ἄδηλον, ὅτι μέντοι ἔνια μαθήματα ᾗ ἀκίνητα καὶ ᾗ χωριστὰ θεωρεῖ, δῆλον. An alternative translation is possible wherein the last clause reads, "it is clear, however, that it considers some mathematical objects *qua* immovable and *qua* separable."

5. Ibid. VI, 1026a13–16: ἡ μὲν γὰρ φυσικὴ περὶ χωριστὰ μὲν ἀλλ' οὐκ ἀκίνητα, τῆς δὲ μαθηματικῆς ἔνια περὶ ἀκίνητα μὲν οὐ χωριστὰ δὲ ἴσως ἀλλ' ὡς ἐν ὕλῃ· ἡ δὲ πρώτη καὶ περὶ χωριστὰ καὶ ἀκίνητα.

6. Ibid. VI, 1026a16–22: ἀνάγκη δὲ πάντα μὲν τὰ αἴτια ἀΐδια εἶναι, μάλιστα δὲ ταῦτα· ταῦτα γὰρ αἴτια τοῖς φανεροῖς τῶν θείων. ὥστε τρεῖς ἂν εἶεν φιλοσοφίαι θεωρητικαί, μαθηματική, φυσική,

Aristotle justifies labeling the science of separable and immovable objects "theological" by appealing to the secondary characteristics of theological objects. In addition to their separability and immovability, they are eternal, divine, and the causes of visible divine objects. Aristotle presumably alludes here to his account in *Metaphysics* Λ, where unmoved mover(s) serve as the final cause of heavenly bodies' motion. Although heavenly bodies are divine, as physical they are movable and, therefore, posterior to immovable divine entities, or theological objects such as the unmoved mover(s).

The account of the theoretical sciences in *Metaphysics* K7 is largely consistent with *Metaphysics* E1.[7] Some of the language Aristotle uses to discuss the objects differs, but in general the categories of the three sciences and the objects' defining characteristics remain consistent. Aristotle defines the theoretical sciences accordingly:

> Physics deals with objects that contain a principle of motion in themselves, and mathematics is theoretical and is a [science] that deals with objects that are permanent but not separable. Therefore there is a [science], distinct from both of these sciences, that deals with the separable and immovable, if there really is a substance of this kind—I mean separable and immovable—as we shall endeavor to prove.[8]

Aristotle here describes mathematical objects as permanent—they remain in place—rather than immovable, but he does not hesitate to label them inseparable. Hence, in both E1 and K7 of the *Metaphysics*, Aristotle demarcates the three theoretical sciences according to the properties of the objects they study. Physics studies objects that are separable and movable, mathematics studies objects that are inseparable—although studied as if they were separable—and immovable, or permanent, and theology studies objects that are separable and immovable.

Aristotle also ranks the three theoretical sciences according to ontological criteria. In *Metaphysics* E1, he calls theology not only "primary" but also

---

θεολογική (οὐ γὰρ ἄδηλον ὅτι εἴ που τὸ θεῖον ὑπάρχει, ἐν τῇ τοιαύτῃ φύσει ὑπάρχει), καὶ τὴν τιμιωτάτην δεῖ περὶ τὸ τιμιώτατον γένος εἶναι.

7. The authenticity of Book K of the *Metaphysics* has been doubted but, even if spurious, it remains consistent with *Metaphysics* E1 in the respects relevant to this study.

8. Aristotle, *Metaphysics* XI, 1064a30–36: ἡ μὲν οὖν φυσικὴ περὶ τὰ κινήσεως ἔχοντ' ἀρχὴν ἐν αὑτοῖς ἐστίν, ἡ δὲ μαθηματικὴ θεωρητικὴ μὲν καὶ περὶ μένοντά τις αὕτη, ἀλλ' οὐ χωριστά. περὶ τὸ χωριστὸν ἄρα ὂν καὶ ἀκίνητον ἑτέρα τούτων ἀμφοτέρων τῶν ἐπιστημῶν ἔστι τις, εἴπερ ὑπάρχει τις οὐσία τοιαύτη, λέγω δὲ χωριστὴ καὶ ἀκίνητος, ὅπερ πειρασόμεθα δεικνύναι.

"the most honorable" science. It is the characteristic of theological objects as divine, "the most honorable genus," that establishes the science that studies them as the most honorable.[9] Similarly, in *Metaphysics* K7, Aristotle declares the following:

> Evidently, then, there are three genera of theoretical sciences: the physical, mathematical, and theological. The highest is the genus of the theoretical [sciences], and of these themselves [viz. the theoretical sciences] the last named [is the highest], because it deals with the most honorable of existing things, and each [science] is called better or worse in accordance with its proper object of study."[10]

In other words, theology is the highest, or most honorable, science because theological objects are the most honorable objects. Physics and mathematics rank below theology, because the objects they study are less honorable than divine, theological ones.

Aristotle does not explicitly rank physics and mathematics in relation to one another, but in *Metaphysics* K7 he suggests that if objects that are both separable and immovable—or theological objects—did not exist then physical objects, being separable and movable, would be primary. He explains, "If, then, natural substances are the first of existing things, physics must be the first of the sciences; but if there is some other nature and substance, separable and immovable, then the science of it must be different from and prior to physics and universal because of its priority."[11] This passage implies that Aristotle ranked physics as prior to and, therefore, more honorable than mathematics, presumably because he considered separability to be a more fundamental characteristic than immovability. Therefore, the separability of physical objects entails the ranking of physics above mathematics, and theology is the most honorable of the theoretical sciences because divine entities are more honorable than physical or mathematical objects. Again, the properties of the objects studied by the sciences determine the sciences' relative ranking.

9. Ibid. VI, 1026a21: τὸ τιμιώτατον γένος

10. Ibid. XI, 1064b1–6: δῆλον τοίνυν ὅτι τρία γένη τῶν θεωρητικῶν ἐπιστημῶν ἔστι, φυσική, μαθηματική, θεολογική. βέλτιστον μὲν οὖν τὸ τῶν θεωρητικῶν γένος, τούτων δ' αὐτῶν ἡ τελευταία λεχθεῖσα· περὶ τὸ τιμιώτατον γάρ ἐστι τῶν ὄντων, βελτίων δὲ καὶ χείρων ἑκάστη λέγεται κατὰ τὸ οἰκεῖον ἐπιστητόν.

11. Ibid. XI, 1064b9–14: εἰ μὲν οὖν αἱ φυσικαὶ οὐσίαι πρῶται τῶν ὄντων εἰσί, κἂν ἡ φυσικὴ πρώτη τῶν ἐπιστημῶν εἴη· εἰ δ' ἔστιν ἑτέρα φύσις καὶ οὐσία χωριστὴ καὶ ἀκίνητος, ἑτέραν ἀνάγκη καὶ τὴν ἐπιστήμην αὐτῆς εἶναι καὶ προτέραν τῆς φυσικῆς καὶ καθόλου τῷ προτέραν.

Ptolemy's Definitions Are Not Aristotle's

At first glance, Ptolemy's account in *Almagest* 1.1 of theoretical philosophy and the three theoretical sciences—physics, mathematics, and theology—seems to be derivative of Aristotle's. In the first sentence, addressing his customary dedicatee, Ptolemy declares, "It seems to me that the legitimate philosophers, Syrus, were entirely right to have distinguished the theoretical [part] of philosophy from the practical."[12] In other words, according to Ptolemy, legitimate philosophers, past and contemporary, maintain a distinction between theoretical and practical philosophy. Ptolemy does not include the productive as a division of philosophy as Aristotle does, but even Aristotle contrasted only the theoretical and practical on occasion, and the omission of the productive was common in post-Hellenistic philosophy. Even more striking is Ptolemy's citation of Aristotle in his division of theoretical philosophy. Ptolemy rarely cites philosophers in his corpus, but in *Almagest* 1.1 he ascribes the trichotomy of theoretical philosophy to Aristotle. Again, he says, "For Aristotle divides the theoretical [part of philosophy], too, very fittingly, into three primary genera, the physical, mathematical, and theological."[13] With this reference to Aristotle, Ptolemy indicates that he is purposefully appropriating Aristotle's schema. The discussion of the sciences that follows, however, reveals a substantial reconceptualization of Aristotelian metaphysics.

Franz Boll, a nineteenth-century philologist and the only scholar besides myself to attempt a complete study of Ptolemy's philosophy, believed that Ptolemy used *Metaphysics* E1 when composing *Almagest* 1.1.[14] As Liba Taub has observed, however, Ptolemy's definitions of the sciences are not Aristotle's from the *Metaphysics*,[15] and it seems unlikely that Ptolemy referred directly to either E1 or K7 of the *Metaphysics* when composing the chapter. While at the heart of Aristotle's definitions of the three theoretical sciences is the pairing of separability and inseparability with movability and immovability, Ptolemy does not define the objects of the sciences according to a pairing of two fundamental contraries.

Considering Ptolemy's departure from Aristotle's definitions of the three theoretical sciences, it is important first to note in what ways Aristotle's

---

12. Ptolemy, *Almagest* 1.1, H4: Πάνυ καλῶς οἱ γνησίως φιλοσοφήσαντες, ὦ Σύρε, δοκοῦσί μοι κεχωρικέναι τὸ θεωρητικὸν τῆς φιλοσοφίας ἀπὸ τοῦ πρακτικοῦ.

13. Ibid., H5: καὶ γὰρ αὖ καὶ τὸ θεωρητικὸν ὁ Ἀριστοτέλης πάνυ ἐμμελῶς εἰς τρία τὰ πρῶτα γένη διαιρεῖ τό τε φυσικὸν καὶ τὸ μαθηματικὸν καὶ τὸ θεολογικόν.

14. Boll, "Studien," 71.

15. Taub, *Ptolemy's Universe*, 19–26.

criteria—the (in)separability and (im)movability of objects—manifest in Ptolemy's account. With regard to theology, Ptolemy describes its object as "the first cause of the first motion of the universe."[16] Most likely Ptolemy identified this first cause with Aristotle's "Prime Mover," portrayed in *Physics* VIII and *Metaphysics* Λ. Ptolemy does not label it the Prime Mover. It is a "first cause" rather than a "first mover," but, as the first cause of the first motion, it is the Prime Mover. As for which motion the Prime Mover causes, in Ptolemy's astronomy the first motion of the universe is the diurnal rotation of the sphere of fixed stars. Ptolemy emphasizes two aspects of the Prime Mover. Again, for Aristotle the defining attributes of theological objects are their separability and immovability. For Ptolemy, this theological entity is distinctively imperceptible and immovable. He describes it as follows:

> the first cause of the first motion of the universe, if one considers it simply, can be thought of as an imperceptible and immovable deity and the species [of theoretical philosophy] concerned with investigating this [can be called] "theology," since this kind of activity, somewhere up in the highest reaches of the cosmos, only can be apprehended by thought and is completely separated from perceptible substances[17]

Ptolemy follows Aristotle in characterizing the object of theology as immovable, but, unlike Aristotle, he does not call it separable, even if it is located far away, in the highest reaches of the heavens. He does not say that it is able to have an absolutely independent existence; he only maintains that it is separate from perceptible substances. Besides its immovability, then, Ptolemy's Prime Mover is distinguished principally by its imperceptibility.

Ptolemy addresses movement in relation to physical and mathematical objects, but, again, he does not discuss whether the objects of the sciences are separable or inseparable. He calls physical objects "ever moving" (ἀεὶ κινουμένη) and identifies them with the perceptible properties of sublunary bodies. He says, "on the other hand, the species [of the theoretical part of

16. Ptolemy, *Almagest* 1.1, H5: τὸ μὲν τῆς τῶν ὅλων πρώτης κινήσεως πρῶτον αἴτιον
Cf. Ptolemy, *Optics* 2.103, where the author appeals to the Prime Mover as an exemplar of an unmoved entity. He calls it "that which moves first" (*quod primo mouet*). Note that Harald Siebert argues that the *Optics* is not by Ptolemy, or even from the second century, but was composed in late antiquity: Siebert, *Die ptolemäische "Optik."*

17. Ptolemy explains here why he calls the first cause a "deity," or god, and the science of this first cause "theology." It seems that, because the first cause is associated with the heavens and is imperceptible, it can be thought of as a god. Ptolemy, *Almagest* 1.1, H5: τὸ μὲν τῆς τῶν ὅλων πρώτης κινήσεως πρῶτον αἴτιον, εἴ τις κατὰ τὸ ἁπλοῦν ἐκλαμβάνοι, θεὸν ἀόρατον καὶ ἀκίνητον ἂν ἡγήσαιτο καὶ τὸ τούτου ζητητικὸν εἶδος θεολογικὸν ἄνω που περὶ τὰ μετεωρότατα τοῦ κόσμου τῆς τοιαύτης ἐνεργείας νοηθείσης ἂν μόνον καὶ καθάπαξ κεχωρισμένης τῶν αἰσθητῶν οὐσιῶν

philosophy] that examines material and ever-moving quality—being concerned with white, hot, sweet, soft, and suchlike—one may call 'physical,' since this sort of substance dwells for the most part among corruptible bodies and below the lunar sphere . . ."[18] Ptolemy characterizes physical objects as mainly sublunary; most of them exist below the lunar sphere. All of them experience perpetual change, and, in the case of sublunary physical objects, this change includes corruption. Four perceptible qualities stand as examples of physical objects: white, hot, sweet, and soft. Ptolemy also defines mathematical objects by way of examples: "and the species [of the theoretical part of philosophy] indicative of the quality concerning forms and movements from place to place, and which serves to investigate shape, number, size, and place, time, and suchlike, one may define as 'mathematical' . . ."[19] Movement from place to place is studied by mathematics, but neither movability nor immovability defines mathematical objects in general for Ptolemy.

Unlike Aristotle, Ptolemy explicitly ranks physics, mathematics, and theology in relation to one another, with regard to both ontological and epistemological considerations. According to Ptolemy, mathematical objects are intermediate between the physical and theological, and he presents two, adjacent, arguments in support of this order:

> this kind of substance falls, as it were, in the middle between the other two since, firstly, it can be conceived of both by means of and independently

18. Ibid., H5: τὸ δὲ τῆς ὑλικῆς καὶ αἰεὶ κινουμένης ποιότητος διερευνητικὸν εἶδος περί τε τὸ λευκὸν καὶ τὸ θερμὸν καὶ τὸ γλυκὺ καὶ τὸ ἁπαλὸν καὶ τὰ τοιαῦτα καταγιγνόμενον φυσικὸν ἂν καλέσειε τῆς τοιαύτης οὐσίας ἐν τοῖς φθαρτοῖς ὡς ἐπὶ τὸ πολὺ καὶ ὑποκάτω τῆς σεληνιακῆς σφαίρας ἀναστρεφομένης

19. Ibid., H5–6: τὸ δὲ τῆς κατὰ τὰ εἴδη καὶ τὰς μεταβατικὰς κινήσεις ποιότητος ἐμφανιστικὸν εἶδος σχήματός τε καὶ ποσότητος καὶ πηλικότητος ἔτι τε τόπου καὶ χρόνου καὶ τῶν ὁμοίων ζητητικὸν ὑπάρχον ὡς μαθηματικὸν ἂν ἀφορίσειε

Ptolemy nowhere explains what he means by the "quality concerning movements," to which he also refers in *Harmonics* 3.3, D94, where he mentions "the quantity and quality of the primary movements" (τὸ ποσὸν καὶ τὸ ποιὸν τῶν πρώτων κινήσεων). Here he contrasts this quality with "material and ever-moving quality" (τῆς ὑλικῆς καὶ αἰεὶ κινουμένης ποιότητος). The quality of movement is different in kind from material quality, to which Ptolemy also refers in *Tetrabiblos* 1.1.2, H4 (τῆς ὑλικῆς ποιότητος), and *Tetrabiblos* 1.2.15, H10 (τὸ ποιὸν τῆς ὕλης). The material qualities are peculiar to matter and include white, sweet, and soft, as well as hot, cold, wet, and dry. As we shall see, *Almagest* 1.1 also suggests that passivity, activity, corruptibility, and incorruptibility are material qualities. In *Geography* 1.1.5, Ptolemy discusses the degree to which geography, in contrast to chorography, is concerned with quality and quantity. He treats likeness (ὁμοιότης) as qualitative, and he identifies shape (σχῆμα) as a form of likeness. Therefore, when Ptolemy refers to the quality of movements, he may mean the shapes and configurations consequent upon movements.

of sense perception, and, secondly, it is an attribute of all beings without exception, both mortal and immortal, for those that are perpetually changing in their inseparable form, it changes with them, while for the eternal, which have an aethereal nature, it keeps their immovable form unmoved.[20]

First, while the object of theology is imperceptible and physical objects are perceptible, mathematical objects, although perceptible, can be thought of both with and without the aid of sense perception. Ptolemy may be suggesting that, while it is possible to observe mathematical objects—as one also observes physical objects—and contemplate them with reference to the sense impressions one has of them, it is also possible to contemplate them independently of sense perception, as if they were imperceptible, like the object of theology. Second, mathematical objects are attributes of all existing things, mortal and immortal, presumably physical and theological.

These two arguments together give incompatible accounts of what theological objects are. The first implies that the theological object under discussion is the Prime Mover, for the relevant object can be thought of only independently of sense perception. In the second argument, aethereal bodies, rather than the Prime Mover, exemplify theological objects. It is possible that in the second argument Ptolemy simply is acknowledging another set of theological objects, in addition to the Prime Mover, but if Ptolemy included both the Prime Mover and aethereal bodies as theological objects then his first argument would be unsound. Again, the argument implies that the object of theology can be contemplated only independently of sense perception. It would be rather odd of Ptolemy to suggest that not only the Prime Mover but also aethereal bodies can be conceived of only independently of sense perception, for the stars and planets are visible, even if the spheres that contain them are imperceptible from the earth. It seems, then, that Ptolemy broadens the scope of theological objects only in his second argument. Perhaps anticipating opposition, he provides more than one argument for this ranking of the objects of the sciences—and, by extension, the sciences themselves—even if the two arguments rest on incompatible premises.

Ptolemy also mentions the trichotomy of theoretical philosophy in the *Harmonics*, his text on music theory. Perhaps because it is an early work of Ptolemy, predating the *Almagest*, the *Harmonics'* account is less nuanced than *Almagest* 1.1, but it is still consistent with it. In *Harmonics* 3.6, Ptolemy distinguishes

---

20. Ptolemy, *Almagest* 1.1, H6: τῆς τοιαύτης οὐσίας μεταξὺ ὥσπερ ἐκείνων τῶν δύο πιπτούσης οὐ μόνον τῷ καὶ δι᾽ αἰσθήσεως καὶ χωρὶς αἰσθήσεως δύνασθαι νοεῖσθαι, ἀλλὰ καὶ τῷ πᾶσιν ἁπλῶς τοῖς οὖσι συμβεβηκέναι καὶ θνητοῖς καὶ ἀθανάτοις τοῖς μὲν ἀεὶ μεταβάλλουσι κατὰ τὸ εἶδος τὸ ἀχώριστον συμμεταβαλλομένην, τοῖς δὲ ἀιδίοις καὶ τῆς αἰθερώδους φύσεως συντηροῦσαν ἀκίνητον τὸ τοῦ εἴδους ἀμετάβλητον.

theoretical from practical philosophy and does not mention the productive. Moreover, he divides theoretical and practical philosophy into three subdivisions each: "For each of the two kinds of principle, that is, the theoretical and the practical, there are three genera, the physical, mathematical, and theological in the case of the theoretical, and the ethical, domestic, and political in the case of the practical . . ."[21] Ptolemy does not define the three theoretical sciences here. He only ranks them. He draws correspondences between the sciences and the musical genera—the enharmonic, chromatic, and diatonic—and, as in *Almagest* 1.1, mathematics is intermediate between physics and theology:

> Thus the enharmonic is to be compared to the physical and the ethical, because of its decrease in magnitude by comparison with the others; the diatonic to the theological and the political, because of the similarity of its order and its majesty to theirs; and the chromatic to the mathematical and the domestic, because of the shared nature of what is intermediate in relation to the extremes. For the mathematical genus is involved to a high degree both in the physical and in the theological, and the domestic shares with the ethical in being private and subordinate, and with the political in being corporate and controlling[22]

Ptolemy does not explain here what mathematics shares with physics and theology that makes it intermediate, but he may be referring to what he articulates in the second argument in *Almagest* 1.1, that mathematical objects inhere in mortal and immortal, sublunary and superlunary, physical and theological entities.

Ptolemy's Aristotelian Definitions

Although Ptolemy's definitions of the sciences are not Aristotle's, they are still Aristotelian.[23] Reconceptualizing the criteria that distinguish the sciences,

---

21. Ptolemy, *Harmonics* 3.6, D98: Καὶ τοίνυν καθ᾽ ἑκατέραν ἀρχήν, τουτέστι τὴν θεωρητικὴν καὶ τὴν πρακτικήν, τριῶν ὄντων γενῶν, ἐπὶ μὲν τῆς θεωρητικῆς τοῦ τε φυσικοῦ καὶ τοῦ μαθηματικοῦ καὶ τοῦ θεολογικοῦ, ἐπὶ δὲ τῆς πρακτικῆς τοῦ τε ἠθικοῦ καὶ τοῦ οἰκονομικοῦ καὶ τοῦ πολιτικοῦ

22. Ibid.: τὸ μὲν οὖν ἐναρμόνιον τῷ τε φυσικῷ καὶ τῷ ἠθικῷ παραβλητέον διὰ τὴν κοινὴν παρὰ τἆλλα τοῦ μεγέθους ἐλάττωσιν, τὸ δὲ διατονικὸν τῷ θεολογικῷ καὶ τῷ πολιτικῷ διὰ τὴν τῆς τάξεως καὶ τῆς μεγαλειότητος ὁμοιότητα, τὸ δὲ χρωματικὸν τῷ τε μαθηματικῷ καὶ τῷ οἰκονομικῷ διὰ τὴν τῆς πρὸς τὰ ἄκρα μεσότητος κοινότητα. τό τε γὰρ μαθηματικὸν γένος ἐπὶ πλεῖστον ἀναστρέφεται κἄν τῷ φυσικῷ κἄν τῷ θεολογικῷ, τό τε οἰκονομικὸν μετέχει καὶ τοῦ ἠθικοῦ κατὰ τὸ ἰδιωτικὸν καὶ ὑποτεταγμένον καὶ τοῦ πολιτικοῦ κατὰ τὸ κοινωνικὸν καὶ ἀρχικόν

23. Liba Taub observes that Ptolemy's definitions are not Aristotle's but "there is something about them which has a familiar ring, which suggests, broadly speaking, some 'Aristotelian' influence": Taub, *Ptolemy's Universe*, 24.

Ptolemy appropriates not only Aristotle's metaphysical terms but also his epistemological concepts, and he reconceives their significance and relationships. Ptolemy, like Aristotle, defines the sciences in accordance with their objects of study, but the criteria he uses to distinguish the objects are not their (in)separability or (im)movability but instead whether and how they are perceptible. In this way, Ptolemy incorporates epistemological criteria at a foundational level to his metaphysics. Whether and how an object presents itself to the human sensory apparatus determines whether the object itself is physical, mathematical, or theological.

Again, Ptolemy's object of theology, the Prime Mover, is imperceptible. Ptolemy's physical and mathematical objects are perceptible, and their description derives from several of Aristotle's texts, including the *De sensu et sensibilibus*, *De insomniis*, *Metaphysics* M, and especially the *De anima*. To reiterate, the qualities Ptolemy lists as exemplifying physical objects are white (λευκόν), hot (θερμόν), sweet (γλυκύ), and soft (ἁπαλόν). Each is perceptible by one sense only: white by sight, hot by touch, sweet by taste, and soft by touch. Because only one sense can perceive each of these qualities, they are, according to Aristotle's theory of perception, special sensibles. Aristotle defines special sensibles in the *De anima*: "I call special whatever cannot be perceived by another sense, and about which it is impossible to be deceived, as for instance sight has color, hearing sound, and taste flavor, while touch has many varieties of object..."[24] For each sense, Aristotle lists at least one pair of contraries that exemplifies the sense's special object: "For every sense seems to be concerned with one pair of contraries, as for instance sight is with white and black, hearing with high and low pitch, and taste with bitter and sweet; but in the field of the touchable there are many pairs of contraries, hot and cold, dry and wet, rough and smooth, and so on for the rest."[25] Among the contraries, Aristotle includes white (λευκόν), hot (θερμόν), and sweet (γλυκύ), each of which Ptolemy lists among the objects of physics. Although Aristotle does not mention softness (ἁπαλόν) as a special sensible, he would no doubt have considered it as such, since it is perceptible only by the sense of touch. A near contemporary of Ptolemy, Alexander of Aphrodisias, the Aristotelian commentator and chair in Peripatetic philosophy at Athens in the late second century and/or early third century, does call softness a special sensible—albeit by

24. Aristotle, *De anima* II, 418a11–14, translation modified from Hamlyn: λέγω δ' ἴδιον μὲν ὃ μὴ ἐνδέχεται ἑτέρᾳ αἰσθήσει αἰσθάνεσθαι, καὶ περὶ ὃ μὴ ἐνδέχεται ἀπατηθῆναι, οἷον ὄψις χρώματος καὶ ἀκοὴ ψόφου καὶ γεῦσις χυμοῦ, ἡ δ' ἁφὴ πλείους [μὲν] ἔχει διαφοράς

25. Ibid. II, 422b23–27: πᾶσα γὰρ αἴσθησις μιᾶς ἐναντιώσεως εἶναι δοκεῖ, οἷον ὄψις λευκοῦ καὶ μέλανος, καὶ ἀκοὴ ὀξέος καὶ βαρέος, καὶ γεῦσις πικροῦ καὶ γλυκέος· ἐν δὲ τῷ ἁπτῷ πολλαὶ ἔνεισιν ἐναντιώσεις, θερμὸν ψυχρόν, ξηρὸν ὑγρόν, σκληρὸν μαλακόν, καὶ τῶν ἄλλων ὅσα τοιαῦτα.

another term, *malakotês*—in his *De anima* 55.23. Hence, the qualities Ptolemy lists as examples of physical objects are not only classifiable as special sensibles, but they also appear as special sensibles in Aristotle's and/or Aristotelian accounts of perception.

Ptolemy's examples of mathematical objects are classifiable as common sensibles in Aristotle's schema. They are perceptible by more than one sense, such as by sight and touch. Again, Ptolemy provides the following examples of mathematical objects: the quality concerning forms (εἴδη) and movements from place to place (μεταβατικὰς κινήσεις) as well as shape (σχῆμα), number (ποσότης), size (πηλικότης), place (τόπος), and time (χρόνος). Aristotle provides two separate, but similar, lists of common sensibles in the *De anima*. In the first, he includes movement (κίνησις), rest (ἠρεμία), number (ἀριθμός), shape (σχῆμα), and magnitude (μέγεθος).[26] In the second, he includes all of the same, except instead of using the term *êremia* for rest he uses *stasis*.[27] Hence, two of Ptolemy's mathematical objects coincide with Aristotle's common sensibles: movements (κινήσεις) and shape (σχῆμα). It is clear that Ptolemy did not use the *De anima* as a direct guide for composing the *Almagest*, but examples of special and common sensibles in the *De anima* stand as physical and mathematical objects, respectively, in *Almagest* 1.1. Moreover, those terms that do not arise in Aristotle's account of special and common sensibles in the *De anima* are still classifiable as such.

The influence of the *De anima* on Ptolemy's *On the Kritêrion and Hêgemonikon* is manifest and lends additional support to my interpretation of Ptolemy's account of physical and mathematical objects in the *Almagest*. As the title indicates, *On the Kritêrion and Hêgemonikon* examines the criterion of truth, the method by which one gains knowledge, as well as the nature and structure of the human soul. I argue in chapter 7 that it is an early—perhaps the earliest—extant text of Ptolemy. It is likely that the epistemological investigations of *On the Kritêrion* influenced Ptolemy's account of the theoretical sciences in *Almagest* 1.1. In *On the Kritêrion*, Ptolemy explains that each perceptual faculty has one object proper to it and, concerning this object, or special sensible, a faculty never errs. Recalling Aristotle's discussion of special sensibles in the *De anima*—as well as Aristotle's dismissal of skeptical challenges to knowledge in, for instance, *Metaphysics* Γ5[28]—Ptolemy asserts, "On its own each of the faculties naturally tells the truth whenever it is concerned with only its own special object and not distracted by having to come to grips with the [other faculties], when, for example, sight is concerned with colors, hearing with

26. Ibid. II, 418a17–18.
27. Ibid. III, 425a16.
28. Aristotle, *Metaphysics* IV, 1010b1–1011a2.

sounds, taste with flavors, smell with odors, touch with [. . .] qualities . . ."[29]
Moreover, Ptolemy recognizes that some objects are common, or perceptible
by more than one sense. His list of common sensibles includes bulk (ὄγκος),
magnitude (μέγεθος), number (πλῆθος), shape (σχῆμα), position (θέσις), arrangement (τάξις), and movement (κίνησις).[30] While the list of mathematical objects in *Almagest* 1.1 corresponds to Aristotle's list of common sensibles
in the *De anima* with regard to only two terms, the list of common sensibles
in *On the Kritêrion* contains three of Aristotle's terms: magnitude (μέγεθος),
shape (σχῆμα), and movement (κίνησις). Ptolemy's appropriation of Aristotle's theory of common sensibles, as well as several of Aristotle's examples of
common sensibles, suggests that he deliberately identifies mathematical objects
with common sensibles in *Almagest* 1.1.

Aristotle provides lists of common sensibles in the *De insomniis* and *De
sensu et sensibilibus* that are slightly different from the lists in the *De anima*,
and these accounts may have influenced Ptolemy's list in *On the Kritêrion*. The
De insomniis records only three common sensibles, each of which Aristotle
includes in the *De anima*: shape (σχῆμα), magnitude (μέγεθος), and motion
(κίνησις).[31] Notably, these three terms are the same three that *On the Kritêrion*
and the *De anima* share. In the *De sensu*, Aristotle provides two distinct lists of
common sensibles. In the first, he includes four examples: magnitude (μέγεθος),
shape (σχῆμα), movement (κίνησις), and number (ἀριθμός).[32] Again, Ptolemy's list in *On the Kritêrion* includes three of these four terms. More interesting is the second list of common sensibles in the *De sensu*, where Aristotle
states, "for magnitude and shape, roughness and smoothness, and, moreover,
the sharpness and bluntness found in solid bodies, are common to all the

29. Ptolemy, *On the Kritêrion*, La16, translation modified from Liverpool-Manchester Seminar on ancient Greek philosophy, in Ptolemaeus, "On the Kriterion" (meeting regularly from
May 1980 to October 1981, the seminar included Henry Blumenthal, Noreen Fox, Pamela Huby,
George Kerferd, A. C. Lloyd, A. A. Long, Mari Nagase, Gordon Neal, and Howard Robinson):
καθ᾽ αὑτὴν μὲν οὖν ἑκάστη τῶν δυνάμεων ὅταν τὸ ἴδιον καὶ οἰκεῖον μόνον ἐπισκοπῇ κατὰ τὸ τῶν
συμπεπλεγμένην ἀπερίσπαστον ἀληθεύειν πέφυκεν, ὡς ὅταν ὄψις μὲν χρώματα, φωνὰς δ᾽ ἀκοή,
γεῦσις δὲ χυμούς, ἀτμοὺς δ᾽ ὄσφρησις, ἁφὴ δὲ † ποιότητας

30. Ibid., La17. Cf. Ptolemy *Optics* 2.2 and 2.13 for a discussion of common sensibles, including corporeity, size, shape, place, movement, and rest (*corpus, magnitudo, figura, situs, motus et
quies*). Note that the list of common sensibles in the *Optics* shares four terms with the first list
in the *De anima*: movement, rest, shape, and magnitude. In addition, Ptolemy's list of mathematical objects in *Almagest* 1.1 coincides with the list of common sensibles in the *Optics* with
respect to four terms: movement, shape, size, and place.

31. Aristotle, *De insomniis* 458b5.

32. Aristotle *De sensu et sensibilibus* 437a9.

senses, or, if not to all, at least to sight and touch."[33] Not only does Aristotle add two dichotomies—roughness and smoothness as well as sharpness and bluntness—but he also specifies that the latter dichotomy is found in objects that are solid, or have bulk (ὄγκοι). This additional example may be significant, for bulk (ὄγκος) is a common sensible in On the Kritêrion. Ptolemy's inclusion of bulk may derive from an Epicurean influence on his theory of matter,[34] or—although the comparison is a loose one, as Aristotle mentions solid bodies but does not call solidity itself a common sensible—Aristotle's discussion of common sensibles in the De sensu, like the De anima and De insomniis, may have influenced the account of common sensibles in On the Kritêrion and, by extension, the list of mathematical objects in Almagest 1.1.

Some late antique and Byzantine philosophers identify mathematical objects with common sensibles in their commentaries on Aristotle's texts. For instance, Philoponus and Sophonias—the sixth-century Neoplatonist and the late thirteenth- to early fourteenth-century Constantinopolitan monk, respectively—examine the nature of mathematical objects in their commentary and paraphrase of Aristotle's De anima. Each of them, likely the latter influenced by the former, defines mathematical objects as forms, abstracted by thought from matter, that are common sensibles.[35] Similarly, in his commentary on Aristotle's Metaphysics, Syrianus—the fifth-century philosopher and head of the Platonic school at Athens—considers the possibility that mathematical objects are common sensibles, abstracted by thought from perceptible objects.[36] It is possible that Ptolemy's identification of mathematical objects with common sensibles influenced these later interpretations of Aristotle's philosophy.

With respect to Ptolemy's first argument in Almagest 1.1 for the intermediate status of mathematical objects—that they are conceivable with and without the aid of sense perception—Ptolemy may have developed it in response to Aristotle's, or an Aristotelian, account of how it is that one examines mathematical objects. Again, in the Metaphysics Aristotle portrays mathematical objects as inseparable but studied as if they were separable, and in Physics II.2

33. Ibid. 442b5–7, translation modified from J. I. Beare in Aristotle, Complete Works: μέγεθος γὰρ καὶ σχῆμα καὶ τὸ τραχὺ καὶ τὸ λεῖον, ἔτι δὲ τὸ ὀξὺ καὶ τὸ ἀμβλὺ τὸ ἐν τοῖς ὄγκοις, κοινὰ τῶν αἰσθήσεών ἐστιν, εἰ δὲ μὴ πασῶν, ἀλλ' ὄψεώς γε καὶ ἁφῆς.

34. An Epicurean influence on Ptolemy's philosophy is apparent in both On the Kritêrion and the Tetrabiblos where Ptolemy portrays the soul as consisting of particles finer than the constituents of body. I discuss this distinction between body and soul in chapters 4 and 7.

35. Philoponus, In Aristotelis libros de anima commentaria 15.57.28–30; Sophonias, In Aristotelis libros de anima paraphrasis 9.28–31. See Harari, "John Philoponus."

36. Syrianus, In Aristotelis metaphysica commentaria 95.13–17. On Philoponus's and Syrianus's theories of mathematical objects, see Mueller, "Aristotle's Doctrine of Abstraction."

Aristotle confirms this characterization of mathematical objects. He explains that the mathematician differs from the natural philosopher in that the former studies figures, solids, lines, and points as if they were independent of physical bodies: "Now the mathematician, though he too treats of these things, nevertheless does not treat each as a limit of a natural body; nor does he consider the attributes indicated as the attributes of such bodies. That is why he separates [them]; for in thought they are separable from motion, and it makes no difference, nor does any falsity result, if they are separated."[37] Correspondingly, in *Metaphysics* M Aristotle addresses the perceptibility of mathematical objects, rather than their inseparability, and he contends that mathematical objects are perceptible but studied in such a way that their perceptibility is irrelevant. He explains, "so too with geometry: just because their objects happen to be perceptible, though not [studied] *qua* perceptible, the mathematical sciences will not be about perceptible objects, nor about other separate objects over and above these."[38] In his polemic against the Platonic conception of mathematical objects as separate, incorporeal, imperceptible entities, Aristotle maintains that mathematical objects are indeed perceptible but studied as if they were imperceptible.

A similar understanding of the nature and study of mathematical objects underlies *Almagest* 1.1. For Ptolemy, mathematical objects are perceptible; they are common sensibles. Yet, the mathematician can contemplate them either in relation to or independently of sense perception. It may be mathematical objects' existence as common sensibles, rather than special sensibles, that admits different means by which they can be contemplated. In the case of special sensibles, cognition of them is irrevocably tied to their perception by a single faculty, whether it be sight, hearing, taste, smell, or touch. In the case of common sensibles, the intellect has the option of considering them with respect to one or more of the senses. The intellect can consider objects perceptible by sight and touch, for example, in relation to both sight and touch, sight independently of touch, or touch independently of sight. Because the intellect can consider these objects independently of sight or touch, I take it that for Ptolemy the intellect also can consider them independently of sight

---

37. Aristotle, *Physics* II, 193b31–35, translation modified from R. P. Hardie and R. K. Gaye in Aristotle, *Complete Works*: περὶ τούτων μὲν οὖν πραγματεύεται καὶ ὁ μαθηματικός, ἀλλ᾽ οὐχ ἡ φυσικοῦ σώματος πέρας ἕκαστον· οὐδὲ τὰ συμβεβηκότα θεωρεῖ ἡ τοιούτοις οὖσι συμβέβηκεν· διὸ καὶ χωρίζει· χωριστὰ γὰρ τῇ νοήσει κινήσεώς ἐστι, καὶ οὐδὲν διαφέρει, οὐδὲ γίγνεται ψεῦδος χωριζόντων.

38. Aristotle, *Metaphysics* XIII, 1078a2–5: οὕτω καὶ τὴν γεωμετρίαν· οὐκ εἰ συμβέβηκεν αἰσθητὰ εἶναι ὧν ἐστί, μὴ ἔστι δὲ ἡ αἰσθητά, οὐ τῶν αἰσθητῶν ἔσονται αἱ μαθηματικαὶ ἐπιστῆμαι, οὐ μέντοι οὐδὲ παρὰ ταῦτα ἄλλων κεχωρισμένων.

and touch, i.e. unaided by and dissociated from sense perception. In this way, the intellect has the capability of divorcing its function from the faculty of sense perception when it contemplates common sensibles, and mathematical objects can be thought of both with and without the aid of sense perception.

Conclusion

Ptolemy appropriates his division of theoretical philosophy into physics, mathematics, and theology from Aristotle, and his definitions of the sciences are, albeit not Aristotle's, Aristotelian. Again, Aristotle distinguishes and ranks the theoretical sciences solely according to ontological criteria. The combination of (in)separability and (im)movability in an object determines which science studies it. Theology is the highest, most honorable science because its objects are divine, and physics is prior to mathematics because separability is a more fundamental property of objects than immovability. Ptolemy adapts criteria from Aristotle's broader corpus to distinguish the objects of the sciences and, by extension, the sciences themselves. Whether and how an object is perceptible determines whether it is studied by physics, mathematics, or theology. An imperceptible object, the Prime Mover, is studied by theology, special sensibles are studied by physics, and common sensibles are studied by mathematics. Moreover, whether the objects studied by the sciences are conceivable with and/or without the aid of sense perception determines the sciences' relative ranking. Mathematics is intermediate between physics and theology, because physical objects are conceivable only with the aid of sense perception, the Prime Mover is conceivable only independently of sense perception, and mathematical objects are conceivable both with and without the aid of sense perception. In this way, Ptolemy undergirds his metaphysics with epistemology. What determines whether an object is in and of itself physical, mathematical, or theological is whether and how it is perceptible to a human being, and how the objects of the sciences are contemplated by human beings determines the sciences' hierarchy.

3

Knowledge or Conjecture

AFTER PTOLEMY defines physics, mathematics, and theology in *Almagest* 1.1, he assesses their epistemic success, whether they produce conjecture, mere guesswork, or whether they have the capacity to generate knowledge, true and skillfully constructed accounts of the objects under consideration. As we saw in the previous chapter, according to Aristotle the three theoretical sciences are types of knowledge, or understanding. To ask whether a type of knowledge yields knowledge is nonsensical, and in fact Aristotle does not evaluate the epistemic efficacy of the three sciences when he defines them in *Metaphysics* E1 and K7. Ptolemy, on the other hand, does not label physics, mathematics, and theology types of knowledge. They are parts of philosophy, and whether a branch of philosophy yields knowledge or conjecture is an apposite question. Ptolemy argues that both physics and theology are conjectural and that mathematics alone yields knowledge. This complex claim—that only mathematics and neither physics nor theology generates knowledge—is unprecedented in the history of philosophy and would have been extraordinarily controversial in the ancient Greek philosophical milieu.

Ptolemy appropriates the terms and concepts of his epistemology from both the Aristotelian and Platonic traditions, but his epistemology is not simply derivative; it is subversive. Ptolemy ridicules attempts by philosophers to study physics and theology, the two theoretical sciences traditionally studied by philosophers. He contends that philosophers are unable to agree on the nature of physical and theological objects and that these fields of inquiry are inherently conjectural. Mathematics alone yields sure and incontrovertible knowledge, and, furthermore, its contribution to physics and theology is epistemically efficacious. Mathematics engenders a good guess at the nature of the Prime Mover, and it reveals the natures of physical objects. Ptolemy casts mathematics as epistemologically the highest part of theoretical philosophy, and he co-opts the fields of inquiry traditionally studied by philosophers for mathematicians.

Ptolemy's Epistemology: Neither Aristotle's nor Plato's

Ptolemy presents his epistemological assessment of physics, mathematics, and theology in the following passage from *Almagest* 1.1:

> From all this we concluded that the first two genera of the theoretical [part of philosophy] should rather be called conjecture than knowledge, the theological because of its complete invisibility and ungraspability, and the physical because of the instability and unclearness of matter, so that, on account of this, [we concluded] never to hope that philosophers will be agreed about them, and that only the mathematical can provide sure and incontrovertible knowledge to its practitioners, if one approaches it rigorously, for its demonstration proceeds by indisputable methods, both arithmetic and geometry[1]

According to Ptolemy, physics and theology are conjectural; one can never know whether the accounts they generate accurately reflect the true natures of physical and theological objects. Conversely, mathematics yields knowledge, accounts that accurately correspond to the phenomena and, moreover, are sure and incontrovertible.

As the Platonic and Aristotelian traditions dominated the second-century philosophical landscape, reference to Plato's and Aristotle's epistemologies of physics and theology reveals how unconventional Ptolemy's denunciation of them is. As is well known, Plato portrays the study of physical, or visible, objects as an intellectual endeavor inferior to the study of metaphysical entities. Whereas the study of the Forms is knowledge, the study of the visible world is opinion, and an account of physical phenomena, such as the cosmology of the *Timaeus*, remains a likely story. Hence, Ptolemy's epistemic evaluation of physics does not contradict Platonic epistemology. According to Aristotle, on the other hand, physics is a division of knowledge alongside mathematics and theology. The characteristics of an object, including its changeability, do not prevent human beings from obtaining knowledge of the universals that enform it. For Aristotle, universals are knowable and unchanging despite their abstraction from changeable bodies.

---

1. Ptolemy, *Almagest* 1.1, H6: ἐξ ὧν διανοηθέντες, ὅτι τὰ μὲν ἄλλα δύο γένη τοῦ θεωρητικοῦ μᾶλλον ἄν τις εἰκασίαν ἢ κατάληψιν ἐπιστημονικὴν εἴποι, τὸ μὲν θεολογικὸν διὰ τὸ παντελῶς ἀφανὲς αὐτοῦ καὶ ἀνεπίληπτον, τὸ δὲ φυσικὸν διὰ τὸ τῆς ὕλης ἄστατον καὶ ἄδηλον, ὡς διὰ τοῦτο μηδέποτε ἂν ἐλπίσαι περὶ αὐτῶν ὁμονοῆσαι τοὺς φιλοσοφοῦντας, μόνον δὲ τὸ μαθηματικόν, εἴ τις ἐξεταστικῶς αὐτῷ προσέρχοιτο, βεβαίαν καὶ ἀμετάπιστον τοῖς μεταχειριζομένοις τὴν εἴδησιν παράσχοι ὡς ἂν τῆς ἀποδείξεως δι᾽ ἀναμφισβητήτων ὁδῶν γιγνομένης, ἀριθμητικῆς τε καὶ γεωμετρίας

Ptolemy's characterization of theology as conjectural, however, is neither Platonic nor Aristotelian, and this aspect of his epistemology is arguably the most subversive.[2] For both Plato and Aristotle, theology yields knowledge as it concerns the highest ontological order. Again, for Plato, the study of metaphysical entities, the Forms, supersedes every other study. In *Metaphysics* M, Aristotle too gestures toward an epistemological hierarchy among the sciences, where the ontological priority of an object determines human beings' success at understanding it. He explains as follows:

> Indeed, the more [that our knowledge] concerns what are prior in definition and simpler, the greater the accuracy (that is, simplicity) [our knowledge] has, so that there is more [accuracy] where there is no magnitude than where there is magnitude, and most of all where there is no movement, though if there is movement [the accuracy] is greatest if it is the primary [movement]; for this is the simplest, and uniform [movement the simplest] of that.[3]

Aristotle places this argument in his book on mathematics, and so it is likely that in it he is addressing the relationship of mathematics to theology and physics. The object permitting the greatest accuracy, which has no magnitude and experiences no movement, is the Prime Mover or unmoved movers of *Metaphysics* Λ8, and their study, theology, is the science that is most accurate. The object whose study is the next most accurate is that which experiences only the primary movement. Aristotle explains in the *Physics* that the primary type of movement is motion from place to place, and the primary motion from place to place is movement in a circle. According to the *De caelo*, the aether, the fifth element of the heavens, experiences only this movement. Therefore, after theology, astronomy—which Aristotle classifies as a type of mathematics—is the most accurate field of inquiry.[4] Whereas theology ranks first and mathematics second, physics is last in the epistemological hierarchy. Physical bodies have magnitude and experience many changes, in addition to motion from place to place. In this way, Aristotle characterizes theology as more accurate

---

2. Liba Taub has observed that Ptolemy's epistemological claim in *Almagest* 1.1 is not Aristotelian: Taub, *Ptolemy's Universe*, 26.

3. Aristotle, *Metaphysics* XIII, 1078a9–13: καὶ ὅσῳ δὴ ἂν περὶ προτέρων τῷ λόγῳ καὶ ἁπλουστέρων, τοσούτῳ μᾶλλον ἔχει τὸ ἀκριβές (τοῦτο δὲ τὸ ἁπλοῦν ἐστίν), ὥστε ἄνευ τε μεγέθους μᾶλλον ἢ μετὰ μεγέθους, καὶ μάλιστα ἄνευ κινήσεως, ἐὰν δὲ κίνησιν, μάλιστα τὴν πρώτην· ἁπλουστάτη γάρ, καὶ ταύτης ἡ ὁμαλή.

4. For Aristotle's classification of astronomy as a type of mathematics, see, for instance, *Physics* II, 194a7–8, where he calls astronomy one of "the more physical of the mathematical sciences" (τὰ φυσικώτερα τῶν μαθημάτων).

than physics and mathematics. Ptolemy's claim, then, that theology is conjectural, contravenes both Plato's and Aristotle's epistemologies.

Platonic Epistemology

Although consistent with neither Plato's nor Aristotle's evaluation of the sciences, Ptolemy's bipartite epistemological claim—that physics and theology are conjectural and mathematics alone yields knowledge—derives from a subversive fusion of Platonic and Aristotelian epistemologies. Ptolemy adapts the Platonic dichotomy between opinion and knowledge to an Aristotelian form of empiricism. Concerning the former, Plato's distinction between opinion and knowledge remained a concern among Platonic philosophers at the time of Ptolemy's writing. Although not only Platonists but also Stoics contrasted opinion and knowledge—as, for instance, Sextus Empiricus, the Pyrrhonian skeptic in the second or early third century, notes in *Adversus mathematicos* 7.151—the similarity in concepts and structure of Alcinous's *Didaskalikos* and *Almagest* 1.1 reveals Ptolemy's philosophy to be a response to the contemporary Platonic tradition and, in particular, the type of Platonic philosophy advanced by Alcinous.

The identity of Alcinous has been a topic of debate for centuries, but John Dillon has narrowed the possibilities to a Platonic philosopher living, most likely, in the first and/or second century CE—in other words, roughly contemporarily with Ptolemy.[5] As was common among middle Platonic authors, Alcinous blends several philosophical influences: Platonism, of course, as well as Aristotelianism and Stoicism. As will become evident, the manner in which Ptolemy practices Platonism is similar to the type of Platonism exercised in Alcinous's *Didaskalikos*. In particular, I argue that Ptolemy structured *Almagest* 1.1 so as to mimic the structure of contemporary philosophical handbooks, like the *Didaskalikos*.

Jaap Mansfeld has observed that *Almagest* 1.1 incorporates structural elements common to ancient introductions to an author or a text. These structural elements became formalized in late antiquity and are known as the *schema isagogicum*, or "introductory scheme." This scheme consists of a set of preliminary questions, "headings" or "main points" (κεφάλαια), that late antique authors, especially, addressed in the introductory sections of their commentaries on particular works or intellectual figures. Late Neoplatonic commentators, for instance, consistently began their commentaries on Aristotle's texts by responding to this scheme, stating the theme or purpose of the work, its position in the Aristotelian corpus, the text's utility, etc. Mansfeld

5. Dillon, "Introduction," xi–xiii.

traces the introductory scheme's roots in the centuries preceding late antiquity, and he reveals its reach across apparently distinct genres. Authors employed the scheme not only when introducing philosophical texts but also in their studies of medicine, patristics, rhetoric, grammar, and mathematics. Mansfeld discovers several of the isagogical headings in *Almagest* 1.1, including the purpose of the author; the utility, aim, and order of the present study; the historical contributions of predecessors; and the qualities required of the student.[6] Thus, Mansfeld demonstrates that *Almagest* 1.1 fits within a far-reaching ancient literary culture, which established practices for how to introduce a text, in whichever field of study.

I suggest, however, that the resemblance of *Almagest* 1.1 to Alcinous's *Didaskalikos* is more particular and genre specific. In the first three chapters, Alcinous defines philosophy and the philosopher, describes the theoretical and practical types of life, and expounds the parts of theoretical and practical philosophy. Correspondingly, in *Almagest* 1.1 Ptolemy identifies the legitimate philosophers, distinguishes the theoretical and practical parts of philosophy, and defines the parts of theoretical philosophy, before portraying the aim of theoretical philosophy, the transformation of one's soul to a divine-like state, a topic that Alcinous discusses later in his handbook and which I discuss in the next chapter. Alcinous and Ptolemy participated not only in the broad literary culture that Mansfeld describes but also in the genre-specific rhetorical culture of philosophical handbooks. Indeed, Ptolemy's imitation of a philosophical handbook would have been apparent to any contemporary philosopher. The question, then, is why Ptolemy decided to appropriate this structure. Why did Ptolemy introduce his astronomical theory with a chapter in the style of a philosophical handbook? His decision could simply have resulted from his embeddedness in the contemporary philosophical context. As a result of reading philosophical handbooks, he may simply have acquired the style. Yet, again, what is interesting is that Ptolemy utilizes the structure not only of ancient prefaces in general but of philosophical handbooks in particular. He deliberately chose to introduce his astronomical models by situating them within the philosophical context of distinguishing the parts of philosophy and defending his choice to spend more of his free time pursuing theoretical rather than practical philosophy. It will become clear later in this chapter that Ptolemy appropriated the style of a philosophical handbook in order to cast *Almagest* 1.1 as a polemic against the discourses of philosophers. Ptolemy appropriates the terms and style of a philosophical handbook in order to subvert the genre and denounce philosophers' attempts to meet the objectives of philosophical inquiry, in this case generating knowledge rather than conjecture.

6. Mansfeld, *Prolegomena mathematica*, 68–69.

First, it is important to note which concepts Ptolemy appropriates from the contemporary Platonic tradition, as evidenced by Alcinous's text. To begin with, in *Didaskalikos* 2.1 Alcinous, like Ptolemy, distinguishes the theoretical and practical domains but omits the productive. He declares, "There are two types of life, the theoretical and the practical. The summation of the theoretical life lies in the knowledge of the truth, while that of the practical life [lies] in the performance of what is dictated by reason. The theoretical life is of primary value; the practical of secondary and involved with necessity . . ."[7] In *Didaskalikos* 3.4, Alcinous divides theoretical philosophy into three sciences:

> Of theoretical [philosophy], that part which is concerned with the immovable and primary causes and such as are divine is called theological; that which is concerned with the motion of the stars, their revolutions and periodic returns, and the constitution of the cosmos [is called] physical; and that which theorizes by means of geometry and the other branches of mathematics [is called] mathematical.[8]

Thus, both Ptolemy and Alcinous define physics, mathematics, and theology.

Alcinous's account of theology most resembles Ptolemy's. Causes that are motionless, primary, and divine are theological, and these characteristics of theological entities could be attributed to an Aristotelian Prime Mover, such as described by Ptolemy in *Almagest* 1.1. Alcinous's account of physics least resembles Ptolemy's. As discussed in the previous chapter, Ptolemy associates physics primarily with the study of the sublunary realm. Alcinous, on the other hand, describes astronomical and cosmological phenomena as the subject matter of physics. The movements of celestial bodies are the objects of astronomy, and the physical constitution of the cosmos, although it may include the sublunary realm, emphasizes the composition of the superlunary region. John Dillon has observed that Alcinous acquired the phrases "the motion of the stars" and "the constitution of the cosmos" from Plato's *Republic* VII, 530a, and *Timaeus* 32c, respectively.[9] No doubt, Alcinous's inclusion of cosmological concepts in his definition of physics derives from a Platonic

---

7. Alcinous, *Didaskalikos* 2.1, translation modified from Dillon, in Alcinous, *Handbook of Platonism*: Διττοῦ δ' ὄντος τοῦ βίου, τοῦ μὲν θεωρητικοῦ τοῦ δὲ πρακτικοῦ, τοῦ μὲν θεωρητικοῦ τὸ κεφάλαιον ἐν τῇ γνώσει τῆς ἀληθείας κεῖται, τοῦ πρακτικοῦ δὲ ἐν τῷ πρᾶξαι τὰ ὑπαγορευόμενα ἐκ τοῦ λόγου. Τίμιος μὲν δὴ ὁ θεωρητικὸς βίος, ἑπόμενος δὲ καὶ ἀναγκαῖος ὁ πρακτικός

8. Ibid. 3.4: Τοῦ δὲ θεωρητικοῦ τὸ μὲν περὶ τὰ ἀκίνητα καὶ τὰ πρῶτα αἴτια καὶ ὅσα θεῖα θεολογικὸν καλεῖται, τὸ δὲ περὶ τὴν τῶν ἄστρων φορὰν καὶ τὰς τούτων περιόδους καὶ ἀποκαταστάσεις καὶ τοῦδε τοῦ κόσμου τὴν σύστασιν φυσικόν, τὸ δὲ θεωρούμενον διὰ γεωμετρίας καὶ τῶν λοιπῶν μαθημάτων μαθηματικόν.

9. Dillon, "Commentary," 60.

tradition of amalgamating the two. More striking, however, is Alcinous's portrayal of the revolutions and periodic returns of celestial bodies as the subject matter of physics. One would expect that they would fall under the category of mathematics, as they concern the mathematically measured movements of heavenly bodies, or astronomical phenomena. Yet, Alcinous offers a very limited account of mathematics in this chapter of the *Didaskalikos*. Although he acknowledges the existence of many types of mathematics, he names only geometry.

In *Didaskalikos* 7.1, Alcinous again defines the three theoretical sciences and offers an expanded account of physics and mathematics:

> next let us discuss the theoretical. We have said earlier that the divisions of this are the theological, physical, and mathematical. The aim of the theological is knowledge of the primary, highest, and originative causes; the [aim] of the physical is to learn what is the nature of the universe, what sort of an animal is man, and what place he has in the cosmos, if God exercises providence over the universe, and if other gods are ranked beneath him, and what is the relation of men to the gods; the [aim] of the mathematical is to examine the nature of two- and three-dimensional being and the phenomena of change and motion.[10]

This account of theology is nearly identical to the one in chapter 3; theology studies primary causes. Physics again concerns cosmological phenomena, such as the nature of the universe, but Alcinous here omits astronomical phenomena. Instead, he frames theological problems, such as the activity and relationships among the gods, as falling within the scope of physics. Mathematics again includes geometry, but, in addition to plane figures, it concerns solids, motion from place to place, and change in general.

Considering that Alcinous includes motion from place to place as an object of mathematics in chapter 7, it is peculiar that in chapter 3 he characterizes the revolutions and periodicities of celestial bodies as the subject matter of physics. One possible explanation for this discrepancy is Alcinous's emphasis in chapter 7. John Dillon has noted the influence of Plato's *Republic* VII and

---

10. Alcinous, *Didaskalikos* 7.1, translation modified from Dillon, in Alcinous, *Handbook of Platonism*: ἑξῆς δὲ περὶ τοῦ θεωρητικοῦ λέγωμεν. Τούτου τοίνυν τὸ μὲν εἴπομεν εἶναι θεολογικόν, τὸ δὲ φυσικόν, τὸ δὲ μαθηματικόν· καὶ ὅτι τοῦ μὲν θεολογικοῦ τέλος ἡ περὶ τὰ πρῶτα αἴτια καὶ ἀνωτάτω τε καὶ ἀρχικὰ γνῶσις, τοῦ δὲ φυσικοῦ τὸ μαθεῖν, τίς ποτ' ἐστὶν ἡ τοῦ παντὸς φύσις καὶ οἷόν τι ζῷον ὁ ἄνθρωπος καὶ τίνα χώραν ἐν κόσμῳ ἔχων, καὶ εἰ θεὸς προνοεῖ τῶν ὅλων καὶ εἰ ἄλλοι θεοὶ τεταγμένοι ὑπὸ τούτῳ, καὶ τίς ἡ τῶν ἀνθρώπων πρὸς τοὺς θεοὺς σχέσις· τοῦ δὲ μαθηματικοῦ τὸ ἐπεσκέφθαι τὴν ἐπίπεδόν τε καὶ τριχῇ διεστηκυῖαν φύσιν, περί τε κινήσεως καὶ φορᾶς ὅπως ἔχει.

the education of the philosopher-king on this chapter, and he has gone so far as to contend that Alcinous's text betrays no independent interest in mathematics.[11] This lack of interest would account for the inconsistency in Alcinous's definitions of astronomy, such that the definitions would be dependent on the context. In *Didaskalikos* 7.3, Alcinous delineates the role of mathematics in the education of the philosopher-king and, in this context, astronomy—alongside arithmetic, geometry, stereometry, and harmonics—is a branch of mathematics. Alcinous defines astronomy "by means of which we will study in the heavens the motions of the stars and the heavens, and the creator of night and day, the months and the years . . ."[12] Although the periodic revolutions of the stars fall within the domain of physics in chapter 3, here they are the subject matter of a branch of mathematics. One is left with the distinct impression that Alcinous did not maintain consistent definitions of the three theoretical sciences, physics and mathematics especially.

More significant to our current purpose is Alcinous's epistemology and his concern with the distinction between opinion and knowledge. According to *Didaskalikos* 4.3, the intellect's contemplation of the objects of intellection is knowledge (ἐπιστήμη) and scientific reason (ἐπιστημονικὸς λόγος). Because this type of intellection concerns principles that are sure and stable, it too is sure and stable. Opinion (δόξα), on the other hand, derives from sense perception and, because it concerns unstable objects, it is only likely (εἰκός). Alcinous explains this dichotomy between knowledge and opinion accordingly:

> This latter [form of reason], too, has two forms: one concerned with the objects of intellection and the other concerned with the objects of sense perception. Of these, that concerning the objects of intellection is science and scientific reason, while that concerning the objects of sense perception is opinion and reason based on opinion. For this reason, scientific [reason] possesses stability and permanence, inasmuch as it concerns principles that are stable and permanent, while the [reason] based on persuasion and opinion [possesses] a high degree of [mere] likelihood, by reason of the fact that it is not concerned with permanent objects.[13]

11. Dillon, "Commentary," 86.

12. Alcinous, *Didaskalikos* 7.3, translation modified from Dillon, in Alcinous, *Handbook of Platonism*: καθ' ἣν ἐν τῷ οὐρανῷ θεασόμεθα ἄστρων τε φορὰς καὶ οὐρανοῦ καὶ τὸν δημιουργὸν νυκτὸς καὶ ἡμέρας, μηνῶν τε καὶ ἐνιαυτῶν

13. Ibid. 4.3: Διττὸς δὲ καὶ οὗτος, ὁ μὲν περὶ τὰ νοητά, ὁ δὲ περὶ τὰ αἰσθητά· ὧν ὁ μὲν περὶ τὰ νοητὰ ἐπιστήμη τέ ἐστι καὶ ἐπιστημονικὸς λόγος, ὁ δὲ περὶ τὰ αἰσθητὰ δοξαστικός τε καὶ δόξα. Ὅθεν ὁ μὲν ἐπιστημονικὸς τὸ βέβαιον ἔχει καὶ μόνιμον, ἅτε περὶ τῶν βεβαίων καὶ μονίμων ὑπάρχων. ὁ δὲ πιθανὸς καὶ δοξαστικὸς πολὺ τὸ εἰκὸς διὰ τὸ μὴ περὶ τὰ μόνιμα εἶναι.

The ontological permanence of an object, whether it is stable or unstable, determines the epistemic security of human beings' reasoning concerning it. This one-to-one correspondence between the properties of an object and the type of cognition human beings have of it, of course, derives from Plato's characterization of opinion and knowledge. In the divided line of *Republic* VI, 509d–511e, for instance, the clarity or obscurity of objects determines the type of apprehension a human being has of them: understanding, thought, belief, or conjecture.

Alcinous distinguishes two types of intellection. The first, he claims in *Didaskalikos* 4.6, exists before the transmigration of the soul into the body, and the second arises after the soul's embodiment. Each type of intellection contemplates a different set of intelligible objects. The former understands the Forms, as Plato conceived them; the second comprehends enmattered forms, an Aristotelian concept. Alcinous explains in *Didaskalikos* 4.7, "and since of intelligible objects some are primary, such as the [transcendent] ideas, and others secondary, such as the forms in matter, which are inseparable from matter, so also intellection will be twofold, the one kind of primary objects, the other of secondary."[14] Intellect studies the Forms through direct thought or intuition; enmattered forms are apprehended through scientific reasoning and remembrance of the intellection of the Forms before embodiment.

Alcinous maintains an absolute distinction between opinion and knowledge. Apprehension of the intelligible world is knowledge obtained through recollection; cognition of the sensible world is opinion. He explains this distinction, "As the intelligible cosmos is the primary object of intellection, and the sensible [cosmos] is a composite, the intelligible cosmos is judged by intellection along with reason, that is to say, not without the aid of reason, and the sensible [cosmos is judged] by opinion-based reason not without the aid of sense perception."[15] Although Alcinous incorporates Aristotelian and Stoic elements into his epistemology—such as enmattered forms and natural concepts, respectively—his overall understanding of opinion, knowledge, and, especially, the latter's acquisition through recollection of the Forms is thoroughly Platonic.

---

14. Ibid. 4.7: καὶ ἐπεὶ τῶν νοητῶν τὰ μὲν πρῶτα ὑπάρχει, ὡς αἱ ἰδέαι, τὰ δὲ δεύτερα, ὡς τὰ εἴδη τὰ ἐπὶ τῇ ὕλῃ ἀχώριστα ὄντα τῆς ὕλης, καὶ νόησις ἔσται διττή, ἡ μὲν τῶν πρώτων, ἡ δὲ τῶν δευτέρων.

15. Ibid. 4.8: Τοῦ νοητοῦ δὲ κόσμου πρώτου ὄντος νοητοῦ, τοῦ δ' αἰσθητοῦ ἀθροίσματος, τὸν μὲν νοητὸν κόσμον κρίνει νόησις μετὰ λόγου, τουτέστιν οὐκ ἄνευ λόγου, τὸν δὲ αἰσθητὸν ὁ δοξαστικὸς λόγος οὐκ ἄνευ αἰσθήσεως.

Ptolemy on Opinion and Knowledge

Several of Ptolemy's texts testify to his appropriation of the Platonic distinction between opinion and knowledge, and I argue that Ptolemy's fusion of this Platonic distinction with an Aristotelian form of empiricism underlies his claim in *Almagest* 1.1 that physics and theology are conjectural and mathematics alone yields knowledge. First of all, Ptolemy explicitly defines opinion and knowledge when advancing an Aristotelian epistemology in *On the Kritêrion and Hêgemonikon*. Similar to how in Aristotle's *De anima* thought requires images, and learning and understanding are not possible in the absence of perception, for Ptolemy the intellect depends on the transmission of sense impressions by means of the faculty of *phantasia*.[16] While for Alcinous one gains knowledge through recollection of the Forms, for Ptolemy human beings develop knowledge through memory of sense impressions and concepts developed in relation to them. The intellect makes inferences about perceived objects, and these inferences are either opinion (δόξα) or knowledge (ἐπιστήμη). Ptolemy indicates that internal *logos*, or reason, takes two forms: "its simple and unanalyzed apprehension is opinion and supposition, and its skillful and firmly grounded [apprehension] is knowledge and understanding."[17] Opinion and knowledge are of the same objects, but their inferences are at different stages of development. Opinion is an inference that is immediate, unanalyzed, and closely tied to initial sense impressions; knowledge judges objects in relation to remembered sense impressions. Moreover, this latter judgment, this re-examination of external objects, is skillful. Ptolemy elaborates on the distinction between opinion and knowledge in the following:

> When the internal *logos* of thought combines with these simple and non-inferential *kriteria*, even [*logos*] can still only form opinions if it concentrates exclusively on its immediate object, but when it makes clear and skillful distinctions it at once enters the state of knowledge; this involves separating and combining the differences and non-differences between actual things, and moving up from particulars to universals and on to the genera and species of the objects before it[18]

16. See Aristotle, *De anima* III, 431a14–17 and 432a3–10.

17. Ptolemy, *On the Kritêrion*, La6: τούτου δ᾽ ἡ μὲν ἁπλῆ καὶ ἀδιάρθρωτος ἐπιβολὴ γίνεται δόξα καὶ οἴησις, ἡ δὲ τεχνικὴ καὶ ἀμετάπιστος ἐπιστήμη καὶ γνῶσις.

18. Ibid., La18, translation modified from Liverpool-Manchester Seminar, in Ptolemaeus, "On the Kriterion": τούτοις δὲ τοῖς ἁπλοῖς καὶ ἀσυλλογίστοις κριτηρίοις ἐπισυναφθεὶς ὁ τῆς διανοίας ἐνδιάθετος λόγος κατὰ μὲν τὴν ἀπολελυμένην ἐπιβολὴν καὶ αὐτὸς ἔτι δοξάζει μόνον, κατὰ δὲ τὴν ἐναργῆ καὶ τεχνικὴν διάκρισιν ἤδη τὴν ἐπιστημονικὴν ἕξιν ἀπολαμβάνει χωρίζων τε καὶ

The intellect can move from opinion to knowledge by means of clear and skill-ful analysis. By comparing the sense impressions of an object to remembered sense impressions, an individual re-examines the object and is able to classify it accurately.

The idea that opinion and knowledge are of the same objects is not unique to Ptolemy. Even Plato suggests as much. In the *Meno*, for instance, Socrates interrogates a slave boy about a mathematical problem, and Socrates's ques-tioning leads the slave boy to reject false opinions and assume true beliefs. Socrates indicates that further questioning would transform these true beliefs into knowledge: "These opinions have now just been stirred up like a dream, but if he were repeatedly asked these same questions in various ways, you know that in the end his knowledge about these things would be as accurate as anyone's."[19] For Plato, one transforms opinion into knowledge through rec-ollection of the Forms from the time before embodiment. Ptolemy, on the other hand, does not portray a belief in the pre-existence of the soul. According to *On the Kritêrion*, knowledge is merely the skillful examination of perceived objects in relation to remembered sense impressions. It is because opinion and knowledge both rely on sense perception that they can be different epistemic states related to the same object, and an individual can utilize skilled reason-ing methods to move from opinion to knowledge.

Ptolemy's use of skillful (τεχνική) judgment, in particular, to distinguish knowledge from opinion may derive from Aristotle's treatment of the rela-tionship between skill and knowledge. In *Metaphysics* A1, Aristotle claims that experience, derived from sense impressions, leads to both craft, or, as I will translate it, skill (τέχνη) and knowledge (ἐπιστήμη), where skill, like knowl-edge, results from judgments about universals rather than particulars. In the following passage, Aristotle explains the relationship between skill and uni-versals by contrasting it with experience: "In reference to the practical, expe-rience seems in no way inferior to skill, and we even see men of experience succeeding more than those who have theory without experience, the cause being that experience is understanding of particulars but skill is [knowledge] of universals, and actions and productions are all concerned with the par-ticular . . ."[20] Skill, like knowledge, is the apprehension of universals, and the

---

συνάγων τάς τε διαφορὰς καὶ τὰς ἀδιαφορίας τῶν ὄντων καὶ ἀνάγων ἀπὸ τῶν κατὰ μέρος ἐπὶ τὰ καθόλου καὶ ἀνωτάτω τά τε γένη καὶ τὰ εἴδη τῶν ὑποκειμένων

19. Plato, *Meno*, 85c–d, translated by Grube: Καὶ νῦν μέν γε αὐτῷ ὥσπερ ὄναρ ἄρτι ἀνακεκίνηνται αἱ δόξαι αὗται· εἰ δὲ αὐτόν τις ἀνερήσεται πολλάκις τὰ αὐτὰ ταῦτα καὶ πολλαχῇ, οἶσθ' ὅτι τελευτῶν οὐδενὸς ἧττον ἀκριβῶς ἐπιστήσεται περὶ τούτων.

20. Aristotle, *Metaphysics* I, 981a12–17: πρὸς μὲν οὖν τὸ πράττειν ἐμπειρία τέχνης οὐδὲν δοκεῖ διαφέρειν, ἀλλὰ καὶ μᾶλλον ἐπιτυγχάνουσιν οἱ ἔμπειροι τῶν ἄνευ τῆς ἐμπειρίας λόγον ἐχόντων

relationships Aristotle constructs among skill, knowledge, and universals seem to have influenced Ptolemy's interpretation of the distinction between opinion and knowledge. According to Ptolemy, the move from opinion to knowledge, from judgments of particulars to universals, is skillful.

In the *Harmonics*, Ptolemy clarifies the relationships of opinion and knowledge with conjecture and skill, respectively. In particular, he describes skill as a cause corresponding to reason: "Of the cause that is in accordance with reason, one aspect is intellect, corresponding to the diviner form, one is skill, corresponding to reason itself, and one is habit, corresponding to nature . . ."[21] Therefore, in the *Harmonics* Ptolemy associates skill with reason, and in *On the Kritêrion*, as explained above, the application of skill—or, as one may interpret it, reason—to sense impressions yields knowledge. Moreover, Ptolemy identifies opinion with conjecture (εἰκασία) in *Harmonics* 3.5, where he divides the intellectual part of the soul into seven species. Two of the seven species are opinion, which he defines as concerned with superficial conjecture (ἐξεπιπολῆς εἰκασίαν), and knowledge, which concerns truth and apprehension (παρὰ τὴν ἀλήθειαν καὶ τὴν κατάληψιν).[22] The association of opinion with conjecture has a precedent in Socrates's discussion of the divided line in *Republic* VII, where conjecture falls within the domain of opinion.[23] If opinion correlates with conjecture, then when Ptolemy calls theology and physics conjectural in *Almagest* 1.1, he associates them with opinion, which, according to *On the Kritêrion*, is a simple and unanalyzed apprehension.

Ptolemy employs two terms in *Almagest* 1.1 to denote knowledge: *katalêpsis epistêmonikê* and *eidêsis*. For Ptolemy, the term *katalêpsis* has lost its Stoic connotation of a firm grasp of some impression and instead denotes any type of apprehension.[24] The adjective *epistêmonikê* indicates that the type of apprehension in this case is one of knowledge. As for *eidêsis*, considering that the

---

(αἴτιον δ' ὅτι ἡ μὲν ἐμπειρία τῶν καθ' ἕκαστόν ἐστι γνῶσις ἡ δὲ τέχνη τῶν καθόλου, αἱ δὲ πράξεις καὶ αἱ γενέσεις πᾶσαι περὶ τὸ καθ' ἕκαστόν εἰσιν

21. Ptolemy, *Harmonics* 3.3, D92: ἐπεὶ δὲ τοῦ κατὰ τὸν λόγον αἰτίου τὸ μέν ἐστιν ὡς νοῦς καὶ παρὰ τὸ θειότερον εἶδος, τὸ δὲ ὡς τέχνη καὶ παρ' αὐτὸν τὸν λόγον, τὸ δὲ ὡς ἔθος καὶ παρὰ τὴν φύσιν

My translations of Ptolemy's *Harmonics* are modifications of Andrew Barker's translation in *Greek Musical Writings*, 270–391.

22. Ptolemy, *Harmonics* 3.5, D96.

23. Plato, *Republic* VII, 534a.

24. See, for instance, *Tetrabiblos* 1.1.2, H4, where Ptolemy contrasts two types of *katalêpsis*, one that is sure and one that is merely possible, which distinguish what we today call astronomy from what we today call astrology. For more on this distinction between astronomy and astrology, see chapter 8.

term derives from the verb *oida*, literally meaning "to have seen," it is fitting that the term Ptolemy uses to denote knowledge reflects the process by which the intellect obtains it: the clear and skillful examination of sense impressions. Given this distinction between knowledge and conjecture, how is it, then, that Ptolemy argues that theology and physics are conjectural and that mathematics alone yields knowledge?

I argue that, for Ptolemy, whether a human being can generate knowledge or only conjecture when studying a science is entirely contingent on the qualities of the objects contemplated. These qualities include two epistemic properties, perceptibility and clarity, as well as one ontological property, stability. The necessity of the objects' perceptibility for knowledge follows from Ptolemy's appropriation of Aristotelian epistemology; the requirements of clarity and stability are consistent with Platonic epistemology. Hence, Ptolemy's criteria derive from his own, eclectic blend of Aristotelian and Platonic epistemologies.

With regard to the first criterion, perceptibility, the object of theology—the Prime Mover—is imperceptible. The intellect has no sense impressions of the Prime Mover to examine, and therefore theology cannot give rise to knowledge. The object of theology is entirely ungraspable (ἀνεπίληπτον)—one can only guess at the Prime Mover's nature—and, consequently, theology is conjectural. Physical and mathematical objects, on the other hand, are perceptible. I argued in the previous chapter that Ptolemy's examples of physical objects are special sensibles, perceptible by only one sense organ, and his examples of mathematical objects are common sensibles, perceptible by more than one sense. Given the perceptibility of physical and mathematical objects, whether they are stable and clear determines whether their respective studies generate knowledge or conjecture. For Ptolemy, stability and clarity go hand in hand. Stable objects are clear objects or, perhaps even more, the stability of an object entails its clear appearance to human sense perception.

In *Almagest* 1.1, Ptolemy contends that physical objects do not meet these criteria of stability and clarity. They are unstable and unclear, and their instability and lack of clarity prevent the intellect from making clear and stable judgments of their sense impressions. It seems that Ptolemy calls physical objects unstable and unclear, because, at least in *Almagest* 1.1, he associates them with sublunary phenomena and, especially, the special sensibles that enform sublunary bodies. It is because sublunary bodies experience manifold changes, including corruption, that human beings' sense impressions of physical objects are unstable and unclear and, thus, preclude the intellect's clear and skillful analysis of them. If the intellect cannot clearly and skillfully examine sense impressions, then it is limited to an unanalyzed notion of a perceived object, and its judgment remains conjecture rather than knowledge. This implicit argument is surprising, for in *On the Kritêrion* Ptolemy, like Aristotle before him, maintains that special sensibles do not admit of error but naturally tell the truth.[25] It seems, then, that in the *Almagest* Ptolemy is overturning the traditional association of special sensibles with truth and instead is casting them as fodder for conjecture. In *Almagest* 1.1, the changeability of sublunary bodies entails that the special sensibles enforming them are unstable, one's impressions of them unclear, and their study conjectural rather than productive of knowledge.

Mathematical objects, on the other hand, are stable and clear. Because the properties of objects determine the type of apprehension one has of them, the study of these stable and clear objects is itself stable and clear. Ptolemy clarifies this epistemological position when he explains his preference for mathematics, and astronomy—the branch of mathematics that studies divine, heavenly objects—in particular:

> we were drawn to cultivate this sort of theory [viz. mathematics], the whole of it as far as we were able, but especially the study of divine and heavenly things, for that alone is engaged in the investigation of the eternal and unchanging; for that reason it too can be eternal and unchanging, which is a proper attribute of knowledge, in its own apprehension, which is neither unclear nor disorderly[26]

Again, the qualities of an object determine the qualities of the field of inquiry examining it. The study of objects that are eternal and unchanging is itself eternal and unchanging, and the type of apprehension that is eternal and unchanging is knowledge. Moreover, like the objects it concerns, knowledge is neither unclear nor disorderly. In this passage, Ptolemy suggests that mathematics in general is productive of knowledge, but, appropriate to the theme of the *Almagest*, he singles out astronomy, the science that examines heavenly bodies' movements and configurations. As we saw in the previous chapter, movements are one example of the subject matter of mathematics. The movements and configurations of heavenly bodies, in particular, belong to a branch of mathematics—namely, astronomy. Furthermore, in this passage astronomy exemplifies mathematics. The movements and configurations of heavenly bodies are eternal, unchanging, and, by implication, they are stable and one's

25. See Ptolemy, *On the Kritêrion*, La16, and Aristotle, *De anima* II, 418a11–12.

26. Ptolemy, *Almagest* 1.1, H6–7: προήχθημεν ἐπιμεληθῆναι μάλιστα πάσης μὲν κατὰ δύναμιν τῆς τοιαύτης θεωρίας, ἐξαιρέτως δὲ τῆς περὶ τὰ θεῖα καὶ οὐράνια κατανοουμένης, ὡς μόνης ταύτης περὶ τὴν τῶν ἀεὶ καὶ ὡσαύτως ἐχόντων ἐπίσκεψιν ἀναστρεφομένης διὰ τοῦτό τε δυνατῆς οὔσης καὶ αὐτῆς περὶ μὲν τὴν οἰκείαν κατάληψιν οὔτε ἄδηλον οὔτε ἄτακτον οὖσαν ἀεὶ καὶ ὡσαύτως ἔχειν, ὅπερ ἐστὶν ἴδιον ἐπιστήμης

sense impressions of them are clear. Therefore, if one approaches mathematics rigorously, one can skillfully examine clear sense impressions of mathematical objects, such as the movements and configurations of celestial bodies, and thereby create knowledge.

The Indisputability of Mathematical Demonstration[27]

Because physics and theology are conjectural, according to Ptolemy philosophers will never be able to agree on the nature of physical and theological objects. Philosophers' attempts to create knowledge in their own areas of study are necessarily futile. Mathematicians, on the other hand, can and do generate knowledge, and for two reasons. First, as we have seen, the stability and clarity of mathematical objects give rise to knowledge; mathematical objects are the only type of object that supplies clear and stable sense impressions, which permit skillful analysis. Second, the method of mathematics, demonstration by means of arithmetic and geometry, makes its study rigorous and its proofs indisputable. This declaration of arithmetical and geometrical demonstrations' indisputability situates mathematical proofs in direct opposition to philosophers' discourses. If philosophers cannot reach consensus because their studies are conjectural, then their accounts of physical and theological objects are necessarily disputable. As indisputable, mathematical demonstrations are epistemically superior to the discourses of philosophers.

The first extant articulation in the Greek corpus of the indisputability of geometrical demonstration lies in the corpus of Hero of Alexandria, a mechanician who lived in the late first century CE.[28] In the proem to the third book of the *Metrica*, Hero discusses the just apportionment of land. Seeking the most exact means of determining the areas of land, Hero explains, "if one wished to divide land according to the given ratio, so that not one grain, so to speak, exceeds or falls short of the given ratio, there would be need of only geometry, in which the fit is fair, justice is in the proportion, and the demonstration concerning these is indisputable, which none of the other crafts or

27. This section is modified from Jacqueline Feke, "Meta-mathematical Rhetoric: Hero and Ptolemy against the Philosophers," *Historia Mathematica* 41, no. 3 (2014): 261–76. Copyright © 2014 by Elsevier.

28. Concerning when he lived, Hero discusses a lunar eclipse visible in both Rome and Alexandria in *Dioptra* 35. Otto Neugebauer dates the eclipse to 62 CE and argues that Hero personally observed it: Neugebauer, "Methode zur Distanzbestimmung," 23. Nathan Sidoli argues that 62 CE may be taken at most as a *terminus post quem* for Hero's activity: Sidoli, "Heron of Alexandria's Date."

sciences professes."[29] According to Hero, geometry is the only craft or science that determines areas precisely. It is successful in this regard because its demonstration is indisputable. Although Ptolemy adds that arithmetical demonstration is indisputable, the language used by Hero and Ptolemy is otherwise identical. A geometrical demonstration (ἀπόδειξις) is indisputable (ἀναμφισβήτητος).

Mathematical demonstrations were held in high regard long before Hero. Aristotle, for instance, famously held the geometrical proof as a paradigm of demonstration in general. In *Nicomachean Ethics* I, 1094b, he observes that geometrical demonstrations are more exact than other studies: "for it is the mark of an educated man to seek as much precision in each genus just so far as the nature of the subject admits; it appears equally foolish to accept probable arguments from a mathematician and to demand demonstrations from a rhetorician."[30] For Aristotle, what makes mathematical demonstrations more precise than other discourses is not their mathematical content but their adherence to the principles of demonstration. The deduction from true and primitive premises, mathematical or not, establishes a demonstration's epistemic security. Aristotle defines "demonstration" in the *Posterior Analytics*:

> By demonstration I mean a scientific deduction, and by scientific I mean one in accordance with which, by having it, we know something. If then knowing is as we posited, it is necessary that demonstrative knowledge spring from things that are true and primitive and immediate and more familiar than and prior to and explanatory of the conclusion, for in this way the principles also will be appropriate to what is being proven. For there will be a deduction even without these things, but there will not be a demonstration, for it will not produce knowledge.[31]

---

29. Hero, *Metrica* 141–42: εἰ δέ τις βούλοιτο κατὰ τὸν δοθέντα λόγον διαιρεῖν τὰ χωρία, ὥστε μηδὲ ὡς εἰπεῖν κέγχρον μίαν τῆς ἀναλογίας ὑπερβάλλειν ἢ ἐλλείπειν τοῦ δοθέντος λόγου, μόνης προσδεήσεται γεωμετρίας· ἐν ᾗ ἐφαρμογὴ μὲν ἴση, τῇ δὲ ἀναλογίᾳ δικαιοσύνη, ἡ δὲ περὶ τούτων ἀπόδειξις ἀναμφισβήτητος, ὅπερ τῶν ἄλλων τεχνῶν ἢ ἐπιστημῶν οὐδεμία ὑπισχνεῖται.

30. Aristotle, *Nicomachean Ethics* I, 1094b23–27: πεπαιδευμένου γάρ ἐστιν ἐπὶ τοσοῦτον τἀκριβὲς ἐπιζητεῖν καθ' ἕκαστον γένος, ἐφ' ὅσον ἡ τοῦ πράγματος φύσις ἐπιδέχεται· παραπλήσιον γὰρ φαίνεται μαθηματικοῦ τε πιθανολογοῦντος ἀποδέχεσθαι καὶ ῥητορικὸν ἀποδείξεις ἀπαιτεῖν.

31. Aristotle, *Posterior Analytics* I, 71b17–25: ἀπόδειξιν δὲ λέγω συλλογισμὸν ἐπιστημονικόν· ἐπιστημονικὸν δὲ λέγω καθ' ὃν τῷ ἔχειν αὐτὸν ἐπιστάμεθα. εἰ τοίνυν ἐστὶ τὸ ἐπίστασθαι οἷον ἔθεμεν, ἀνάγκη καὶ τὴν ἀποδεικτικὴν ἐπιστήμην ἐξ ἀληθῶν τ' εἶναι καὶ πρώτων καὶ ἀμέσων καὶ γνωριμωτέρων καὶ προτέρων καὶ αἰτίων τοῦ συμπεράσματος· οὕτω γὰρ ἔσονται καὶ αἱ ἀρχαὶ οἰκεῖαι τοῦ δεικνυμένου. συλλογισμὸς μὲν γὰρ ἔσται καὶ ἄνευ τούτων, ἀπόδειξις δ' οὐκ ἔσται· οὐ γὰρ ποιήσει ἐπιστήμην.

According to Aristotle, a demonstration produces knowledge because its premises, or principles, are true, primitive, immediate, and explanatory of the conclusion.[32]

Aristotle never calls a demonstration or its principles indisputable (ἀναμφισβήτητος), but philosophers in the second century CE did employ this terminology. In his commentary on Aristotle's *Metaphysics*, for instance, Alexander of Aphrodisias paraphrases Aristotle's criteria for the principles of demonstrations: "he means that all who demonstrate reduce their arguments to this ultimate principle, as immediately familiar and clear and indisputable…"[33] Sextus Empiricus rejects the indisputability of premises in *Adversus mathematicos* 8.353:

> Moreover, the premises of the demonstration which the Laconian mentions are either disputed and untrustworthy or indisputable and trustworthy. But if they are disputed and untrustworthy, the demonstration out of these [premises] evidently will be untrustworthy for the constructive reasoning of anything. And that they are trustworthy and indisputable is an object of prayer rather than truth.[34]

The mandate for indisputable premises has a precedent as early as Diogenes of Apollonia, the late fifth-century BCE philosopher known for revising Milesian cosmological theory. Diogenes Laertius, in the early third century CE, quotes Diogenes of Apollonia: "It seems to me that one beginning any account ought to present a principle that is indisputable and an explanation that is simple and august."[35] Characterizations of discourses by a host of terms implying indisputability—such as unmoved by persuasion (ἀμετάπειστος), unchangeable (ἀμετάπτωτος), and infallible (ἀμετάπταιστος)[36]—abound in ancient Greek philosophical texts, but Hero and Ptolemy make a more specific claim. For

---

32. On Aristotle's theory of demonstration, see especially Lloyd, "Theories and Practices of Demonstration."

33. Alexander, *In Aristotelis metaphysica commentaria* 271.7–9, translation modified from Madigan, in Alexander of Aphrodisias, *On Aristotle Metaphysics 4*: λέγων πάντας τοὺς ἀποδεικνύντας ἐπὶ ἐσχάτην ταύτην ἀρχὴν ὡς αὐτόθεν γνώριμον καὶ ἐναργῆ καὶ ἀναμφισβήτητον τὸν λόγον ἀνάγειν

34. Sextus Empiricus, *Adversus mathematicos* 8.353: καὶ μὴν τὰ λήμματα ἧς λέγει ἀποδείξεως ὁ Λάκων ἤτοι ἀμφισβητεῖται καὶ ἄπιστά ἐστιν ἢ ἀναμφισβήτητά ἐστι καὶ πιστά. ἀλλ᾽ εἰ μὲν ἀμφισβητεῖται καὶ ἄπιστά ἐστιν, πάντως καὶ ἡ ἐξ αὐτῶν ἀπόδειξις ἄπιστος γενήσεται πρὸς τήν τινος κατασκευήν. τὸ δὲ πιστὰ αὐτὰ εἶναι καὶ ἀναμφισβήτητα εὐχὴ μᾶλλόν ἐστιν ἢ ἀλήθεια. Cf. 1.157.

35. Diogenes Laertius, *Lives* 9.57: "λόγου παντὸς ἀρχόμενον δοκεῖ μοι χρεὼν εἶναι τὴν ἀρχὴν ἀναμφισβήτητον παρέχεσθαι, τὴν δ᾽ ἑρμηνείαν ἁπλῆν καὶ σεμνήν."

36. G.E.R. Lloyd lists several examples, with citations, in "Pluralism of Greek 'Mathematics,'" 306.

them, demonstrations in their entirety, premises and conclusions, have the potential to be indisputable, and the type of demonstration that is indisputable is arithmetical and/or geometrical.

In the second century CE, Nicomachus, the Neopythagorean philosopher, and Galen associated mathematical discourses, arithmetical or geometrical, with indisputability. In *Introduction to Arithmetic* 1.23.4, Nicomachus describes an arithmetical method (ἔφοδος) that clearly and indisputably (σαφέστατα καὶ ἀναμφιλέκτως) sets before the mind the distinctions between the limited and unlimited, the comprehensible and incomprehensible, the fine and the ugly. Discussing the principles of demonstrations in *De animi cuiuslibet peccatorum dignotione et curatione*, Galen envisions how a master builder would address the question of whether a body moves or remains still in a void: "For I know that he uses in his demonstrations principles of this sort, clear and indisputably agreed on by all."[37] Like Hero and Ptolemy, Galen laments the disagreements of philosophers, and he holds mathematicians as a paradigm of consensus.[38] In his commentary on Hippocrates's *On Diet in Acute Diseases*, he observes that individuals who create geometrical demonstrations "not only persuade those individuals who are learning, but also in their personal opinion they hold them as most true. Therefore, they say that they have used clear and indisputable proofs in geometrical demonstrations."[39] Hence, Galen acknowledges his familiarity with the claim of mathematicians, like Hero and Ptolemy, that geometrical demonstrations are indisputable.

Searching the *Thesaurus Linguae Graecae*,[40] I have found only ten more cases in the Greek corpus that make some reference to the indisputability of geometrical and/or arithmetical demonstration. In the third century CE, the Christian bishop Gregory of Nyssa declares geometry in general indisputable (ἀναμφισβήτητον);[41] in the fourth century CE, the mathematician Theon of Alexandria calls arithmetical and geometrical demonstrations indisputable (ἀναμφισβητήτων) when commenting on *Almagest* 1.1;[42] active in the late

---

37. Galen, *De animi cuiuslibet peccatorum dignotione et curatione* 5.98.14–99.1: τοιαύταις γὰρ ἀρχαῖς εἰς τὰς ἀποδείξεις οἶδα χρώμενον αὐτὸν ἐναργέσι τε καὶ ἀναντιλέκτως ὑπὸ πάντων ὁμολογουμέναις.

38. See Galen, *De placitis Hippocratis et Platonis* 8.1.16–20; *De libris propriis liber* 11 (*Scripta Minora* 2.115.21–117.20); *De propriorum animi cuiuslibet affectuum dignotione et curatione* 5.42, 93.

39. Galen, *In Hippocratis de victu acutorum commentaria iv* 15.440.4–7: οὐ μόνον γὰρ αὐτοὺς τοὺς μανθάνοντας πείθουσιν, ἀλλὰ καὶ παρὰ τοῖς ἰδιώταις δόξαν ἔχουσιν ὡς ἀληθέσταται. γραμμικαῖς οὖν ἀποδείξεσι κεχρῆσθαί φασι τοὺς ἐναργῶς τι καὶ ἀναμφιλέκτως δείξαντας.

40. *Thesaurus Linguae Graecae: A Digital Library of Greek Literature*, www.tlg.uci.edu.

41. Gregorius Thaumaturgus, *In Origenem oratio panegyrica* 8.19–20.

42. Theon, *Commentaria in Ptolemaei syntaxin mathematicam* 323.17–18.

fourth and early fifth centuries CE, Synesius—who studied under Hypatia, the daughter of Theon of Alexandria—explains that astronomy proceeds to its demonstrations not disputably (οὐκ ἀμφισβητησίμως), for it uses geometry and arithmetic;[43] in the fifth century CE, Proclus claims that Euclid improved on the work of his predecessors, such as Eudoxus and Theaetetus, by reforming their propositions into irrefutable demonstrations (ἀνελέγκτους ἀποδείξεις);[44] in the sixth century CE, Simplicius describes geometrical demonstration as proceeding indisputably (ἀναμφιλέκτως) in his commentary on Aristotle's *De caelo*;[45] the late antique anonymous author of the *Prolegomena to the Almagest* maintains that Ptolemy's goal in the text is to make the hypothesis of the heavens' uniform circular motion consistent with the phenomena and to accomplish this goal by means of geometrical and irrefutable (ἀναντιρρήτοις) demonstrations.[46] In the Byzantine tradition, several scholars call geometrical demonstrations irrefutable (ἀναντίρρητοι): Michael Psellus, the eleventh-century Constantinopolitan statesman, rhetor, and teacher; pseudo-Alexander of Aphrodisias, possibly Michael of Ephesus, the Aristotelian commentator who flourished under the patronage of the Byzantine princess Anna Comnena in the early twelfth century; Theodorus Metochites, the late thirteenth- to early fourteenth-century intellectual who is known for his paraphrases of Aristotle's writings and who composed *Stoicheiôsis astronomikê*, an introduction to Ptolemaic astronomy; and Nicephorus Gregoras, Metochites's student.[47] What, then, distinguishes the instantiation of this claim in Hero's and Ptolemy's texts? Again, the terminology of Hero and Ptolemy is identical, and they both make the claim while casting mathematics as superior to competing discourses. For Hero, the statesman—the distributor of goods, honors, and portions of land—should turn to geometry. For Ptolemy, physics and theology are conjectural, but mathematics furnishes knowledge because its objects lend themselves to a study that is eternal, unchanging, clear, orderly, and—when practiced rigorously, by means of arithmetical or geometrical demonstration—indisputable.

---

43. Synesius, *Ad Paeonium de dono astrolabii* 4.10–12.

44. Proclus, *In primum Euclidis elementorum librum commentarii* 68.9–10. Cf. 3.9, 11.21–22, 70.17.

45. Simplicius, *In de caelo* 3.1.562.31–32.

46. Acerbi, Vinel, and Vitrac, "Prolégomènes à l'Almageste," 76.

47. Michael Psellus, *De omnifaria doctrina* 127.12–13; pseudo-Alexander of Aphrodisias, *In Aristotelis metaphysica commentaria* 441.9–10; Theodorus Metochites, *Stoicheiôsis astronomikê* 1.3.4.42–43; Nicephorus Gregoras, *Antirrhetica priora* 2.6,337.16–17. Cf. Nicephorus Gregoras, *Astrolabica B* 226.17–18.

Mathematics' Contribution to Physics and Theology

For Ptolemy, the study of physical and theological objects is not entirely fruitless. Although physics and theology are conjectural when independently studied, the contribution of mathematics to their study is epistemically efficacious. One cannot have knowledge of physical and theological objects, but at least one can guess well, and mathematics ensures that this guess is a good one. Ptolemy explains as follows:

> [mathematics] can work in the domains of the other [two parts of the theoretical part of philosophy] no less than they. For this is the best [part of philosophy] to pave the way for theology, as it is the only one able to guess well at [the nature] of the immovable and separated activity, from its familiarity with the attributes of substances that are, on the one hand, perceptible, both moving and being moved, and, on the other hand, eternal and unchanging, [I mean the attributes] having to do with motions and the arrangements of movements; [mathematics] contributes to physics not accidentally, for nearly every peculiar attribute of material substance is made apparent from the peculiar qualities of its motion from place to place[48]

Mathematics lends its epistemic capability to theology and physics to a limited degree. By means of the application of mathematics, one can make a good guess at the nature of theological and physical objects.

Concerning theology, it is because mathematical objects—specifically astronomical objects, meaning the movements and configurations of heavenly bodies—have certain characteristics in common with the Prime Mover that mathematics furnishes a good guess at its nature.[49] Again, when Ptolemy declares his preference for studying mathematics and astronomy in particular, he describes the objects of astronomy as divine, eternal, and unchanging. While aethereal bodies are mostly unchanging, in that the only change they experience is motion from place to place, the subject matter of astronomy, the

---

48. Ptolemy, *Almagest* 1.1, H7: πρὸς δὲ τὰς ἄλλας οὐχ ἧττον αὐτῶν ἐκείνων συνεργεῖν. τό τε γὰρ θεολογικὸν εἶδος αὕτη μάλιστ᾽ ἂν προοδοποιήσειε μόνη γε δυναμένη καλῶς καταστοχάζεσθαι τῆς ἀκινήτου καὶ χωριστῆς ἐνεργείας ἀπὸ τῆς ἐγγύτητος τῶν περὶ τὰς αἰσθητὰς μὲν καὶ κινούσας τε καὶ κινουμένας, ἀιδίους δὲ καὶ ἀπαθεῖς οὐσίας συμβεβηκότων περί τε τὰς φορὰς καὶ τὰς τάξεις τῶν κινήσεων· πρός τε τὸ φυσικὸν οὐ τὸ τυχὸν ἂν συμβάλλοιτο· σχεδὸν γὰρ τὸ καθόλου τῆς ὑλικῆς οὐσίας ἴδιον ἀπὸ τῆς κατὰ τὴν μεταβατικὴν κίνησιν ἰδιοτροπίας καταφαίνεται

49. For a different interpretation of the reasoning underlying Ptolemy's claim that the application of mathematics provides a good guess at the nature of theological activity, see Taub, *Ptolemy's Universe*, 26–29.

movements of the heavenly bodies, is absolutely unchanging. Celestial bodies' movements give the appearance of irregularity, but in fact they are regular and eternal. Astronomical objects' divinity, eternity, and lack of change characterize the Prime Mover as well. One can make a good guess at the nature of the Prime Mover, then, by means of an analogy, from astronomical objects to the theological.

Ptolemy's argument for the contribution of astronomy to theology is Platonic in character, for in *Republic* VII Plato puts forward a similar claim. Socrates maintains that astronomy is necessary in the education of the philosopher-king because, if studied correctly, it guides one's mind toward the study of metaphysical reality. Celestial bodies are, according to Socrates, the most beautiful and exact of visible bodies. They imitate the Forms to a more perfect degree than any other component of the visible realm and, in this way, their attributes participate in the Forms to a greater degree than do other visible bodies. Socrates portrays celestial bodies as relatively perfect in the following passage:

> We should consider these decorations in the heavens to be the most beautiful and most exact of such [i.e., visible] things, seeing that they have been embroidered on a visible surface, but [we should consider] their motions to fall far short of the true ones, motions that are really fast or slow as measured in true numbers, that trace out true geometric figures, that are all in relation to one another, and that are the true motions of the things carried along in them, and these, of course, must be grasped by reason and thought, but not by sight[50]

Even though celestial bodies are not absolutely perfect, in that their characteristics, such as their movements and configurations, are not exact, compared with the other components of the visible world they are more perfect. Moreover, they can serve as a conceptual model of the Forms. Socrates proclaims, "Therefore, we should use the embroidery in the heavens as a model in the study of these other things . . ."[51] Celestial bodies are comparatively good images of the Forms and, consequently, their study is propaedeutic to the study of the Forms.

---

50. Plato, *Republic* VII, 529c–d, translation modified from Grube, rev. Reeve: ταῦτα μὲν τὰ ἐν τῷ οὐρανῷ ποικίλματα, ἐπείπερ ἐν ὁρατῷ πεποίκιλται, κάλλιστα μὲν ἡγεῖσθαι καὶ ἀκριβέστατα τῶν τοιούτων ἔχειν, τῶν δὲ ἀληθινῶν πολὺ ἐνδεῖν, ἃς τὸ ὂν τάχος καὶ ἡ οὖσα βραδυτὴς ἐν τῷ ἀληθινῷ ἀριθμῷ καὶ πᾶσι τοῖς ἀληθέσι σχήμασι φοράς τε πρὸς ἄλληλα φέρεται καὶ τὰ ἐνόντα φέρει, ἃ δὴ λόγῳ μὲν καὶ διανοίᾳ ληπτά, ὄψει δ' οὔ

51. Ibid., 529d: Οὐκοῦν, εἶπον, τῇ περὶ τὸν οὐρανὸν ποικιλίᾳ παραδείγμασι χρηστέον τῆς πρὸς ἐκεῖνα μαθήσεως ἕνεκα

Similarly, Alcinous articulates an argument for the propaedeutic value of mathematics in *Didaskalikos* 7.4. After listing the divisions of mathematics that Plato establishes in Book VII of the *Republic*—arithmetic, geometry, stereometry, astronomy, and harmonics—Alcinous addresses the merits of harmonics and astronomy:

> We will cultivate also music, referring the sense of hearing to the same objects; for even as the eyes are naturally suited to astronomy, so is the sense of hearing to harmony; and even as in applying our minds to astronomy we turn from visible objects to invisible and intelligible substance, so in listening to harmonious sound we in the same way transfer our attention from audible things to what is contemplated by the mind itself; whereas if we do not approach these studies in this way, our view of them will be imperfect and unproductive and of no account. For one must pass swiftly from visible things and audible things to those things which may be seen only by the rational activity of the soul.[52]

Alcinous discusses the ability of astronomy to direct one's mind toward contemplation of metaphysical reality, and Ptolemy, influenced by the contemporary Platonic tradition, employs a similar argument for astronomy's contribution to theology.

Still, Ptolemy's argument differs in its content. What is relevant in Ptolemy's argument is not celestial bodies per se but rather their movements, and, for Ptolemy, the characteristics that astronomical and metaphysical objects share are not differentiated by an ontological hierarchy. According to Ptolemy, astronomical objects and the Prime Mover are eternal, divine, and unchanging in exactly the same way. Ptolemy, then, adapts the Platonic argument for the propaedeutic value of astronomy to an Aristotelian ontological system. According to *Almagest* 1.1, one can make a good guess at the nature of the Prime Mover not by contemplating the Prime Mover alone but only by inferring its nature from astronomical objects. A mathematician studies the movements of heavenly bodies, these movements are eternal, divine, and

52. Alcinous, *Didaskalikos* 7.4, translation modified from Dillon, in Alcinous, *Handbook of Platonism*: Καὶ μουσικῆς δὲ ἐπιμελησόμεθα, ἐπὶ τὰ αὐτὰ τὴν ἀκοὴν ἀναφέροντες· ὡς γὰρ πρὸς ἀστρονομίαν ὄμματα συνέστηκεν, οὕτως ἀκοὴ πρὸς τὸ ἐναρμόνιον· καὶ ὥσπερ ἀστρονομίᾳ τὸν νοῦν προσέχοντες ἀπὸ τῶν ὁρωμένων ἐπὶ τὴν ἀόρατον καὶ νοητὴν οὐσίαν ποδηγούμεθα, οὕτω καὶ τῆς ἐναρμονίου φωνῆς κατακούοντες ἀπὸ τῶν ἀκουστῶν ἐπὶ τὰ αὐτῷ τῷ νῷ θεωρούμενα κατὰ ταὐτὰ μεταβαίνομεν· ὡς εἰ μὴ οὕτω μετίοιμεν ταῦτα τὰ μαθήματα, ἀτελής τε καὶ ἀνόνητος καὶ οὐδενὸς λόγου ἀξία ἡ περὶ τούτων σκέψις γένοιτ' ἂν ἡμῖν. Δεῖ γὰρ ὀξέως ἀπὸ τῶν ὁρατῶν καὶ ἀπὸ τῶν ἀκουστῶν μεταβαίνειν ἐπ' ἐκεῖνα, ἃ ἔστιν ἰδεῖν μόνῳ τῷ τῆς ψυχῆς λογισμῷ.

unchanging, and, by way of inference, one makes a good guess at the nature of the Prime Mover, that it too is eternal, divine, and unchanging.

Still, because Ptolemy calls the Prime Mover completely ungraspable, the problem remains of how good this guess is. Ptolemy seems to suggest that the guess mathematics provides is at least well reasoned. It is impossible to know whether it accurately describes the Prime Mover, but the inference from astronomical objects furnishes a reasonable method by which to construct an account of the Prime Mover's properties, and, because the method is reasonable, it is the best guess and it permits consensus. In other words, one cannot know whether the guess is accurate, but a reasonable method produces a good guess, one that allows for a consensus view. Thus, the guess furnished by way of mathematics stands as an improvement on the contributions of philosophers, who cannot agree on the nature of either physical or theological objects.

Ptolemy suggests that mathematics has even more success contributing to physics, for mathematics contributes to physics not accidentally but essentially. Ptolemy echoes this sentiment in *Geography* 1.1.9, wherein he asserts that mathematics reveals the nature of the heavens and the earth:

> These things belong to the highest and finest theory: to exhibit to human apprehension through the mathematical [sciences] the heavens themselves in their nature, since they can be seen revolving about us, and the earth [in its nature] through an image, since the true [earth], being enormous and not surrounding us, cannot be inspected by any one person either as a whole or part by part.[53]

In the *Geography*, the contribution of mathematical sciences reveals the physical nature of the heavens and earth. In *Almagest* 1.1, Ptolemy illustrates the contribution of mathematics to element theory. In this way, he shifts from characterizing physics as concerned with what are special, or peculiar, to a single faculty of sense perception, the special sensibles, to what are peculiar to material substance, meaning the fundamental qualities of the elements, sublunary as well as superlunary. This shift in focus does not negate his earlier definition of physical objects as special sensibles. Rather, Ptolemy broadens the scope of what objects fall within the domain of physics. Special sensibles and the elements' fundamental properties are both peculiar to material substance and, therefore, they are the objects of physics. Ptolemy explains that the observation of

---

53. Ptolemy, *Geography* 1.1.9, translation modified from Berggren and Jones, *Ptolemy's "Geography"*: Ἃ τῆς ἀνωτάτω καὶ καλλίστης ἐστὶ θεωρίας, ἐπιδεικνύναι διὰ τῶν μαθημάτων ταῖς ἀνθρωπίναις καταλήψεσι τὸν μὲν οὐρανὸν αὐτόν, ὡς ἔχει φύσεως, ὅτι δύναται φαίνεσθαι περιπολῶν ἡμᾶς, τὴν δὲ γῆν διὰ τῆς εἰκόνος, ὅτι τὴν ἀληθινὴν καὶ μεγίστην οὖσαν καὶ μὴ περιέχουσαν ἡμᾶς, οὔτε ἀθρόαν οὔτε κατὰ μέρος ὑπὸ τῶν αὐτῶν ἐφοδευθῆναι δυνατόν.

elements' movements reveals these fundamental qualities: "as the corruptible [is distinguished] from the incorruptible by its straight or circular [motion], and the heavy from the light, and the passive from the active, by [its motion] toward the center or away from the center."[54] Whether a body moves rectilinearly or circularly indicates whether its elements are corruptible or incorruptible, and, if it moves rectilinearly, whether it moves toward or away from the center of the cosmos indicates whether the elements composing it are heavy and passive or light and active, respectively.

Liba Taub has observed that Ptolemy's conception of the contribution of mathematics to element theory is consistent with Aristotle's theory of motion in the *Physics*.[55] I would add that it also finds a precedent in the *De caelo*,[56] where Aristotle describes the natural, rectilinear motion of the four elements: earth, water, air, and fire. Fire, being light, naturally rises to the circumference of the sublunary realm. Earth, being heavy, naturally falls to the center. Water and air, intermediate between earth and fire, rise or fall until they reach their respective natural places, in between the realms of earth and fire. Moreover, in *On Generation and Corruption* Aristotle identifies the hot and cold as active principles.[57] The addition of either heat or cold causes the generation and corruption of the elements, or their transmutation into one another. Aristotle's fifth element, the aether, does not undergo generation or corruption. Characterized by neither heaviness nor lightness, it moves circularly for all eternity. In this way, it is always active. Ptolemy's assertion, then, that the movement of an object reveals its physical nature is consistent with the element theory of Aristotle's *Physics*, *De caelo*, and *On Generation and Corruption*.

Ptolemy applies geometry to element theory not only in *Almagest* 1.1 but also in *Almagest* 1.7, *Planetary Hypotheses* 2.3, and two of his lost works, *On the Elements* and *On Weights*. In *Almagest* 1.7, he depicts the motion of the four elements in geometrical terms in order to situate natural rectilinear motion within a spherical cosmos:

> there is no up or down in the cosmos with respect to itself, any more than one could contrive such in a sphere, but rather the proper and natural motion of the compound bodies in it is such that light and rarefied bodies drift outward toward the periphery, but seem to move in the direction that

54. Ptolemy, *Almagest* 1.1, H7: ὡς τὸ μὲν φθαρτὸν αὐτὸ καὶ τὸ ἄφθαρτον ἀπὸ τῆς εὐθείας καὶ τῆς ἐγκυκλίου, τὸ δὲ βαρὺ καὶ τὸ κοῦφον ἢ τὸ παθητικὸν καὶ τὸ ποιητικὸν ἀπὸ τῆς ἐπὶ τὸ μέσον καὶ τῆς ἀπὸ τοῦ μέσου.

55. Taub, *Ptolemy's Universe*, 28.

56. See especially Aristotle, *De caelo* I, 268b17–24, 269a17–b17, and IV, 311a15–28. Alan Bowen notes Taub's neglect of the *De caelo* in his "Review of *Ptolemy's Universe*," 140.

57. Aristotle, *On Generation and Corruption* II, 329b23–31.

is "up" for each [observer] since the overhead direction for all of us, which is also called "up," points toward the surrounding [i.e., aether or limit of the cosmos], but heavy and dense bodies are carried toward the middle and the center [of the cosmos], but seem to fall downward because, again, the direction that is for all of us toward our feet, called "down," also points toward the center of the earth[58]

In *Planetary Hypotheses* 2.3, Ptolemy also describes the sublunary elements as rising and falling. The four elements move rectilinearly, and the fifth element, the aether, moves uniformly and circularly. Ptolemy applied geometry to element theory in *On the Elements* and *On Weights* as well. These two books are no longer extant, but, in his commentary on the *De caelo*, Simplicius attests to Ptolemy's theory of natural motion. Whereas Ptolemy asserts in *Planetary Hypotheses* 2.3 that the four elements rest in their natural places, in *On the Elements* Ptolemy reportedly argued that elements either rest or move circularly when in their natural places and move rectilinearly only when displaced from their natural places.[59] In *On Weights*, so Simplicius maintains, Ptolemy contended that neither air nor water has weight in its natural place.[60] Despite minor discrepancies among these individual accounts, Ptolemy consistently describes element theory in geometrical terms. Furthermore, he applies mathematics to the physics of composite bodies in the *Harmonics*, *Tetrabiblos*, and *Planetary Hypotheses*. In the *Harmonics* he applies harmonics to psychology; in the *Tetrabiblos* he applies astronomy to astrology; in the *Planetary Hypotheses* he applies astronomy to cosmology. I examine Ptolemy's application of mathematics to these fields of inquiry in subsequent chapters.

Conclusion

Ptolemy modeled *Almagest* 1.1 on contemporary philosophical handbooks. In the chapter, he discusses the topics appropriate to a handbook's introductory chapters—including the definition of legitimate philosophers, the distinction

---

58. Ptolemy, *Almagest* 1.7, H22–23: τοῦ μὲν κάτω ἢ ἄνω μηδενὸς ὄντος ἐν τῷ κόσμῳ πρὸς αὐτήν, καθάπερ οὐδὲ ἐν σφαίρᾳ τις ἂν τὸ τοιοῦτον ἐπινοήσειεν, τῶν δὲ ἐν αὐτῷ συγκριμάτων τὸ ὅσον ἐπὶ τῇ ἰδίᾳ καὶ κατὰ φύσιν ἑαυτῶν φορᾷ τῶν μὲν κούφων καὶ λεπτομερῶν εἰς τὸ ἔξω καὶ ὡς πρὸς τὴν περιφέρειαν ἀναριπιζομένων, δοκούντων δὲ εἰς τὸ παρ' ἑκάστοις ἄνω τὴν ὁρμὴν ποιεῖσθαι διὰ τὸ καὶ πάντων ἡμῶν τὸ ὑπὲρ κεφαλῆς, ἄνω δὲ καλούμενον καὶ αὐτό, νεύειν ὡς πρὸς τὴν περιέχουσαν ἐπιφάνειαν, τῶν δὲ βαρέων καὶ παχυμερῶν ἐπὶ τὸ μέσον καὶ ὡς πρὸς τὸ κέντρον φερομένων, δοκούντων δὲ εἰς τὸ κάτω πίπτειν διὰ τὸ καὶ πάντων πάλιν ἡμῶν τὸ πρὸς τοὺς πόδας, καλούμενον δὲ κάτω, καὶ αὐτὸ νεύειν πρὸς τὸ κέντρον τῆς γῆς

59. Simplicius, *In de caelo* 1.2.20.10–25. For an analysis of this passage, see chapter 8.
60. Ibid. 4.4.710.14–711.9.

between theoretical and practical philosophy, and the divisions of theoretical philosophy—while at the same time he undermines attempts by philosophers to generate knowledge in their domains of inquiry—namely, physics and theology. Ptolemy fuses Aristotelian and Platonic epistemologies to put forward a new, radical epistemology: one in which mathematics reigns supreme. He argues that philosophers will never reach consensus on the nature of physical and theological objects, because these fields of inquiry are, on their own, conjectural. Mathematics, on the other hand, yields sure and incontrovertible knowledge, and its contributions to physics and theology are epistemically efficacious. Mathematics enables a good guess at the nature of the Prime Mover, and it reveals the nature of physical objects. In this way, Ptolemy bolsters the value of mathematics and co-opts the fields of inquiry traditionally studied by philosophers for mathematicians. This epistemological polemic is one of two in *Almagest* 1.1. As we will see in the following chapter, Ptolemy argues that mathematics is not only the one type of theoretical philosophy that generates knowledge but it is also the only part of philosophy—theoretical or practical—that furnishes the ultimate goal of ancient virtue ethics: the good life. If one desires not only knowledge but also a life well lived, then one must study mathematics.

4

Mathematics and the Good Life[1]

THE POLEMIC in *Almagest* 1.1 is two part. As we saw in the previous chapter, Ptolemy argues that mathematics is superior to physics and theology with respect to the epistemological objective of obtaining knowledge, as opposed to mere conjecture. In addition, Ptolemy argues that mathematics is the only part of philosophy capable of transforming an individual's soul into a virtuous, or excellent, condition. If one desires a good life, then one must study mathematics. Ptolemy asserts the ethical supremacy of mathematics near the end of *Almagest* 1.1, but the argument has its beginning in the first lines of the text, where he distinguishes theoretical and practical philosophy. Indeed, the entirety of *Almagest* 1.1 is a defense of theoretical over practical philosophy. The exhortation to live the contemplative life over the life of action was advanced by many ancient Greek philosophers, including Aristotle, and Ptolemy adapts this genre to his own purpose.[2] He not only defends theoretical philosophy in general but mathematics in particular, and his description of the relationship between theoretical and practical philosophy establishes mathematics as the only route to the good life. In this way, Ptolemy transforms ancient virtue ethics and champions the mathematician's way of life as the best way of life.

---

1. This chapter is modified from Jacqueline Feke, "Ptolemy's Defense of Theoretical Philosophy," *Apeiron: A Journal for Ancient Philosophy and Science* 45, no. 1 (January 2012): 61-90. Copyright © 2012 by Walter de Gruyter. https://www.degruyter.com/view/j/apeiron.2012.45 .issue-1/apeiron-2011-0011/apeiron-2011-0011.xml.

2. For instance, Aristotle defends the contemplative life in the *Protrepticus*. See Hutchinson and Johnson, "Authenticating Aristotle's *Protrepticus*."

The Distinction between
Theoretical and Practical Philosophy

To unpack how Ptolemy constructs the argument for the mathematical way of life, one must look first to the distinction between theoretical and practical philosophy he articulates at the beginning of *Almagest* 1.1. Ptolemy launches his defense of theoretical philosophy in the very first line of the text. He states, "It seems to me that the legitimate philosophers, Syrus, were entirely right to have distinguished the theoretical [part] of philosophy from the practical."[3] I mentioned in chapter 2 that Ptolemy appropriates the distinction between theoretical and practical philosophy from Aristotle. What I did not mention is that Ptolemy also reveals himself here to be engaging in a contemporary debate over the relationship between theoretical and practical philosophy, one which may have evolved from post-Hellenistic attempts to synthesize Aristotle's and Plato's accounts. Aristotle's texts establish an absolute distinction between the theoretical and practical, but Plato's corpus does not consistently distinguish them. At times, Plato's interlocutors suggest that they are completely distinct, as in *Gorgias* 500c–d and *Statesman* 258d–e. One might read *Republic* VI, 500d–e, on the other hand, as a portrayal of the dependency of the practical on the theoretical. Socrates argues here that the philosopher should apply what he sees when contemplating the divine to shaping his and others' characters:

> And if he should come to be compelled to put what he sees there into people's characters, whether into a single person or into a populace, instead of shaping only his own, do you think he will be a poor craftsman of moderation, justice, and the whole of popular virtue? . . . And when the majority realize that what we are saying about the philosopher is true, will they be harsh with him or mistrust us when we say that the city will never find happiness until its outline is sketched by painters who use the divine model?[4]

---

3. Ptolemy, *Almagest* 1.1, H4: Πάνυ καλῶς οἱ γνησίως φιλοσοφήσαντες, ὦ Σύρε, δοκοῦσί μοι κεχωρικέναι τὸ θεωρητικὸν τῆς φιλοσοφίας ἀπὸ τοῦ πρακτικοῦ.

4. Plato, *Republic* VI, 500d–e, trans. Grube, rev. Reeve: Ἂν οὖν τις, εἶπον, αὐτῷ ἀνάγκη γένηται ἃ ἐκεῖ ὁρᾷ μελετῆσαι εἰς ἀνθρώπων ἤθη καὶ ἰδίᾳ καὶ δημοσίᾳ τιθέναι καὶ μὴ μόνον ἑαυτὸν πλάττειν, ἆρα κακὸν δημιουργὸν αὐτὸν οἴει γενήσεσθαι σωφροσύνης τε καὶ δικαιοσύνης καὶ συμπάσης τῆς δημοτικῆς ἀρετῆς; [. . .] Ἀλλ' ἐὰν δὴ αἴσθωνται οἱ πολλοὶ ὅτι ἀληθῆ περὶ αὐτοῦ λέγομεν, χαλεπανοῦσι δὴ τοῖς φιλοσόφοις καὶ ἀπιστήσουσιν ἡμῖν λέγουσιν ὡς οὐκ ἄν ποτε ἄλλως εὐδαιμονήσειε πόλις, εἰ μὴ αὐτὴν διαγράψειαν οἱ τῷ θείῳ παραδείγματι χρώμενοι ζωγράφοι;

Molding the city according to the divine model, the philosopher shapes the city into its most happy and just state. After contemplating the divine, the philosopher is able to rule the city in the best possible way. In the terms of second-century philosophy, practical philosophy is dependent on theoretical philosophy.

The analogy of the cave at *Republic* VII, 514a–17a, reiterates this dependency. The individual who has left the cave and seen the sun is compelled to return to the cave. Correspondingly, the philosopher who has contemplated the Forms and obtained knowledge is qualified and compelled to rule Kallipolis. At *Republic* VII, 540a–b, Socrates explains that, at the age of fifty, individuals who have succeeded thus far in both practical and scientific matters (ἐν ἔργοις τε καὶ ἐπιστήμαις) must proceed to study the Forms and contemplate the Good. Thereafter, they use the Good as the paradigm for the city they are compelled to rule:

> and once they've seen the good itself, they must each in turn put the city, its citizens, and themselves in order, using it as their model. Each of them will spend most of his time with philosophy, but, when his turn comes, he must labor in politics and rule for the city's sake, not as if he were doing something fine, but rather something that has to be done[5]

The rulers of Kallipolis mold the city according to the divine model they have contemplated. Again, practical philosophy appears to depend on theoretical philosophy.

The discrepancy between Plato's and Aristotle's texts on whether practical philosophy is dependent on or independent of theoretical philosophy seems to have produced a debate, for which we have evidence from the second century CE. In addition to Ptolemy's *Almagest*, the commentary on the *Nicomachean Ethics* by Aspasius—a Peripatetic philosopher active in the first half of the second century—attests to the debate. Aspasius's commentary is the earliest commentary on any of Aristotle's texts—indeed, any philosophical text—to survive in large part. In the first lines he writes, "The treatment of ethics and especially politics is prior to theoretical philosophy in respect to necessity but subsequent in respect to value."[6] Later in the text, when defining

---

5. Ibid. VII, 540a–b, translation modified from Grube, rev. Reeve: καὶ ἰδόντας τὸ ἀγαθὸν αὐτό, παραδείγματι χρωμένους ἐκείνῳ, καὶ πόλιν καὶ ἰδιώτας καὶ ἑαυτοὺς κοσμεῖν τὸν ἐπίλοιπον βίον ἐν μέρει ἑκάστους, τὸ μὲν πολὺ πρὸς φιλοσοφίᾳ διατρίβοντας, ὅταν δὲ τὸ μέρος ἥκῃ, πρὸς πολιτικοῖς ἐπιταλαιπωροῦντας καὶ ἄρχοντας ἑκάστους τῆς πόλεως ἕνεκα, οὐχ ὡς καλόν τι ἀλλ' ὡς ἀναγκαῖον πράττοντας

6. Aspasius, *In ethica Nicomachea commentaria* 1.2–4, translation modified from Konstan, in Aspasius, *On Aristotle's "Nicomachean Ethics"*: Ἡ περὶ τὰ ἤθη πραγματεία καὶ μάλιστα ἡ

action (πρᾶξις), he classifies ethics and politics as practical: "but activity in accord with a practical science is also called 'action.' In general, all those arts are called 'active' which have no other product apart from an action, for example dancing and flute-playing. But more particularly the political and ethical arts are called active, and activities that concern what is noble and shameful are called actions."[7] Evaluating the relationship between theoretical and practical philosophy, Aspasius argues that ethics and politics, arts representative of practical philosophy, are prior or subsequent to theoretical philosophy in respect to necessity or value.

Ptolemy and Aspasius seem to be engaging in a common debate over the relationship between theoretical and practical philosophy. Of note is that both Ptolemy and Aspasius discuss theoretical and practical "philosophy" rather than "thought" (διάνοια), as Aristotle does. In addition, both Ptolemy and Aspasius use the term "prior" (πρότερος) when addressing the relationship between theoretical and practical philosophy. According to Aspasius, they are prior (προτέρα) or subsequent to one another in various respects but remain independent. Hence, Aspasius adheres to what one might call the Aristotelian view, where theoretical and practical philosophy are independent of one another. For Ptolemy, the point of concern is the potential dependency of practical philosophy on theoretical philosophy—that before (πρότερον) practical philosophy is practical, it is theoretical. Ptolemy addresses this potential dependency and then asserts a great difference separating them. Therefore, Ptolemy admits the possible legitimacy of what one might call the Platonic view—that practical philosophy is dependent on theoretical philosophy— but thereafter he appears to champion the Aristotelian view, that they are independent.

Ptolemy takes a similar approach to a philosophical debate in *On the Kritêrion and Hêgemonikon*. As in the *Almagest*, he confronts a debate over whether one thing is reducible to another. In *On the Kritêrion*, the debate concerns the relationship between the soul and the body and, specifically, whether the soul is of the same or of a different kind as the body. At first, Ptolemy gives the appearance of dismissing such debates. Before defining the soul and body, he eschews quibbles over terminology. He calls them "battles over words" (φωνομαχίαις), of which some are "meddlesome tattle" (ἀδολεσχίας

πολιτικὴ [ἠθικὴ] κατὰ μὲν τὸ ἀναγκαῖον προτέρα ἐστὶ τῆς θεωρητικῆς φιλοσοφίας, κατὰ δὲ τὸ τίμιον ὑστέρα.

7. Ibid. 3.19–23: λέγεται δὲ πρᾶξις καὶ ἡ κατὰ πρακτικὴν ἐπιστήμην ἐνέργεια. πρακτικαὶ δὲ λέγονται μὲν κοινῶς πᾶσαι, ὧν μὴ ἔστι ποίημά τι ἄλλο παρὰ τὴν πρᾶξιν, οἷον ὀρχηστικὴ καὶ αὐλητική, ἰδίως δὲ ἥ τε πολιτικὴ καὶ ἠθικὴ πρακτικαὶ λέγονται καὶ πράξεις αἱ περὶ τὸ καλὸν καὶ αἰσχρὸν ἐνέργειαι.

πολυπράγμονος),[8] and he explains that these debates are useful only inasmuch as they include investigations of the objects signified by the debated terminology. Like Galen—who also denounced terminological disputes and claimed that they should be the subject of grammarians and orators rather than physicians and natural philosophers[9]—Ptolemy aims to utilize terms that reflect the nature of the objects signified. He explains, "Such a proposal removes dispute about nomenclature from our discussions but keeps our enquiry into the things underlying the words intact and free from distractions."[10] Ptolemy reveals that he does not rebuke all philosophical debates. He distinguishes debates about the nature of things from mere disputes over words and, dismissing the latter, he concentrates on the former.

In keeping with this emphasis on ontology rather than terminology, Ptolemy asserts the following when defining the soul in contrast to the body:

> this is not the place to bother whether we ought also to call this part "body"; as we have said, we are not at present discussing the names to give to the natural objects before us; what we are investigating is the difference between these things, a difference which we recognize as being unchangeable in reality even if one alters the nomenclature a thousand times, or at one time says that the soul is incorporeal, following those who lay it down that what is known by sense perception is to be called "body," and at another time that it is body, following those who define body as that which can act and be acted upon. The difference between the natures of the said entities being of this kind, there would be universal agreement that it is by the soul that we think and not by the body. And we would also acknowledge that it is by the soul, not the body, that we make our sensory and all other movements, if we took note too of the quantitative aspect of what happens when they separate[11]

8. Ptolemy, *On the Kritêrion*, La9, trans. Liverpool-Manchester Seminar, in Ptolemaeus, "On the Kriterion."

9. Galen, *On Critical Days* 9.789. R. J. Hankinson and Ben Morison disagree on how to interpret this passage: Hankinson, "Usage and Abusage," 171; Morison, "Language," 130–31.

10. Ptolemy, *On the Kritêrion*, La10, trans. Liverpool-Manchester Seminar, in Ptolemaeus, "On the Kriterion.": καὶ ἡ τοιαύτη πρόθεσις ἐξελοῦσα τὸ περὶ τὰς κατηγορίας ἐριστικὸν τῶν διαλεγομένων ἀκεραίαν καὶ ἀπερίσπαστον αὐτῶν συντηρεῖ τὴν περὶ τῶν ὑποκειμένων ἐπίσκεψιν.

11. Ibid., La11–12, translation modified from Liverpool-Manchester Seminar, in Ptolemaeus, "On the Kriterion": εἰ δὲ καὶ ταύτην σῶμα δεῖ καλεῖν, οὐ πολυπραγμονητέον νῦν· οὐ γὰρ τὰ ὀνόματα τῶν ὑποκειμένων φύσεων ζητοῦμεν ἐπὶ τοῦ παρόντος, ὡς ἔφαμεν, ἀλλὰ τὴν ἐν αὐταῖς διαφοράν, ἣν ἔργῳ κατανενοήκαμεν ἀμετάστατον οὖσαν, κἂν μυριάκις τις ἀντιστρέφῃ τὰς ὀνομασίας αὐτῶν, ἢ νῦν μὲν τὴν ψυχὴν ἀσώματον εἶναι φάσκη κατὰ τοὺς νομοθετοῦντας σῶμα καλεῖσθαι τὸ αἰσθήσει γνώριμον, νῦν δὲ σῶμα κατὰ τοὺς τὸ ποιῆσαι καὶ παθεῖν οἷόν τε σῶμα ὁριζομένους. ἀλλ' ἐπειδή γε

In this passage, Ptolemy explains that philosophers disagree on whether the soul and body are alike or different in kind, and he insists that the soul is different from the body, as it causes human beings to think, while the body produces all other movements in human beings. This same pattern of recognizing a debate over the reducibility of one thing to another, and then asserting a difference (διαφορά) between them, also underlies Ptolemy's treatment of the theoretical and the practical in *Almagest* 1.1. He acknowledges the debate over whether practical philosophy is dependent on theoretical philosophy, and he asserts a great difference between the two.

The difference is how one attains virtues in relation to each type of philosophy. Ptolemy gives an account of this difference as follows: "in the first place, it is possible for many people to possess some of the moral virtues even without being taught, whereas it is impossible to obtain the theory of the universe without instruction; furthermore, one derives most benefit in the first case [i.e., the practical part of philosophy] from continuous practice in actual affairs, but in the other [i.e., the theoretical part of philosophy] from making progress in the theories."[12] With respect to practical philosophy, some of the moral virtues are attainable without instruction, and they derive mainly from continuous activity; with respect to theoretical philosophy, apprehension of the universe depends on instruction and most benefit consists in progress in the theories.

Franz Boll and Liba Taub have recognized the influence of Aristotle's *Nicomachean Ethics* II, 1103a, on Ptolemy's differentiation of theoretical and practical philosophy.[13] Here, Aristotle divides virtue in two: "Virtue being, as we have seen, of two kinds, intellectual and moral, intellectual virtue is for the most part both produced and increased by instruction, and therefore requires experience and time; whereas moral virtue is the product of habit, and has indeed derived its name, with a slight variation of form, from that word."[14]

τοιαύτη τίς ἐστιν ἡ τῶν εἰρημένων φύσεων διαφορά, τὸ μὲν τῇ ψυχῇ διανοεῖσθαι καὶ μὴ τῷ σώματι, πᾶς ἂν ὁμολογήσειεν. ὅτι δὲ καὶ τάς τε αἰσθητικὰς καὶ τὰς ἄλλας πάσας κινήσεις τῇ ψυχῇ καὶ οὐ τῷ σώματι ποιούμεθα, κατανοήσαιμεν ἂν εἰ καὶ κατὰ τὸ ποσὸν ἐπιβάλοιμεν αὐτῶν τῇ διαλύσει

12. Ptolemy, *Almagest* 1.1, H4, translation modified from Toomer: οὐ μόνον διὰ τὸ τῶν μὲν ἠθικῶν ἀρετῶν ἐνίας ὑπάρξαι δύνασθαι πολλοῖς καὶ χωρὶς μαθήσεως, τῆς δὲ τῶν ὅλων θεωρίας ἀδύνατον εἶναι τυχεῖν ἄνευ διδασκαλίας, ἀλλὰ καὶ τῷ τὴν πλείστην ὠφέλειαν ἐκεῖ μὲν ἐκ τῆς ἐν αὐτοῖς τοῖς πράγμασι συνεχοῦς ἐνεργείας, ἐνθάδε δ' ἐκ τῆς ἐν τοῖς θεωρήμασι προκοπῆς παραγίγνεσθαι.

13. Boll, "Studien," 70–71; Taub, *Ptolemy's Universe*, 19–20.

14. Aristotle, *Nicomachean Ethics* II, 1103a14–18, translation modified from Rackham: Διττῆς δὴ τῆς ἀρετῆς οὔσης, τῆς μὲν διανοητικῆς τῆς δὲ ἠθικῆς, ἡ μὲν διανοητικὴ τὸ πλεῖον ἐκ διδασκαλίας ἔχει καὶ τὴν γένεσιν καὶ τὴν αὔξησιν, διόπερ ἐμπειρίας δεῖται καὶ χρόνου, ἡ δ' ἠθικὴ ἐξ ἔθους περιγίνεται, ὅθεν καὶ τοὔνομα ἔσχηκε μικρὸν παρεκκλῖνον ἀπὸ τοῦ ἔθους.

Like Ptolemy's practical philosophy, which depends on continuous activity, Aristotle's moral virtue requires habit; like Ptolemy's theoretical philosophy, which progresses by instruction, so, too, Aristotle's intellectual virtue relies on instruction.

Just as Ptolemy did not take his definitions of the three theoretical sciences from the *Metaphysics*, neither is it likely that he simply reinterpreted the *Nicomachean Ethics*. Aristotle's distinction between the moral and intellectual virtues led to a tradition, shared by Peripatetics and Platonists alike, of distinguishing virtues according to their attainment by either instruction or habit. For example, in his commentary on the *Nicomachean Ethics* Aspasius points to learning as the source of intellectual virtues and training or habit as the source of moral virtues. When he comments specifically on *Nicomachean Ethics* II, 1103a, he affirms Aristotle's distinction.[15] Similarly, in *Didaskalikos* 24.4 Alcinous distinguishes the rational and affective parts of the soul by arguing that the former is cultivated through instruction (διδασκαλία) and the latter through habitual practice (τῆς ἔθους ἀσκήσεως). For influences on this passage, John Dillon points to both *Nicomachean Ethics* II, 1103a14–18, and *Republic* VII, 518d–e.[16] In the latter, Socrates characterizes education as the craft that turns the soul toward the Good, and he depicts habit and practice as what cultivate the virtues of the soul that are akin to the body. Correspondingly, in *Phaedo* 82a–b Socrates remarks that popular and political virtue (τὴν δημοτικὴν καὶ πολιτικὴν ἀρετήν) is developed from habit and care (ἔθους τε καὶ μελέτης), and in *Phaedo* 82d–83c he states that philosophy teaches the lovers of learning (φιλομαθεῖς) that their souls, imprisoned in their bodies, guide them to see the intelligible and invisible. Thus, Aristotle's distinction between intellectual and moral virtue has a precedent in Plato's writings, and Aristotelian commentators and middle Platonic philosophers appropriated these sets of ideas.

What is interesting in Ptolemy's account is not only that he distinguishes between the intellectual and moral virtues according to whether they are acquired by instruction or habit but that he also applies this distinction to the differentiation of theoretical and practical philosophy. Although it may be spurious, Plutarch's *De liberis educandis* makes a similar move when it distinguishes the parts of philosophy according to their dependence on teaching or habit. The text divides philosophy into three categories—physics, logic, and ethics—and it distinguishes logic and ethics according to their relation

---

15. Aspasius, *In ethica Nichomachea commentaria* 25.18–26, 37.2–38.6.
16. Dillon, "Commentary," 151.

to learning and habit.[17] Consequently, Ptolemy's application of the distinction between sets of virtues to the differentiation of parts of philosophy appears in the Platonic tradition, but what is unusual, and perhaps unique, with Ptolemy's account is his specific blending of Aristotelian and Platonic trends. Ptolemy uses instruction and habit to distinguish theoretical and practical philosophy.

Virtues

Because Ptolemy distinguishes theoretical and practical philosophy according to how one cultivates virtues in each domain, his delineation of virtues in *Harmonics* 3.5 is manifestly relevant. In this chapter, Ptolemy presents three alternative accounts of the soul. Each portrait of the soul is tripartite, and each of the parts of the soul has a number of species. In his Platonic account of the soul, the rational part (λογιστικόν) has seven species, the spirited part (θυμικόν) has four species, and the appetitive part (ἐπιθυμητικόν) has three species.[18] In this account, the soul's species are species of virtue. The appetitive part of the soul has three species of virtue: moderation (σωφροσύνη), self-control (ἐγκράτεια), and shame (αἰδώς). The spirited part of the soul has four species of virtue: gentleness (πραότης), fearlessness (ἀφοβία), courage (ἀνδρεία), and steadfastness (καρτερία). The rational part of the soul has seven species of virtue: acuteness (ὀξύτης), cleverness (εὐφυΐα), shrewdness (ἀγχίνοια), judgment (εὐβουλία), wisdom (σοφία), prudence (φρόνησις), and experience (ἐμπειρία). By providing a list of virtues, Ptolemy follows a typically Hellenistic trend, which has a precedent in Aristotle's corpus.

By ascribing virtues to distinct parts of the soul, as opposed to the soul in its entirety, Ptolemy joins the middle Platonic tradition. Salvatore R. C. Lilla has noted that several Platonists assign one particular virtue to each part of the soul. The middle Platonists include Alcinous; Philo, the first-century CE Jewish philosopher; and Clement, the Christian philosopher who lived from the late second to the early third century CE. Neoplatonists such as Plotinus and Porphyry maintain this principle.[19] Lilla notes that these several

17. Pseudo-Plutarch, *De liberis educandis* 2a10–11: καλῶ δὲ λόγον μὲν τὴν μάθησιν, ἔθος δὲ τὴν ἄσκησιν. Xenocrates, Aristotle's contemporary, divided philosophy into physics, ethics, and logic. See Sextus Empiricus, *Adversus mathematicos* 7.16. This classification scheme became normative in Hellenistic philosophy.

18. Plato uses the terms λογιστικόν, θυμοειδές, and ἐπιθυμητικόν in *Republic* IV, 440e–441a. The use of θυμικόν instead of θυμοειδές for the spirited part of the soul seems to have been common in the second century. Alcinous, too, uses θυμικόν in *Didaskalikos* 17.4 and 29.1.

19. Lilla, *Clement of Alexandria*, 80.

philosophers assign prudence (φρόνησις) to the rational part (λογιστικόν) of the soul, courage (ἀνδρεία) to the spirited part (θυμοειδές), and moderation (σωφροσύνη) to the appetitive part (ἐπιθυμητικόν). I would add to Lilla's list pseudo-Andronicus's De passionibus, from the late Hellenistic or early Imperial period. Pseudo-Andronicus, too, assigns prudence (φρόνησις) to the rational part (λογιστικόν) and moderation (σωφροσύνη) to the appetitive part (ἐπιθυμητικόν), but he allocates gentleness (πραότης) as well as courage (ἀνδρεία) to the spirited part (θυμοειδές).[20] Although these philosophers differ on which virtue(s) they ascribe to the spirited part, Ptolemy's account coheres with both sets, as he assigns several virtues to each part of the soul. Indeed, it is this point that distinguishes Ptolemy's account. The allocation of several virtues to individual parts of the soul has a precedent in the Magna Moralia, for instance, and even Apuleius—the second-century Roman rhetorician, popularizer of philosophy, and author of the Metamorphoses—assigns two virtues, prudence and wisdom, to the rational soul,[21] but it seems that the convention of assigning more than one virtue to each part of the soul fell into disfavor. Not only do the above Platonists ascribe only one virtue to each part but—in a polemic against Chrysippus, the third head of the Stoa—Galen too assigns a single virtue to each of the soul's parts and argues that only one virtue can belong to each part.[22] Ptolemy, on the other hand, ascribes not one, not two, but several virtues to each part of the soul.

In addition to listing the species of virtue, Ptolemy provides short definitions of them. The closest philological match to these definitions is found in pseudo-Plato's Definitions, composed during the fourth century BCE and most likely developed incrementally over the course of centuries.[23] Franz Boll first drew attention to the potential influence of the Definitions on Ptolemy's list of virtues, and Ingemar Düring later agreed that not only does a great deal of terminological overlap exist between the two texts, but, moreover, this overlap cannot be accidental.[24] Every single virtue that Ptolemy lists in the Harmonics appears in the Definitions. Many of them have their own definitions, but even those that are not listed separately are still included in the definitions of other terms. For example, acuteness (ὀξύτης) does not have its own definition, but it—like cleverness (εὐφυία), which is defined in the Definitions—is used to

---

20. Pseudo-Andronicus, De passionibus 2.1.3, 4.4.1–5.1.2, 6.1.1.

21. Pseudo-Aristotle, Magna Moralia I, 1185b1–13; Apuleius, De Platone et eius dogmate 2.9.234.

22. Galen, De placitis 7.1.24–32.

23. D. S. Hutchinson discusses the Definitions' authorship in the introduction to the Definitions in Plato, Complete Works, 1677–78.

24. Boll, "Studien," 106–8; Düring, Ptolemaios und Porphyrios, 271.

define shrewdness (ἀγχίνοια), another of Ptolemy's virtues of the rational part of the soul.[25]

Furthermore, several of the definitions pseudo-Plato and Ptolemy provide are nearly identical. According to Ptolemy, wisdom (σοφία) has to do with the theoretical (θεωρητικόν), and according to pseudo-Plato it is "non-hypothetical knowledge; knowledge of what always exists; knowledge which contemplates [θεωρητική] the cause of beings."[26] Similarly, both Ptolemy and pseudo-Plato define gentleness (πραότης) in relation to anger (ὀργή)[27] and courage (ἀνδρεία) in relation to dangers (κίνδυνοι).[28] In addition, both Ptolemy and pseudo-Plato use the phrase "endurance of hardships" (ὑπομονὴ πόνων) in their accounts of steadfastness (καρτερία). Ptolemy defines it as "the endurance of hardships,"[29] and pseudo-Plato describes it as "endurance of pain for the sake of what is fine; endurance of hardships for the sake of what is fine."[30] Concerning the virtues Ptolemy associates with the appetitive part of the soul, both Ptolemy and pseudo-Plato define moderation (σωφροσύνη) in relation to pleasures (ἡδοναί),[31] they characterize self-control (ἐγκράτεια) as a type of enduring (ὑπομονή),[32] and they use forms of the term "avoidance" (εὐλάβεια) when defining shame (αἰδώς).[33] Thus, a considerable amount of textual overlap exists between Ptolemy's definitions of the virtues and the definitions of these terms by pseudo-Plato. It is possible, of course, that Peripatetic and Stoic definitions of virtues influenced Ptolemy's account,[34] but these textual correspondences between Ptolemy's and pseudo-Plato's texts suggest a Platonic source. Ptolemy may have used a Platonic handbook, such as the *Definitions*, when composing this section of the *Harmonics*, or perhaps he simply drew these definitions from his earlier education. Either way, it is

---

25. Pseudo-Plato, *Definitions* 412e4–5.

26. Ibid. 414b5–6, trans. Hutchinson, in Plato, *Complete Works*: ἐπιστήμη ἀνυπόθετος· ἐπιστήμη τῶν ἀεὶ ὄντων· ἐπιστήμη θεωρητικὴ τῆς τῶν ὄντων αἰτίας.

27. Ptolemy, *Harmonics* 3.5, D97; pseudo-Plato, *Definitions* 412d6–7.

28. Ptolemy, *Harmonics* 3.5, D97; pseudo-Plato, *Definitions* 412a3–7.

29. Ptolemy, *Harmonics* 3.5, D97, trans. Barker, in *Greek Musical Writings*: ταῖς ὑπομοναῖς τῶν πόνων.

30. Pseudo-Plato, *Definitions* 412c1–2: ὑπομονὴ λύπης ἕνεκα τοῦ καλοῦ· ὑπομονὴ πόνων ἕνεκα τοῦ καλοῦ.

31. Ptolemy, *Harmonics* 3.5, D97; pseudo-Plato, *Definitions* 411e6–7.

32. Ptolemy, *Harmonics* 3.5, D97; pseudo-Plato, *Definitions* 412b3.

33. Ptolemy, *Harmonics* 3.5, D97; pseudo-Plato, *Definitions* 412c8–9.

34. Boll emphasizes the possible influence of pseudo-Andronicus's *De passionibus*, in addition to pseudo-Plato's *Definitions*, on Ptolemy's definitions of the virtues: Boll, "Studien," 106–8.

fitting that Ptolemy should use Platonic definitions of virtues when delineating species in a Platonic account of the soul.

After Ptolemy delineates the parts and species of the soul in *Harmonics* 3.5, in the following chapter, *Harmonics* 3.6, he names the genera to which the species belong and the two principles (ἀρχαί) to which the genera belong. The two principles are the theoretical and the practical, and the genera are the three theoretical and three practical parts of philosophy. Ptolemy explains, "For each of the two kinds of principle, that is, the theoretical and the practical, there are three genera, the physical, mathematical, and theological in the case of the theoretical, and the ethical, domestic, and political in the case of the practical..."[35] As discussed in chapter 2, Ptolemy's three theoretical sciences have a precedent in Aristotle's corpus, such as in *Metaphysics* E1 and K7. The three practical sciences, though they recall *Nicomachean Ethics* VI, 1141b23ff., appear only in scholastic interpretations such as Alcinous's *Didaskalikos* 3.3.[36] Of note is that in *Harmonics* 3.6 Ptolemy portrays theoretical and practical philosophy as absolutely distinct. Therefore, the distinction between theoretical and practical philosophy in *Almagest* 1.1, as explored thus far, is consistent with Ptolemy's portrayal of theoretical and practical philosophy in *Harmonics* 3.6. The remainder of *Almagest* 1.1, however, reveals that the relationship Ptolemy posits between theoretical and practical philosophy is more complex than it first appears.

How to Order Actions

After Ptolemy distinguishes theoretical and practical philosophy at the beginning of *Almagest* 1.1, he defends his preference for the former:

> Hence, we thought it fitting for them [i.e., the legitimate philosophers], on the one hand, to order their actions according to the applications of the *phantasiai* of these in such a way as never to forget, even in ordinary affairs, to strive for a fine and well-ordered state, and, on the other hand, with leisure to devote the most time to the instruction of theories, being many and fine, but especially those particularly called mathematical.[37]

35. Ptolemy, *Harmonics* 3.6, D98: Καὶ τοίνυν καθ᾽ ἑκατέραν ἀρχήν, τουτέστι τὴν θεωρητικὴν καὶ τὴν πρακτικήν, τριῶν ὄντων γενῶν, ἐπὶ μὲν τῆς θεωρητικῆς τοῦ τε φυσικοῦ καὶ τοῦ μαθηματικοῦ καὶ τοῦ θεολογικοῦ, ἐπὶ δὲ τῆς πρακτικῆς τοῦ τε ἠθικοῦ καὶ τοῦ οἰκονομικοῦ καὶ τοῦ πολιτικοῦ

36. Dillon, "Commentary," 60.

37. Ptolemy, *Almagest* 1.1, H4–5: ἔνθεν ἡγησάμεθα προσήκειν ἑαυτοῖς τὰς μὲν πράξεις ἐν ταῖς αὐτῶν τῶν φαντασιῶν ἐπιβολαῖς ῥυθμίζειν, ὅπως μηδ᾽ ἐν τοῖς τυχοῦσιν ἐπιλανθανώμεθα τῆς πρὸς τὴν καλὴν καὶ εὔτακτον κατάστασιν ἐπισκέψεως, τῇ δὲ σχολῇ χαρίζεσθαι τὸ πλεῖστον εἰς τὴν τῶν

In other words, given that progress in theoretical philosophy depends on instruction and the attainment of moral virtues requires activity, it is preferable to spend more time engaged in theoretical rather than practical philosophy. Moreover, it is important to address how one should guide one's actions even in ordinary affairs, as is stated, as well as in practical philosophy, as is implied. According to Ptolemy, one should order one's actions according to the applications of *phantasiai*.

Ptolemy's phrase—"to order their actions according to the applications of the *phantasiai* of these" (τὰς μὲν πράξεις ἐν ταῖς αὐτῶν τῶν φαντασιῶν ἐπιβολαῖς ῥυθμίζειν)—is peculiar and requires unpacking. Prescribing the ordering of actions, Ptolemy uses the verb *rhuthmizein*. This verb occurs in only one other instance in Ptolemy's corpus: *Tetrabiblos* 1.3. Ptolemy dedicates that chapter to defending the usefulness of astrology, and in the relevant passage he defends the utility of foreknowledge: "For, in the first place, we should consider that even with events that will necessarily take place their unexpectedness is very apt to cause excessive panic and delirious joy, while foreknowledge accustoms and orders [ἐθίζει καὶ ῥυθμίζει] the soul by experience of distant events as though they were present, and prepares it to accept with peace and stability whatever comes."[38] Knowledge of the future accustoms and orders the soul, and these processes prepare the individual for future action.[39] Ptolemy's collocation of the verbs to accustom (ἐθίζειν) and to order (ῥυθμίζειν) does not seem to be accidental. "Accustoms" (ἐθίζει) recalls the significance of habit for the development of moral virtues, and the association of accustoming with ordering the soul, specifically in preparation for future events, relates to *Almagest* 1.1, where "to order" (ῥυθμίζειν) seems to carry the same connotation. Ptolemy uses the verb when prescribing the ordering of future actions.

Ptolemy also uses the term *phantasiai*. To extract what this term means in *Almagest* 1.1, one must look to *On the Kritērion and Hēgemonikon*. Among the components of Ptolemy's criterion of truth is *phantasia*, which he describes as the mediator between sense perception and the intellect. He explains, "To the faculty of sense perception belong both sense organs and

θεωρημάτων πολλῶν καὶ καλῶν ὄντων διδασκαλίαν, ἐξαιρέτως δὲ εἰς τὴν τῶν ἰδίως καλουμένων μαθηματικῶν.

38. Ptolemy, *Tetrabiblos* 1.3.5, H15, translation modified from Robbins: πρῶτον μὲν γὰρ δεῖ σκοπεῖν, ὅτι καὶ ἐπὶ τῶν ἐξ ἀνάγκης ἀποβησομένων τὸ μὲν ἀπροσδόκητον τούς τε θορύβους ἐκστατικοὺς καὶ τὰς χαρὰς ἐξοιστικὰς μάλιστα πέφυκε ποιεῖν, τὸ δὲ προγινώσκειν ἐθίζει καὶ ῥυθμίζει τὴν ψυχὴν τῇ μελέτῃ τῶν ἀπόντων ὡς παρόντων καὶ παρασκευάζει μετ' εἰρήνης καὶ εὐσταθείας ἕκαστα τῶν ἐπερχομένων ἀποδέχεσθαι.

39. Cf. Posidonius on the benefits of prefiguring a future event in one's mind and habituating oneself to it, as discussed by Galen: *De placitis* 4.7.7–8.

phantasia. The sense organs are the bodily instruments through which contact is made with perceptible things, while *phantasia* is the impression and transmission to the intellect, whose retention and memory of the things transmitted we call conception . . ."[40] Hence, *phantasia* has a dual technical meaning for Ptolemy: (1) a sense impression, and (2) the transmission of sense impression(s) to the intellect. Ptolemy uses the plural form, *phantasiai*, later in the text when discussing the transmission of sense impressions (αἱ τῶν φαντασιῶν διαδόσεις).[41] The plural form conveys the first meaning; *phantasiai* are sense impressions, which are transmitted to the intellect. In *Almagest* 1.1, the term *phantasiai* conveys the same meaning.

As for what Ptolemy means by the abstruse phrase "according to the applications of the *phantasiai* of these" (ἐν ταῖς αὐτῶν τῶν φαντασιῶν ἐπιβολαῖς), one might look to a similar phrase in *Ars Rhetorica* by Cassius Longinus, the third-century CE rhetorician and philosopher who taught at Athens and spent his final years as the principal adviser to the rulers of Palmyra. Juxtaposing memory with recollection, he defines the former as the preservation of *phantasiai* (σωτηρία φαντασιῶν) and the latter as the application of *phantasia* (ἐπιβολὴ φαντασίας).[42] This translation is further supported by reference to Epicurus, the only other philosopher to use a phrase similar to "applications of the *phantasiai*" before Ptolemy. In the *Letter to Herodotus*, Epicurus describes mental images (φαντασμοί) as applications of the mind (ἐπιβολὰς τῆς διανοίας), which are criteria of truth.[43] In Epicurus's materialist theory of perception and cognition, these appearances relay to the mind the external objects being perceived. Ptolemy's and Cassius Longinus's employment of the phrase "application(s) of *phantasia(i)*" has its origin in Epicureanism, but by their time it had become common intellectual currency. Still, *phantasia* is for Ptolemy a component of the criterion of truth, and in the *Almagest* sense impressions, *phantasiai*, are applied. They are applied to a certain purpose: the ordering of actions.

The text indicates that one should perform this ordering when engaged in ordinary affairs and, I would argue, practical philosophy as well. The use of applications of *phantasiai* to order actions may extend from Ptolemy's employment of virtues to distinguish theoretical and practical philosophy. In *Harmonics* 3.5, Ptolemy defines wisdom (σοφία) as the virtue having to do

---

40. Ptolemy, *On the Kritêrion*, La5, trans. Liverpool-Manchester Seminar, in Ptolemaeus, "On the Kriterion": τῆς τε γὰρ αἰσθητικῆς δυνάμεως αἰσθητήρια μέν ἐστι τὰ τοῦ σώματος ὄργανα, δι' ὧν ἅπτεται τῶν αἰσθητῶν, φαντασία δ' ἡ τύπωσις καὶ διάδοσις ἡ ἐπὶ τὸν νοῦν, οὗ τὴν κατοχὴν καὶ μνήμην τῶν διαδοθέντων καλοῦμεν ἔννοιαν

41. Ptolemy, *On the Kritêrion*, La16.

42. Cassius Longinus, *Ars Rhetorica* 572.9–12.

43. Diogenes Laertius, *Lives* 10.51. Cf. ibid. 10.31, 10.50, and 10.147.

with the theoretical and prudence (φρόνησις) as the virtue having to do with the practical.[44] In *Almagest* 1.1, when Ptolemy contrasts theoretical and practical philosophy and mandates how one should order actions when engaged in practical philosophy, he also may have in mind this juxtaposition of wisdom and prudence.

In *Nicomachean Ethics* VI, 1142a, Aristotle explains that prudence is not a type of knowledge because it has to do with the thing to be done (πρακτόν); it has to do with the particular, which is the object of perception rather than knowledge. Moreover, he calls prudence a form of perception.[45] This relation of prudence to perception influenced middle Platonic philosophy, as in Plutarch's *De animae procreatione in Timaeo* 1025d–e and Alcinous's *Didaskalikos* 2.2. In the latter, Alcinous labels the state of the soul when engaged in contemplation (θεωρία) as prudence—perhaps because of the influence of *Phaedo* 79d[46]—and, contrasting the practical with the theoretical life, he remarks, "Action, on the other hand, and the practical [life], being pursued through the body, are subject to external hindrance and would be engaged in when circumstances demand by practicing the transferal to human affairs of the visions of the theoretical life."[47] In other words, visions in accordance with the theoretical life guide actions in the practical life. For an influence on this passage, John Dillon points to *Republic* VI, 500d–e, quoted above in relation to Plato's conception of the dependency of the practical on the theoretical.[48] Again, Socrates argues that the philosopher should transform the populace by putting what he sees (ὁρᾷ) of the ordered and divine into human beings' characters (εἰς ἀνθρώπων ἤθη). Both Aristotle's identification of prudence with perception and Socrates's prescription to apply what one sees when contemplating the divine seem to inform Alcinous's portrayal of how the theoretical life may impact the practical.

Alcinous's text seems to convey a similar meaning to Ptolemy's. Alcinous transfers visions of the contemplative life to human affairs, and Ptolemy applies *phantasiai* in his ordering of actions. Given this similarity, two questions arise: (1) Why does Ptolemy employ the term *phantasiai* rather than

---

44. Ptolemy, *Harmonics* 3.5, D97.

45. Aristotle, *Nicomachean Ethics* VI, 1142a29–30: ἀλλ' αὕτη μᾶλλον αἴσθησις ἢ φρόνησις, ἐκείνης δ' ἄλλο εἶδος.

46. Dillon, "Commentary," 55.

47. Alcinous, *Didaskalikos* 2.2, translation modified from Dillon, in Alcinous, *Handbook of Platonism*: Ἡ μέντοι πρᾶξις καὶ τὸ πρακτικὸν διὰ τοῦ σώματος περαινόμενα κωλυθῆναί τε δύναται καὶ πράττοιτο ἂν ἀπαιτούντων τῶν πραγμάτων, ἃ κατὰ τὸν θεωρητικὸν βίον ὁρᾶται, μελετῆσαι εἰς ἀνθρώπων ἤθη.

48. Dillon, "Commentary," 56.

"visions" or "that which is seen" (ἃ ὁρᾶται), which Alcinous uses? (2) Although Alcinous's visions are of the contemplative life, of what are Ptolemy's *phantasiai*, or, in other words, what does "these" (αὐτῶν) refer to in the phrase "according to the applications of the *phantasiai* of these" (ἐν ταῖς αὐτῶν τῶν φαντασιῶν ἐπιβολαῖς)"? The answer to the first question is apparent from Ptolemy's theory of perception and cognition. *Phantasia* is not simple perception but rather the impression of what is perceived into the soul and its transmission to the intellect. The relation of Ptolemy's phrase to Cassius Longinus's text highlights the particular significance of *phantasiai*. Again, Cassius Longinus defines recollection as the application of *phantasia* (ἐπιβολὴ φαντασίας). Ptolemy, then, uses *phantasiai* rather than visions because he is invoking a theory of memory. Unlike Cassius Longinus, Ptolemy does not define recollection (ἀνάμνησις) in his texts, but he does define memory, and he does so in conjunction with his definition of *phantasia*.[49] Again, in *On the Kritērion* Ptolemy states, "*phantasia* is the impression and transmission to the intellect, whose retention and memory of the things transmitted we call conception . . ."[50] Ptolemy's choice to define *phantasia* in relation to memory derives from his appropriation of Aristotle's theory of memory. According to Aristotle's *On Memory*, sense perception is of what is present to the individual, while memory is of the past. Impressions of what were previously thought or perceived are the objects of memory, and these impressions are *phantasiai*. Hence, by invoking *phantasiai* rather than visions, Ptolemy indicates that one should apply impressions that one remembers.

In response to the second question, I would argue that "these" (αὐτῶν) in "according to the applications of the *phantasiai* of these" (ἐν ταῖς αὐτῶν τῶν φαντασιῶν ἐπιβολαῖς) refers to "theories" (θεωρήμασι) in the previous sentence.[51] My contention that *phantasiai* may represent not only images but also theories—only metaphorically related to images—is, of course, controversial. One might take a more orthodox position and instead ascribe "these" (αὐτῶν) to "affairs" (πράγμασι) in the clause preceding the one containing "theories" (θεωρήμασι). For instance, Toomer translates the phrase as "under the impulse of our actual ideas [of what is to be done]."[52] The interpolated "[of what is to

---

49. The only occurrence of the term ἀνάμνησις in Ptolemy's corpus is in *Tetrabiblos* 4.5.8, H310, where it has a technical, astrological meaning.

50. Ptolemy, *On the Kritērion*, La5, trans. Liverpool-Manchester Seminar, in Ptolemaeus, "On the Kriterion": φαντασία δ' ἡ τύπωσις καὶ διάδοσις ἡ ἐπὶ τὸν νοῦν, οὗ τὴν κατοχὴν καὶ μνήμην τῶν διαδοθέντων καλοῦμεν ἔννοιαν

51. I must thank Alain Bernard for bringing to my attention the possible connection between these two terms in *Almagest* 1.1.

52. Toomer, in Ptolemy, *Almagest*, 35.

be done]" does not refer directly to the "affairs" (πράγμασι). It interpolates the term *prakton*, or "what is to be done," into the text, rather than drawing on the potential antecedents in the previous sentence. Still, Toomer's translation indicates that he interpreted the *phantasiai* to represent some phenomena related to action. His translation of *phantasiai* as "actual ideas," however, reveals that he did not have a strong sense of Ptolemy's psychological theory, and, consequently, his translation here should not be taken as authoritative. If Ptolemy meant the *phantasiai* to represent some phenomena related to action, whether what is to be done or the actual affairs, why would he prescribe the application of the *phantasiai*? The application, itself, implies that the *phantasiai* represent objects different in kind from the actions being ordered. In other words, impressions of actions do not order actions; impressions of higher-order phenomena, such as theories, order actions.

In her analysis of Aristotle's *On Memory*, Martha Nussbaum champions a claim similar to mine in the case of Aristotle's psychology. According to Nussbaum, Aristotle's *phantasiai* may occasionally represent phenomena that are only metaphorically pictorial. At *On Memory* 450b11ff., for instance, Aristotle discusses the nature of what one remembers. He repeatedly mentions the contemplation (θεωρεῖν) of images, and at 450b26 he juxtaposes an object of contemplation, or theory (θεώρημα), with an image (φάντασμα): "just in the same way we have to conceive that the image within us is both something in itself and relative to something else. In so far as it is regarded in itself, it is only an object of contemplation or an image . . ."[53] Nussbaum contends that Aristotle most likely did not consider the object of contemplation in this passage and the lines that follow to be pictorial.[54] Her interpretation is controversial, as Aristotle repeatedly contends that all thought depends on *phantasiai*, and the conventional interpretation supposes that *phantasiai* are always pictorial representations.[55] Nussbaum, however, aims to recast *phantasiai* as selective and interpretive data, which may or may not be pictorial. Whether or not Aristotle conceived of *phantasiai* as only pictorial, the ambiguity in his corpus on the nature of *phantasiai* may have influenced Ptolemy's theory that *phantasiai* are not only juxtaposed with but, an even stronger correlation, representative of theories.

---

53. Aristotle, *On Memory* 450b24–26, translation modified from J. I. Beare, in Aristotle, *Complete Works*: οὕτω καὶ τὸ ἐν ἡμῖν φάντασμα δεῖ ὑπολαβεῖν καὶ αὐτό τι καθ' αὑτὸ εἶναι καὶ ἄλλου [φάντασμα]. ἡ μὲν οὖν καθ' αὑτό, θεώρημα ἢ φάντασμά ἐστιν

54. Nussbaum, *Aristotle's "De Motu Animalium,"* 250.

55. See Aristotle, *De anima* III, 432a10–14, and *On Memory* 449b30–450a1, where Aristotle cites *De anima*.

Although Aristotle's theory of cognition significantly shaped ancient theories of *phantasia*, the Stoic tradition had just as much influence on the concept's development. The Stoics unified the mind and cast it as a purely rational faculty. This unification allowed them to expand the faculty of *phantasia* to encompass both perceptual and nonperceptual impressions, including abstract thought.[56] Brad Inwood has examined evidence for a Stoic theory of propositional *phantasiai*, but he ultimately dismissed it as inconsistent with Stoic psychological theory.[57] Inwood's argument is compelling, but what is relevant to this discussion is the possibility that a theory could have evolved—perhaps with a foundation in the Stoic theory of *phantasia* and/or Aristotle's ambiguous portrayal of nonpictorial *phantasiai*—where *phantasiai* represent not only images and thoughts but also theories. This conceptual evolution is certainly speculative, but it would support the relation, syntactic as well as semantic, of "these" (αὐτῶν) in Ptolemy's "according to the applications of the *phantasiai* of these" (ἐν ταῖς αὐτῶν τῶν φαντασιῶν ἐπιβολαῖς) to "theories" (θεωρήμασι) in the previous sentence. Furthermore, it would establish a strong correspondence between Ptolemy's and Alcinous's texts, such that Ptolemy's applications of *phantasiai* of theories to the ordering of actions would convey a similar meaning to Alcinous's transferal of visions of the contemplative life to human affairs.

Resembling the Divine

Whether or not the *phantasiai* in *Almagest* 1.1 represent theories, the reason Ptolemy provides for why one should order actions in this way indicates that theoretical philosophy is relevant to the ordering. Again, Ptolemy claims that one should order actions in order to strive, even in ordinary affairs, for a fine and well-ordered state (τὴν καλὴν καὶ εὔτακτον κατάστασιν). He explains what he means by, and how one achieves, this fine and well-ordered state near the end of *Almagest* 1.1. Promulgating a theory common in ancient Greek philosophy and rampant in the Platonic tradition, Ptolemy champions the *telos*, or objective, of becoming godlike:

> With regard to virtuous conduct in actions and character, [mathematics], above all, could make clear-sighted men; from the constancy, good order, commensurability, and calm that are contemplated in the case of the divine, it, on the one hand, makes its followers lovers of this divine beauty, and, on

56. See Long, "Stoic Psychology," 572–76.
57. Inwood, *Ethics and Human Action*, 11–13, 55–60.

the other hand, accustoms and, as it were, reforms their natures to a similar state of the soul.[58]

Ptolemy embraces the *telos* of *homoiôsis theôi* broadly construed. He advocates transforming the soul to a state similar to the divine. The means by which one achieves this goal is mathematics, and astronomy in particular. Ptolemy reaffirms this view in *Almagest* 4.9, where he calls the goal, or promise (ἐπαγγελία), of astronomy divine (θεία).[59]

For early endorsements of *homoiôsis theôi*, one may look to Plato's *Theaetetus* 176b and *Republic* VI, 500c–d.[60] Both texts maintain that the philosopher should transform himself into a condition as similar to the divine as is possible for human beings, where contemplation of the divine makes one like the divine. In the second century, several Platonic philosophers appropriated this objective. The anonymous commentator on the *Theaetetus* confirms this goal of human beings at 7.18–19; in *On Plato and His Doctrine* 2.23.252, Apuleius affirms that the end of wisdom is the emulation of the gods;[61] in *Introduction to the Dialogues of Plato* 6.25–29, Albinus asserts that the theoretical life, referring to the study of physics and ethics, and the practical life, which involves the practice of politics and economics, together yield a becoming like a god (τὸ ὁμοιωθῆναι θεῷ). Alcinous dedicates an entire chapter, *Didaskalikos* 28, to *homoiôsis theôi*. Tracing the doctrine through several of Plato's texts, he quotes the *Theaetetus*, *Phaedo*, and *Laws*. Thus, the concept of *homoiôsis theôi* was common in the second century, and in *Almagest* 1.1 Ptolemy joins this tradition.

Plato's *Timaeus* influenced Ptolemy's ethical statement more obviously than did the *Theaetetus* and *Republic*, for Ptolemy claims that one attains a state of the soul similar to the divine through the contemplation of astronomical objects. Correspondingly, Timaeus explains how individuals may organize the revolutions of their souls according to the order of the universe:

> the god invented sight and gave it to us so that we might observe the orbits of intelligence in the heavens and apply them to the revolutions of our own

---

58. Ptolemy, *Almagest* 1.1, H7: πρός γε μὴν τὴν κατὰ τὰς πράξεις καὶ τὸ ἦθος καλοκαγαθίαν πάντων ἂν αὕτη μάλιστα διορατικοὺς κατασκευάσειεν ἀπὸ τῆς περὶ τὰ θεῖα θεωρουμένης ὁμοιότητος καὶ εὐταξίας καὶ συμμετρίας καὶ ἀτυφίας ἐραστὰς μὲν ποιοῦσα τοὺς παρακολουθοῦντας τοῦ θείου τούτου κάλλους, ἐνεθίζουσα δὲ καὶ ὥσπερ φυσιοῦσα πρὸς τὴν ὁμοίαν τῆς ψυχῆς κατάστασιν.

59. Ibid. 4.9, H328.

60. Liba Taub traces the roots of Ptolemy's statement to three of Plato's texts: the *Symposium*, *Republic*, and *Theaetetus*: Taub, *Ptolemy's Universe*, 31–33.

61. Cf. Apuleius *De Platone et eius dogmate* 2.5.227, where Apuleius defines virtue as a condition of the mind that renders one concordant with oneself, calm, and constant.

understanding, for there is a kinship between them, [even though our revolutions are] disturbed, [whereas the universal orbits are] undisturbed, and so once we have come to know them and to share in the ability to make correct calculations according to nature, we should stabilize the straying revolutions within ourselves by imitating the completely unstraying revolutions of the god.[62]

Because a kinship exists between the immortal part of a human soul and the heavens, one can model the soul's revolutions after the order of the heavens. In this way, individuals stabilize their souls' revolutions and return their souls to their proper order. According to Ptolemy, the study of mathematics and, in particular, the contemplation of the constancy, good order, commensurability, and calm of astronomical objects makes the individual a lover of divine beauty and transforms his soul to a state similar to the one contemplated.

Strikingly, three terms Ptolemy uses to signify the divine qualities one contemplates—constancy, good order, and commensurability (ὁμοιότητος καὶ εὐταξίας καὶ συμμετρίας)—indicate that the objects contemplated are distinctly mathematical in kind. Examining the relation of the fine and good to mathematics in *Metaphysics* M, Aristotle states, "The chief forms of the beautiful are order, commensurability, and definiteness [τάξις καὶ συμμετρία καὶ τὸ ὡρισμένον], which are what the mathematical sciences show to the highest degree."[63] In addition, Alexander of Aphrodisias, the second-century Peripatetic philosopher, consistently uses the term "well-ordered" when describing the movement of heavenly bodies. The most explicit instances of the term's astronomical connotation are in his *De fato*, where, according to Charles Genequand, the influence of heavenly bodies' movements on sublunary generation and corruption is a central notion in the theory of providence.[64] Alexander states the following in the *De fato*: "It is for this reason, too, that men say that the first causes of the coming-to-be of each thing in accordance with nature (that is, the divine and their well-ordered revolution) are also causes of fate. For the beginning of all coming-to-be is the divine, in their motion, being

62. Plato, *Timaeus* 47b–c, translation modified from Donald J. Zeyl in Plato, *Complete Works*: θεὸν ἡμῖν ἀνευρεῖν δωρήσασθαί τε ὄψιν, ἵνα τὰς ἐν οὐρανῷ τοῦ νοῦ κατιδόντες περιόδους χρησαίμεθα ἐπὶ τὰς περιφορὰς τὰς τῆς παρ' ἡμῖν διανοήσεως, συγγενεῖς ἐκείναις οὔσας, ἀταράκτοις τεταραγμέναις, ἐκμαθόντες δὲ καὶ λογισμῶν κατὰ φύσιν ὀρθότητος μετασχόντες, μιμούμενοι τὰς τοῦ θεοῦ πάντως ἀπλανεῖς οὔσας, τὰς ἐν ἡμῖν πεπλανημένας καταστησαίμεθα. Cf. Plato, *Timaeus* 44d, 90a–d.

63. Aristotle, *Metaphysics* XIII, 1078a36–b2: τοῦ δὲ καλοῦ μέγιστα εἴδη τάξις καὶ συμμετρία καὶ τὸ ὡρισμένον, ἃ μάλιστα δεικνύουσιν αἱ μαθηματικαὶ ἐπιστῆμαι.

64. Genequand, "Quelques aspects," 115.

in one type of position or another to things on [earth]."[65] The divine things, whose well-ordered revolution determines the fate of sublunary bodies, are manifestly heavenly bodies. Hence, the terms Ptolemy lists as exemplars of divine qualities connote mathematical and, in particular, astronomical entities. For Ptolemy, the contemplation of astronomical objects—the movements and configurations of celestial bodies, rather than the celestial bodies in and of themselves—makes the individual a lover of divine beauty. Transforming his soul to a state similar to the divine, heavenly one, an individual imposes onto his soul mathematical and, in particular, astronomical qualities: constancy, good order, and commensurability.

The term "good order" (εὐταξία) appears again in *Harmonics* 3.4, and it is the only other instance of the term in Ptolemy's corpus outside of *Almagest* 1.1. Its use here is particularly illuminating. Defining the power of *harmonia*, which I examine in chapter 5, Ptolemy asserts the following:

> Let this be enough to show that the power of *harmonia* is a form of the cause corresponding to reason, the form that concerns itself with the commensurability of movements, and that the theoretical science of *harmonia* is a form of mathematics, the [form] concerned with the ratios of differences between things that are heard, this [form] itself contributing to the good order [εὐταξίαν] that arises out of the theory and understanding to people habituated [ἐθιζομένοις] in it.[66]

As in *Almagest* 1.1, Ptolemy proclaims that mathematical study leads to good order. In the *Almagest*, Ptolemy emphasizes the psychological benefits of astronomy, and here in the *Harmonics* he recommends astronomy's cousin science, harmonics. In taking not only astronomy but also harmonics as capable of organizing the soul's parts, Ptolemy follows the *Timaeus*.[67]

Continuing on from the passage above, Ptolemy argues that the power of *harmonia* is present in all objects that have in them a principle of movement, but it is present to the greatest extent in those objects that have a more

65. Alexander, *De fato* 169.23–26, translation modified from Sharples, *On Fate*: διὸ καὶ τὰ πρῶτα τῆς κατὰ φύσιν ἑκάστοις γενέσεως αἴτια (ἔστιν δὲ ταῦτα <τὰ> θεῖα καὶ ἡ τούτων εὔτακτος περιφορὰ) καὶ τῆς εἱμαρμένης αἴτια λέγουσιν. πάσης γὰρ γενέσεως ἀρχὴ ἡ τῶν θείων κατὰ τὴν κίνησιν ποιὰ σχέσις πρὸς τὰ τῇδε. Cf. 203.22–3.

66. Ptolemy, *Harmonics* 3.4, D94–95: Ὅτι μὲν οὖν ἥ τε τῆς ἁρμονίας δύναμις εἶδός ἐστι τοῦ παρὰ τὸν λόγον αἰτίου τὸ περὶ τὰς τῶν κινήσεων συμμετρίας, καὶ ἡ θεωρητικὴ ταύτης ἐπιστήμη μαθηματικῆς ἐστιν εἶδος τὸ περὶ τοὺς λόγους τῶν ἀκουστῶν διαφορῶν, καὶ αὐτὸ διατεῖνον ἐπὶ τὴν ἐκ τῆς θεωρίας καὶ παρακολουθήσεως περιγινομένην τοῖς ἐθιζομένοις εὐταξίαν, διὰ τούτων ὑποτετυπώσθω.

67. See Plato, *Timaeus* 47c-d.

complete and rational nature (τελειοτέρας καὶ λογικωτέρας φύσεως).[68] In other words, every natural body is characterized to some degree by a ratio, in its movements and in the configuration of its matter, and the more complete and rational an object is in its movements and form, the more it is characterized by harmonic ratios. According to Ptolemy, the most complete and rational objects are, in addition to musical pitches, heavenly bodies and human souls:

> [the power of *harmonia* is found] in the [movements] that are engaged most closely with forms. These, as we said, are the [movements] of things that are more complete and more rational in their natures, as among divine things are the [movements] of the heavenly bodies and among mortal things the [movements] of human souls, most particularly, since it is only to each of these of the aforementioned things that there belong not only the primary and most complete movement, that in respect of place, but also the characteristic of being rational.[69]

In their movements and relations, musical pitches, heavenly bodies, and human souls exhibit harmonic ratios. Because human souls are corruptible, human beings need to work to maintain the proper relations among their souls' parts. The harmonic form of the human soul entails the objective of maintaining this harmonic form, and in fact Ptolemy identifies formal and teleological causes in *Harmonics* 3.3: "Since all things, then, have as their first principles matter and movement and form, matter corresponding to the underlying and the out of which, movement to the cause and the by which, and form to the end and the for the sake of which ..."[70] Human beings can reform their souls to a harmonic structure, return them to their proper form, a good order, by contemplating the movements of musical pitches and/or heavenly bodies.

Returning to the *Almagest*, Ptolemy uses a form of the term "good order" (εὐταξία) twice in *Almagest* 1.1. In the first instance, he claims, as mentioned above, that one should order one's actions by applying *phantasiai* so as never to forget even in ordinary affairs and practical philosophy to strive for a fine and well-ordered state (τὴν καλὴν καὶ εὔτακτον κατάστασιν). The second instance is the one just examined, where Ptolemy lists "good order" (εὐταξία) among

---

68. Ptolemy, *Harmonics* 3.4, D95.

69. Ibid.: ἐπὶ δὲ τῶν ἐν τοῖς εἴδεσι τὸ πλεῖστον ἀναστρεφομένων. αὗται δέ εἰσιν αἱ τῶν τελειοτέρων, ὡς ἔφαμεν, καὶ λογικωτέρων τὰς φύσεις, ὡς ἐπὶ μὲν τῶν θείων αἱ τῶν οὐρανίων, ἐπὶ δὲ τῶν θνητῶν αἱ τῶν ἀνθρωπίνων μάλιστα ψυχῶν, ὅτι μόνοις ἑκατέροις τῶν εἰρημένων μετὰ τῆς πρώτης καὶ τελειοτάτης κινήσεως, τουτέστι τῆς κατὰ τόπον, ἔτι καὶ τὸ λογικοῖς εἶναι συμβέβηκεν.

70. Ptolemy, *Harmonics* 3.3, D92: Τῶν ὄντων τοίνυν ἁπάντων ἀρχαῖς κεχρημένων ὕλῃ καὶ κινήσει καὶ εἴδει, τῇ μὲν ὕλῃ κατὰ τὸ ὑποκείμενον καὶ ἐξ οὗ, τῇ δὲ κινήσει κατὰ τὸ αἴτιον καὶ ὑφ' οὗ, τῷ δὲ εἴδει κατὰ τὸ τέλος καὶ τὸ οὗ ἕνεκεν

the qualities of astronomical objects and explains that a well-ordered state of the soul results from the contemplation of astronomical objects. Comparison of these passages reveals how the *telos* of resembling the divine relates to theoretical and practical philosophy. The *telos* is achievable by means of theoretical philosophy—specifically mathematics, astronomy and/or harmonics—and one should keep this *telos* in mind when engaged in practical philosophy as well as in ordinary affairs.

That one achieves the *telos* by means of theoretical philosophy is perhaps surprising, for Ptolemy introduces it with a nod to actions and character, or, presumably, practical philosophy. Again, he asserts, "With regard to virtuous conduct in actions and character, [mathematics], above all, could make clear-sighted men . . ."[71] Ptolemy posits here a particular relationship between theoretical and practical philosophy. The contemplation of astronomical objects and the resulting transformation of the soul relate to virtuous conduct in actions and character. The way they relate is that theoretical study prepares an individual for action. Ptolemy claims that mathematics, more than any other part of philosophy, makes one clear-sighted (διορατικός). I contend that this clear-sightedness relates to the *phantasiai* mentioned earlier in *Almagest* 1.1. Again, I argue that Ptolemy's use of *phantasiai* may derive in part from Aristotle's definition of prudence (φρόνησις) as a type of perception in *Nicomachean Ethics* VI, 1142a, and this interpretation is supported further by the correspondence of Ptolemy's text to Alcinous's prescription to transfer to human affairs the visions of the contemplative life. For Ptolemy, clear-sightedness enhances one's application of *phantasiai* to actions.

Clear-sightedness has this effect because of the relationship between sight and the soul's faculty of thought. In *On the Kritêrion and Hêgemonikon*, Ptolemy discusses the relationship of the senses to the faculty of thought, and regarding four of the five senses—sight, hearing, smell, and taste—he argues, "moreover, of these some are more easily moved and more valuable, i.e., sight and hearing, and because they are located above the others they are extended more toward the soul's faculty of thought . . ."[72] Because the senses of sight and hearing are located in a human being's head higher than the senses of taste and smell, they are physically closer to the soul's faculty of thought, the most valuable of the human soul's faculties and the *hêgemonikon*, the ruling part

---

71. Ptolemy, *Almagest* 1.1, H7: πρός γε μὴν τὴν κατὰ τὰς πράξεις καὶ τὸ ἦθος καλοκαγαθίαν πάντων ἂν αὕτη μάλιστα διορατικοὺς κατασκευάσειεν

72. Ptolemy, *On the Kritêrion*, La20–21: τούτων τ' αὖ τὰς μὲν μᾶλλον εὐκινητοτέρας καὶ τιμιωτέρας, ὄψιν καὶ ἀκοήν, ἀνωτέρας τε οὔσας τῶν ἄλλων μᾶλλον τετάσθαι πρὸς τὸ διανοητικὸν τῆς ψυχῆς

of the soul in regard to both living and living well.[73] Before pronouncing the faculty of thought the *hêgemonikon*, Ptolemy states:

> If we give the name *hêgemonikon* to what is the best absolutely and the most valuable, it will be located in the brain. We have given sufficient proof that the faculty of thought has a higher degree of value and divinity, both in power and in substance and both in the universe and in us, and also that its place is the highest position, the heavens in the cosmos and the head in man.[74]

In other words, the faculty of thought is more valuable and more divine than the other faculties of the human soul, just as the heavens are divine and more valuable than the sublunary realm. The reference to a faculty of thought in the heavens may refer to the Prime Mover of *Almagest* 1.1 or the universal animal (*al-ḥayawân al-kullî*) or celestial souls in the *Planetary Hypotheses*.[75]

Although in a human being the faculty of thought is the *hêgemonikon* in regard to both living and living well, sight and hearing assist in the latter purpose. In *On the Kritêrion*, Ptolemy ascribes the *telos* of living well to sight and hearing:

> If a second prize has to be awarded to one of the other means toward the end of living well, the prize would go elsewhere than to the faculty of thought and what is around the heart would not even be runner-up. It would go rather to the senses, and, if not to all of them, then only to those that contribute most to assist thought in its consideration and judging of actual things, i.e., hearing and sight, which are themselves positioned near the head and the brain, above the other [senses], and near neighbors, because of their special relationship, to the first and chief cause of living well.[76]

73. Ibid., La22.

74. Ibid., translation modified from Liverpool-Manchester Seminar, in Ptolemaeus, "On the Kriterion": ἐὰν μὲν γὰρ τὸ βέλτιστον ἁπλῶς καὶ τιμιώτατον καλῶμεν ἡγεμονικόν, ἐν ἐγκεφάλῳ τοῦτο ἔσται. δέδεικται γὰρ ἡμῖν ἱκανῶς, ὅτι τὸ διανοητικὸν καὶ δυνάμει καὶ οὐσίᾳ τιμιώτερον καὶ θειότερόν ἐστιν ἔν τε τῷ παντὶ καὶ ἐν ἡμῖν, καὶ ὅτι τόπος αὐτοῦ τὰ ἀνωτάτω, τοῦ κόσμου μὲν ὁ οὐρανός, ἀνθρώπου δὲ ἡ κορυφή.

75. See chapter 8. On the aethereal, and therefore more divine, composition of the human soul's faculty of thought, see chapter 7.

76. Ptolemy, *On the Kritêrion*, La23, translation modified from Liverpool-Manchester Seminar, in Ptolemaeus, "On the Kriterion": εἰ δὲ καὶ δευτερεῖά τινι τῶν ἄλλων τῶν πρὸς τὸ τοῦ εὖ ζῆν τέλος δοτέον, ἕτερον ἂν εἴη τὸ μετὰ τὸ διανοητικὸν δεύτερον καὶ οὐδ' ὡς τὸ περὶ τὴν καρδίαν. μᾶλλον γὰρ αἱ αἰσθήσεις καὶ, εἰ μὴ πᾶσαι, μόναι γε αἱ πρὸς τὸ θεωρεῖν καὶ κρίνειν τὰ πράγματα πλεῖστον τῇ διανοίᾳ συμβαλλόμεναι, τοῦτ' ἐστιν ἀκοή τε καὶ ὄψις, αἵ καὶ αὐταὶ κατὰ τὴν κορυφὴν

According to *On the Kritêrion*, sight and hearing assist the faculty of thought in causing a human being to live in a good way. Therefore, clear-sightedness is integral to a good life, and, as such, it is a means to virtuous conduct in actions and character, or success in practical philosophy, as implied in *Almagest* 1.1.

What is it that the clear-sighted men see? If the "sightedness" is literal, then the passage in *Almagest* 1.1 suggests that they see the celestial bodies' movements and configurations. The benefit to virtuous conduct of this literal clear-sightedness would result from the close relationship between sight and the faculty of thought. Literally seeing in a clear way would entail metaphorically seeing how to act. How the simple observation of celestial bodies' movements results in virtuous conduct, however, is unclear. Therefore, the sightedness, like the subsequent seeing how to act, may be metaphorical. With astronomy, one does more than simply observe celestial bodies' movements and configurations; the mathematician constructs hypotheses that account for them. In this metaphorical reading, what the men clearly see are mathematical theories, accounts of the movements and configurations of celestial bodies that reveal their constancy, good order, and commensurability, which are apparent not from simple observation but only from mathematical study. These mathematical theories are the catalysts that, when contemplated, make an individual a lover of divine, heavenly beauty and transform his soul to a similar state.

This metaphorical interpretation of "clear-sightedness" connects this passage more firmly to Ptolemy's employment of *phantasiai* earlier in *Almagest* 1.1. If the clear-sighted men see mathematical theories, then the application of *phantasiai* of theories would indeed order an individual's actions. Representations of mathematical theories of constant, well-ordered, and commensurable phenomena would allow one to order one's actions in a manner concordant with these divine properties, and they would ensure that the individual keep in mind the *telos* of resembling the divine, which results from these qualities' contemplation. Indeed, Ptolemy indicates that fidelity to this *telos* is the goal he has in mind for ordering actions; he proclaims that one should apply *phantasiai* to the ordering of actions so as never to forget, even in ordinary affairs, to strive for a fine and well-ordered state.

In addition, the divine-like state itself, like clear-sightedness, prepares the individual for future action. Evidence for this interpretation lies in *Harmonics* 3.7, wherein Ptolemy discusses how listening to music affects the human soul. As a melody modulates, the soul—consisting of the same ratios as exist among musical pitches—experiences various conditions in relation to the

καὶ τὸν ἐγκέφαλόν εἰσι τετάγμεναι τῶν γε ἄλλων ἀνωτάτω καὶ γειτνιῶσαι δι᾿ οἰκειότητα τῷ πρώτῳ τοῦ καλῶς ζῆν αἰτιωτάτῳ.

activities of the melody. With reference to this theory, Ptolemy concludes the following:

> I suppose it was because he understood this fact that Pythagoras advised people that when they arose at dawn, before setting off on any activity, they should apply themselves to music and to soothing melody, so that the disorder of their souls resulting from arousal out of sleep should first be transformed into a pure and settled state and an orderly gentleness, and so make their souls well attuned and concordant for the actions of the day.[77]

In this passage, Ptolemy employs his psychological theory to explain the legendary habits of Pythagoras and his followers. Similarly, Aristoxenus, the fourth-century BCE music theorist, wrote that the Pythagoreans used music to purify the soul,[78] and in the first century CE Quintilian, the Roman rhetorician, wrote, "On awakening, it was the Pythagoreans' custom to arouse their souls with the sound of the lyre, so that they might be more alert for action, and before going to sleep they soothed their minds by means of this same music in order to calm them down, in case too turbulent thoughts might still inhabit them."[79] Hence, Quintilian, like Ptolemy after him, portrays the Pythagoreans as listening to music every morning in order to prepare their souls for action. He depicts the morning music as arousing, or exciting, a soul, but Ptolemy claims that Pythagoras's music transforms the soul into a pure and settled state. Similarly to Ptolemy, Porphyry in the third century cites the novelist Antonius Diogenes, who depicts Pythagoras as singing paeans at dawn in order to calm the soul (καθημεροῦν τὴν ψυχήν).[80] Therefore, in *Harmonics* 3.7 Ptolemy relates a common story about Pythagoras and, moreover, he explains Pythagoras's legendary practice. According to Ptolemy, Pythagoras advised people to settle their souls into a state (κατάστασις) aligned by harmonic ratios in order to prepare for ensuing actions. I suggest that the *Harmonics'* relation of the soul's harmonic state to success in future actions underlies the *Almagest's* relation of theoretical and practical philosophy. The reformation of one's soul

77. Ptolemy, *Harmonics* 3.7, D100: ὅπερ οἶμαι καὶ τὸν Πυθαγόραν κατανενοηκότα παραινεῖν ἅμα ἕω διαναστάντας, πρὶν ἄρξασθαί τινος ἐνεργείας, μούσης ἅπτεσθαι καὶ μελῳδίας προσηνοῦς, ὅπως τὸ ἀπὸ τῆς διεγέρσεως τῶν ὕπνων περὶ τὰς ψυχὰς ταραχῶδες, πρότερον εἰς κατάστασιν εἰλικρινῆ καὶ πραότητα τεταγμένην μεταβαλόν, εὐαρμόστους αὐτὰς καὶ συμφώνους ἐπὶ τὰς ἡμερησίους πράξεις παρασκευάζῃ.

78. Aristoxenus, *Fragmenta* 26.

79. Quintilian, *Institutio Oratoria* 9.4.12, trans. Riedweg, in *Pythagoras: His Life*: Pythagoreis certe moris fuit et cum evigilassent animos ad lyram excitare, quo essent ad agendum erectiores, et cum somnum peterent ad eandem prius lenire mentes, ut si quid fuisset turbidiorum cogitationum componerent.

80. Porphyry, *Vita Pythagorae* 32.1–9. Cf. Iamblichus, *De vita Pythagorica* 25.110–14.

to a divine-like state is the product of theoretical philosophy, specifically mathematics, and the divine-like state—while it is an objective in and of itself—has the additional benefit of enabling virtuous conduct in actions, the domain of practical philosophy.

Thus, despite his initial effort to assert the independence of theoretical and practical philosophy, to posit a great difference between them, Ptolemy ultimately portrays practical philosophy as dependent on theoretical philosophy. Furthermore, the great difference between them that Ptolemy articulates loses its force when one brings the *Harmonics* to bear on the *Almagest*. Again, in *Almagest* 1.1, Ptolemy allows that some moral virtues may benefit from learning, even though some are possible without it, but he portrays theoretical philosophy as subject to instruction only. In *Harmonics* 3.4, however, it is the people who are habituated (ἐθιζομένοις) in theoretical philosophy—and in harmonics, a mathematical science, in particular—who attain good order in their souls. Ptolemy's distinction between theoretical and practical philosophy, then, is not absolute. Instruction in theoretical philosophy prepares one to act in practical philosophy, and habit, the tool of practical philosophy, is integral to theoretical philosophy.

This pattern—of asserting a great distinction and then dissolving it—also appears in *On the Kritêrion and Hêgemonikon*. I mentioned above how Ptolemy distinguishes soul and body. Later in the text, however, he reveals his materialism. The soul consists of fine particles that scatter when released from the body: "the soul is so constituted as to scatter immediately to its proper elements, like water or breath released from a container, because of the preponderance of finer particles ... the body, on the other hand, although it stays in the same state for a considerable time because of the thicker consistency of its matter ...".[81] The soul consists of finer particles than the particles constituting the body, and these finer particles are so small that they are imperceptible. Correspondingly, Ptolemy affirms in *Tetrabiblos* 3.12 that the body is more material (ὑλικώτερον) than the soul.[82] Therefore, Ptolemy's more detailed expositions of soul and body reveal that he considered them to differ merely in degree rather than kind. They produce different effects in human beings, but they both are material. They consist of particles, such that the soul's particles are smaller in size than the body's. Adhering to a similar structure of exposition, Ptolemy asserts in *Almagest* 1.1 a great difference between theoretical and

81. Ptolemy, *On the Kritêrion*, La12, translation modified from Liverpool-Manchester Seminar, in Ptolemaeus, "On the Kriterion": καθ' ἥν ἡ μὲν ψυχὴ διὰ τὴν ὑπερβολὴν τῆς λεπτομερείας καθάπερ ὕδωρ ἢ πνεῦμα τοῦ συνέχοντος ἀνεθὲν εὐθὺς εἰς τὰ οἰκεῖα στοιχεῖα πέφυκε χωρεῖν [. . .] τὸ δὲ σῶμα διὰ τὸ τῆς ὕλης παχυμερὲς ἐπιδιαμένον συχνὸν χρόνον ἐν ταῖς αὐταῖς καταστάσεσιν

82. Ptolemy, *Tetrabiblos* 3.12.1, H224.

practical philosophy, which establishes their independence from one another, but he goes on to portray practical philosophy as dependent on theoretical philosophy.

Conclusion

Thus, in *Almagest* 1.1 Ptolemy engages in a contemporary debate over the relationship between theoretical and practical philosophy. He first asserts that the two are independent, differentiated by the manner in which one attains virtues in each domain, whether by instruction or continuous activity. Thereafter, he diminishes the distinction by revealing how they relate. Theoretical philosophy, specifically mathematics, transforms the soul. The study of astronomical objects—the movements and configurations of heavenly bodies—reveals their constancy, good order, commensurability, and calm. Mathematicians, aided by habit, come to appreciate these qualities and transform their souls into a fine and well-ordered state. Organizing their actions in accordance with astronomical theories, they never forget their ultimate objective, the divine-like condition of the soul. The study of mathematics is crucial to obtaining this good life. To gain knowledge, to conduct oneself well in any and all aspects of life, and to achieve the highest objective, to resemble the divine, one must study mathematics.

This ethical objective, to live well by striving to resemble the divine, is Ptolemy's explicit motivation for studying mathematics. If we take him at his word, then Ptolemy dedicated his life to mathematics because he believed its study transformed his soul into an excellent condition, one that rendered his life the best possible. Ptolemy's single-minded dedication to mathematics is evident throughout his corpus. He composed numerous mathematical texts and even in his studies of physical phenomena he applies mathematics. In the following chapters I delve deeper into Ptolemy's studies and applications of mathematics and I examine how his metaphysics, epistemology, and ethics inform his studies of the mathematical and physical sciences, including harmonics, astronomy, psychology, astrology, and cosmology.

5

Harmonia

PTOLEMY DECLARES in *Almagest* 1.1 that the study of the constancy, good order, commensurability, and calm of astronomical objects makes the mathematician a lover of this divine beauty and transforms his soul to a similar state. His soul becomes constant, well ordered, commensurable, and calm; but besides their possession of, or ability to acquire, these four characteristics, what do astronomical objects and human souls have in common? What makes the stars' movements and configurations not just similar to but, even more, exemplars for human souls? Ptolemy reveals in the *Harmonics* that the heavens and human souls share the same formal structure. They are characterized by the same set of ratios, the ones studied by harmonics, the mathematical science that investigates the relations among musical pitches. In other words, harmonic ratios describe the relations among not only musical pitches but also the parts of the human soul and the movements and configurations of heavenly bodies. Moreover, Ptolemy explains why it is that these various objects are characterized by these same ratios. It is because of *harmonia*, a concept that I argue is the crux of Ptolemy's ethical theory. It is because *harmonia* operates most fully in human souls, the heavens, and musical scale systems that human beings' mathematical study of the latter two sets of phenomena has the power to shape and transform their souls.

Harmonia: What and Where It Is

To understand what *harmonia* is, one must first understand Ptolemy's general causal framework. It is well known that Aristotle advanced four causes: the material, formal, moving (or efficient), and final.[1] Ptolemy, on the other hand, propounded a total of three principles (ἀρχαί): matter, form, and movement.

---

1. Aristotle also calls the causes "principles" (ἀρχαί). See especially *Metaphysics* XII, 1070b22–27, where he explains that a principle is a cause, and he refers to "four causes and principles" (αἰτίαι δὲ καὶ ἀρχαὶ τέτταρες). At *Metaphysics* V, 1013a17, he says "all causes are principles" (πάντα γὰρ τὰ

He mentions these three in *Almagest* 1.1, for instance, where he says, "As all things are composed of matter and form and movement, each of which cannot be observed, but only conceived of, in its substratum by itself and without the others . . ."[2] In the *Harmonics*, he posits the same three principles and defines them: "Since all things, then, have as their first principles matter and movement and form, matter corresponding to the underlying and the out of which, movement to the cause and the by which, and form to the end and the for the sake of which . . ."[3] According to Ptolemy, matter is what underlies a body, like Aristotle's material cause; movement is the principle that produces some end, like Aristotle's efficient cause; form is the end itself, identifying Aristotle's formal and final causes. The union of formal and final causes is not entirely novel, as it has a precedent even in Aristotle's corpus. In *Physics* II, 198a24–26, for instance, Aristotle acknowledges that the formal and final causes often coincide. For Ptolemy, however, it is not simply the case that in particular hylomorphic bundles the formal and final principles are identical; rather, they are identical as a rule. The formal is the final.

Ptolemy further elaborates on this causal framework by positing three fundamental kinds of cause (αἴτιον), which are species of the moving principle. Again, these causes produce forms, or ends, in hylomorphic bundles. The three causes are nature, reason, and god, which Ptolemy describes as follows: "Now, as causes at the highest level are understood threefold, one corresponding to nature and only being, one corresponding to reason and only being well, and one corresponding to god and being well and eternally . . ."[4] Ptolemy draws on the ontology of the three theoretical sciences—physics, mathematics, and

---

αἴτια ἀρχαί). For Ptolemy, principles and causes are not identical; the causes are species of the moving principle.

2. Ptolemy, *Almagest* 1.1, H5: πάντων γὰρ τῶν ὄντων τὴν ὕπαρξιν ἐχόντων ἔκ τε ὕλης καὶ εἴδους καὶ κινήσεως χωρὶς μὲν ἑκάστου τούτων κατὰ τὸ ὑποκείμενον θεωρεῖσθαι μὴ δυναμένου, νοεῖσθαι δὲ μόνον, καὶ ἄνευ τῶν λοιπῶν

3. Ptolemy, *Harmonics* 3.3, D92: Τῶν ὄντων τοίνυν ἁπάντων ἀρχαῖς κεχρημένων ὕλῃ καὶ κινήσει καὶ εἴδει, τῇ μὲν ὕλῃ κατὰ τὸ ὑποκείμενον καὶ ἐξ οὗ, τῇ δὲ κινήσει κατὰ τὸ αἴτιον καὶ ὑφ' οὗ, τῷ δὲ εἴδει κατὰ τὸ τέλος καὶ τὸ οὗ ἕνεκεν

4. Ibid.: καὶ μὴν τῶν αἰτίων τῶν ἀνωτάτω τριχῶς λαμβανομένων, τοῦ μὲν παρὰ τὴν φύσιν καὶ τὸ εἶναι μόνον, τοῦ δὲ παρὰ τὸν λόγον καὶ τὸ εὖ εἶναι μόνον, τοῦ δὲ παρὰ τὸν θεὸν καὶ τὸ εὖ καὶ ἀεὶ εἶναι

Ptolemy's three causes may derive from the Stoic active principle—both reason and god—and nature, which often is equated with god in Stoic philosophy. For an account of the Stoic passive and active principles, see Diogenes Laertius, *Lives* 7.134. For an analysis of the principles, as well as their relation to nature, in Stoicism see especially Gourinat, "Stoics on Matter." Ptolemy separates what are identified in Stoicism. Although nature, reason, and god are all causes, they are distinct causes.

theology—to demarcate the three causes. Nature is the physical cause, which preserves an entity's being, or existence. God is the theological cause of an object's existence for eternity. Reason is the cause intermediate between nature and god—similarly to how mathematics is intermediate between physics and theology—which joins with the others to produce an object's good existence. For divine entities, this good existence is for eternity; for mortal beings, it is fleeting.

As discussed in chapter 2, Ptolemy portrays mathematics as a science intermediate between physics and theology in *Almagest* 1.1, and he presents two arguments in support of this hierarchy. In the second argument he maintains, concerning the subject matter of mathematics, that "it is an attribute of all beings without exception, both mortal and immortal, for those that are ever changing in their inseparable form, it changes with them, while for the eternal, which have an aethereal nature, it keeps their immovable form unchanged."[5] Ptolemy argues that mathematical objects inhere in both physical bodies—which in *Almagest* 1.1 are mainly sublunary—and in aethereal bodies, which this argument treats as theological, as opposed to physical, entities. Because sublunary bodies are corruptible, when physical bodies change so do the mathematical objects inhering in them. Superlunary bodies, however, are incorruptible, and so the mathematical objects of the heavens remain eternally unchanged. In *Harmonics* 3.3, reason has a similar role as an intermediary. It joins with nature and god to make all objects—physical, mathematical, and theological—exist in a good way.

After defining the three causes, Ptolemy classifies *harmonia* as a species of reason:

> the [cause] corresponding to *harmonia* is not to be identified as corresponding to nature—for it does not preserve being in the underlying—nor as corresponding to god, since it is not the primary cause of eternal being, but, clearly, [it is to be identified] as corresponding to reason, which falls between the other causes mentioned and joins with each in helping to complete well [being]; it is with the gods always, as they are always the same, but [it is] not with all natural things, nor in every way, for the opposite reason.[6]

5. Ptolemy, *Almagest* 1.1, H6: ἀλλὰ καὶ τῷ πᾶσιν ἁπλῶς τοῖς οὖσι συμβεβηκέναι καὶ θνητοῖς καὶ ἀθανάτοις τοῖς μὲν ἀεὶ μεταβάλλουσι κατὰ τὸ εἶδος τὸ ἀχώριστον συμμεταβαλλομένην, τοῖς δὲ ἀϊδίοις καὶ τῆς αἰθερώδους φύσεως συντηροῦσαν ἀκίνητον τὸ τοῦ εἴδους ἀμετάβλητον.

6. Ptolemy, *Harmonics* 3.3, D92: τὸ κατὰ τὴν ἁρμονίαν οὔτε παρὰ τὴν φύσιν θετέον—οὐ γὰρ τὸ εἶναι περιποιεῖ τοῖς ὑποκειμένοις—οὔτε παρὰ τὸν θεόν, ἐπεὶ μηδὲ τοῦ ἀεὶ εἶναι πρῶτόν ἐστιν αἴτιον, ἀλλὰ δηλονότι παρὰ τὸν λόγον, ὃς μεταξὺ τῶν εἰρημένων αἰτίων πίπτων ἑκατέρῳ

Harmonia is a rational cause that joins with nature and god to make harmonic objects exist in a good way.

As a rational cause, *harmonia* produces good forms, or ends, in the objects it moves. While classifying what type of cause *harmonia* is, Ptolemy remarks:

> we must not accept that *harmonia* is that which underlies—for it is one of the active things and not one of the passive things—nor that it is the end, since on the contrary it produces some end, such as good melody, good rhythm, good order, and good conduct, but that it is the cause, which preserves the proper form in that which underlies.[7]

As a subspecies of moving principle, *harmonia* produces some end in the objects it moves and, because *harmonia* is a rational cause, these ends are good—e.g., good melody, good rhythm, good order, and good conduct. Ptolemy notes that reason, including harmonic reason, is the cause of order and proportion: "For reason, simply and in general, is productive of order and proportion . . ."[8] What makes the melodies, rhythms, orders, and conducts good that *harmonia* moves is the production of formal structures, specifically harmonic ratios, in the objects' movements and configurations. These ratios are the ones most readily perceptible in music, and most of them are multiples (where the greater term is a multiple of the smaller; e.g., 2:1 and 4:1) or epimorics (where the greater term is equal to the smaller term plus one; e.g., 3:2, 4:3, and 9:8).

Although it is a rational cause, *harmonia* does not exist separately from material bodies. For Ptolemy, it exists in the entities it moves. The ontological status of *harmonia* becomes clear when Ptolemy pluralizes it to "*harmoniai*" in *Harmonics* 3.10. At the beginning of the chapter, he states, "Let these definitions have been laid down sufficiently concerning the circular movement itself, considered in respect of both *harmoniai* [ἀμφοτέρας τὰς ἁρμονίας] and of the configurations that generally are called 'concordant' and 'discordant.'"[9] Following Ingemar Düring's interpretation, I suggest that in this passage Ptolemy

συναπεργάζεται τὸ εὖ, τοῖς μὲν θεοῖς ἀεὶ συνὼν ὡς ἂν ἀεὶ τοῖς αὐτοῖς οὖσι, τοῖς δὲ φυσικοῖς οὔτε πᾶσιν, οὔτε πάντως διὰ τοὐναντίον.

7. Ibid.: τὴν ἁρμονίαν οὔτε ὡς τὸ ὑποκείμενον ἀποδεκτέον—τῶν ποιητικῶν γάρ τί ἐστι καὶ οὐδέ τι τῶν παθητικῶν—οὔτε ὡς τὸ τέλος, ἐπειδήπερ αὕτη τοὐναντίον ἀπεργάζεταί τι τέλος, οἷον ἐμμέλειαν, εὐρυθμίαν, εὐνομίαν, εὐκοσμίαν, ἀλλ᾽ ὡς τὸ αἴτιον, ὃ τῷ ὑποκειμένῳ περιποιεῖ τὸ οἰκεῖον εἶδος.

8. Ibid.: ὁ μὲν γὰρ λόγος ἁπλῶς καὶ καθόλου τάξεώς ἐστι καὶ συμμετρίας περιποιητικός

9. Ptolemy, *Harmonics* 3.10, D104: Τὰ μὲν οὖν παρ᾽ αὐτὴν τὴν ἐγκύκλιον κίνησιν θεωρούμενα κατ᾽ ἀμφοτέρας τὰς ἁρμονίας καὶ τὰ κοινῶς καλούμενα σύμφωνά τε καὶ διάφωνα τῶν σχημάτων ἐπὶ τοσοῦτον διωρίσθω.

refers to two sets of *harmonia*: that which exists in music and that which exists among astronomical objects.[10] These *harmoniai* are not different with respect to the ratios, as the same ratios exist among musical pitches and heavenly phenomena. Rather, the two sets of *harmonia* are distinguished ontologically. One *harmonia* exists among musical pitches and another among astronomical objects. Ptolemy again refers to multiple *harmoniai* in *Harmonics* 3.5. Delineating the parts of the human soul, he describes one part as "concerned with the things that can produce and the things that participate in *harmoniai* . . ."[11] Again, the existence of *harmonia* in several sets of objects entails the existence of multiple *harmoniai*.

All objects that have in them a principle of movement—that is, all natural bodies—are moved to some degree by *harmonia*. In other words, every natural body is characterized to some degree by a ratio, in its movements and in the configuration of its matter. The more complete and rational an object is, the more it is moved by *harmonia* and the more it is characterized by harmonic ratios. Ptolemy explains the effects of *harmonia* when discussing the *dynamis harmonikê*, to which Ptolemy ascribes three meanings: (1) the capacity to be moved by *harmonia*, (2) the formal characteristics and function of an object within a harmonic system, and (3) the capacity of human beings to understand and practice harmonics, the science of harmonic ratios. In the following passage, the power of *harmonia*—which I take to be a periphrasis of *dynamis harmonikê*—denotes the capacity of objects to be moved by *harmonia*:

> We also must insist that this sort of power must necessarily be present to some extent in all things that have in themselves a principle of movement, just as must the other [powers], but especially and to the greatest extent in those that share in a more complete and rational nature, because of the suitability of the way in which they were generated, and in these alone can it be revealed as maintaining fully and distinctly, to the highest degree, the likeness of the ratios that produce the suitability and attunement in the differing species. For in general, each of the things administered by nature has a share in some ratio in both its movements and in its underlying materials[12]

10. Düring, *Harmonielehre*, 104, n. 19.

11. Ptolemy, *Harmonics* 3.5, D97: παρὰ τὰ ποιητικὰ καὶ τὰ μετέχοντα τῶν ἁρμονιῶν

12. Ibid. 3.4, D95: προσπαραμυθητέον δ᾽ ὅτι καὶ τὴν τοιαύτην δύναμιν ἀναγκαῖον μὲν ἂν εἴη καὶ πᾶσι τοῖς ἀρχὴν ἐν αὑτοῖς ἔχουσι κινήσεως καθ᾽ ὁσονοῦν ἐνυπάρχειν, ὥσπερ καὶ τὰς ἄλλας, μάλιστα δὲ καὶ τὸ πλεῖστον τοῖς τελειοτέρας καὶ λογικωτέρας φύσεως κεκοινωνηκόσι διὰ τὴν οἰκειότητα τῆς γενέσεως, ἐν οἷς καὶ μόνοις καταφαίνεσθαι δύναται, διόλου τε καὶ σαφῶς συντηροῦσα, ὡς ἔνι μάλιστα, τὴν ὁμοιότητα τῶν τὸ πρόσφορον καὶ ἡρμοσμένον ἐν τοῖς διαφέρουσιν εἴδεσι ποιούντων λόγων. καθόλου μὲν γὰρ ἕκαστον τῶν φύσει διοικουμένων λόγου τινὸς κεκοινώνηκεν ἔν τε ταῖς κινήσεσι καὶ ταῖς ὑποκειμέναις ὕλαις

All natural objects possess some ratio in their movements and in the configuration of their underlying materials. The more complete and rational (i.e., affected by reason, one of the three causes) an object is, the more it is moved by *harmonia*. In the most complete and rational objects, the formal structures not only approximate harmonic ratios but, even more, they instantiate them exactly.

According to Ptolemy, in addition to musical scale systems, human souls and heavenly bodies are the most complete and rational of all entities:

> but [the power of *harmonia*] is not found in the movements that alter the matter itself, since, because of [matter's] inconstancy, neither its quality nor quantity is capable of being defined, but [it is found] in the [movements] that are engaged most closely with forms. These, as we said, are the [movements] of things that are more complete and more rational in their natures, as among divine things are the [movements] of the heavenly bodies and among mortal things the [movements] of human souls, most particularly, since it is only to each of these of the aforementioned things that there belong not only the primary and most complete movement, that in respect of place, but also the characteristic of being rational.[13]

In addition to musical scale systems, the heavens and human souls are the most complete and rational. As such, they are characterized by harmonic ratios in the configuration of their parts and in their movements, and they experience only one type of movement: circular motion, which Ptolemy considers the primary and most complete kind of movement.

Ascribing circular motion to musical scale systems, Ptolemy contends with the fact that musical pitches seem to progress linearly:

> then it is [verified] also by the fact that all of the circuits of the aethereal bodies are circular and regular and that the periodic recurrences of the harmonic *systêmata* have the same features. Inasmuch as the order and pitch of the notes seem to advance, as it were, along a straight line, but their function and their relation to one another, which constitute their special character, are accomplished and enclosed within one and the same circuit in accordance with the principle of circular motion[14]

---

13. Ibid. 3.4, D95: ἀλλ' ἐπὶ μὲν τῶν αὐτῆς τῆς ὕλης ἀλλοιωτικῶν κινήσεων οὐ συνορᾶται, μήτε τοῦ ποιοῦ τοῦ κατ' αὐτήν, μήτε τοῦ ποσοῦ διὰ τὴν ἀστασίαν ὁρίζεσθαι δυναμένου, ἐπὶ δὲ τῶν ἐν τοῖς εἴδεσι τὸ πλεῖστον ἀναστρεφομένων. αὗται δέ εἰσιν αἱ τῶν τελειοτέρων, ὡς ἔφαμεν, καὶ λογικωτέρων τὰς φύσεις, ὡς ἐπὶ μὲν τῶν θείων αἱ τῶν οὐρανίων, ἐπὶ δὲ τῶν θνητῶν αἱ τῶν ἀνθρωπίνων μάλιστα ψυχῶν, ὅτι μόνοις ἑκατέροις τῶν εἰρημένων μετὰ τῆς πρώτης καὶ τελειοτάτης κινήσεως, τουτέστι τῆς κατὰ τόπον, ἔτι καὶ τὸ λογικοῖς εἶναι συμβέβηκεν.

14. Ibid., 3.8, D100–101: ἔπειτα καὶ τὸ τάς τε τῶν αἰθερίων περιόδους ἐγκυκλίους τε πάσας εἶναι καὶ τεταγμένας καὶ τὰς τῶν ἁρμονικῶν συστημάτων ἀποκαταστάσεις ὁμοίως ἔχειν. ἐπειδήπερ

Musical pitches give the appearance of progressing linearly, but their functions and relations to one another reveal that they actually are enclosed within a circuit. Hence, homophones, the notes that mark octaves, sound as though they are in unison even though they are one or more octaves apart. In the heavens, every moving body moves circularly, but individual musical pitches and parts of the human soul do not. It is together with, in relation to the other parts of their respective systems, that they complete movement in a circle. Therefore, it is the functional relationship among individual notes and parts of the soul that reveals the circular motion of their resultant systems.

Human souls are unique in that not only are they moved by *harmonia*, but they also have the ability to understand and produce harmonic ratios. Again, when Ptolemy delineates the parts of the human soul in *Harmonics* 3.5, he labels one part as "concerned with the things that can produce and the things that participate in *harmoniai* . . ."[15] It is essential to human nature not only to participate in *harmonia*—to be moved by *harmonia* in the soul's movements and internal configuration—but also to produce harmonic ratios externally. While the heavens are immutable, musical pitches exist in the sublunary realm of generation and corruption. They occur naturally but human beings also have the ability to create them, to sing or play instruments and thereby bring musical pitches and their relations into existence. Accordingly, Ptolemy maintains that harmonics, as well as mathematics in general, is not simply the contemplation of mathematical objects but also their exhibition:

> [Reason] makes correct the order in things that are heard, which in particular we call "melodiousness," through the theoretical discovery of proportions by means of the intellect, through their practical exhibition by means of skill, and through experience in following them by means of habit. When we consider this—that reason in general also discovers what exists well, renders in practice what it has apprehended, and brings the underlying into conformity with it by habituation—it is to be expected that the general science [ἐπιστήμην] that [encompasses] the species [of science] corresponding to [forms of] reason, which is properly called "mathematics" [μαθηματικήν], is not only a theoretical grasp of beautiful things, as some people would suppose, but it jointly includes their exhibition and practice, which arise out of the very [act of] understanding.[16]

ἡ μὲν τάξις καὶ τάσις τῶν φθόγγων ἐπ᾽ εὐθείας ἂν ὥσπερ δοκοίη προκόπτειν, ἡ δὲ δύναμις καὶ τὸ πῶς ἔχειν πρὸς ἀλλήλους, ὅπερ ἐστὶν ἴδιον αὐτῶν, περαίνεταί τε καὶ συγκλείεται πρὸς μίαν καὶ τὴν αὐτὴν περίοδον κατὰ τὸν τῆς ἐγκυκλίου κινήσεως λόγον

15. Ibid. 3.5, D97: παρὰ τὰ ποιητικὰ καὶ τὰ μετέχοντα τῶν ἁρμονιῶν

16. Ibid. 3.3, D92–93: κατορθοῖ δὲ τὴν ἐν τοῖς ἀκουστοῖς τάξιν, ἣν ἐμμέλειαν ἰδίως καλοῦμεν, διά τε τῆς θεωρητικῆς τῶν συμμετρίων εὑρέσεως παρὰ τὸν νοῦν, καὶ διὰ τῆς χειρουργικῆς αὐτῶν

Ptolemy here expands the domain of mathematics. It is a science that is not solely contemplative. Because some mathematical objects, such as harmonic ratios, can be created, it is the duty of the mathematician not only to understand them but also to engage in practical activities with respect to them. He must exhibit and experience harmonic ratios. As we saw in the previous chapter, Ptolemy complicates the definition of theoretical philosophy. Although the theoretical sciences remain distinct from the practical parts of philosophy, among the former, mathematics, at least, has practical aspects.

The exhibition and experience referred to in this passage could simply consist in the creation of harmonic ratios among musical pitches, such as by means of the monochord, but Ptolemy also could be making reference here to his ethical theory. After all, human beings can exhibit harmonic ratios not only among musical pitches but also in their souls. Because human souls are corruptible, the parts of the human soul do not retain their proper configuration. When the soul's parts are in their proper configuration, they are attuned, but they can and do enter a condition contrary to their nature, a lack of attunement. Ptolemy explains: "and common in both genera [i.e., in music and in the human soul] is the attunement of their parts, when each is in the [condition] in conformity to nature, and the lack of attunement, when in the [condition] contrary to nature."[17] Human beings need to work to maintain the proper relations among their souls' parts. They can attune their souls by reestablishing the harmonic ratios among the parts of the soul, and they can exhibit this proper attunement through their good conduct.

Moreover, when Ptolemy describes the theoretical and practical components of mathematics in *Harmonics* 3.3 he mentions habit, which evokes the manner by which individuals acquire moral virtues. As we saw in the previous chapter, Ptolemy follows Aristotle in maintaining that human beings acquire the moral virtues by means of continuous practice, or habit, and in *Harmonics* 3.4 he proclaims that "the theoretical science of [*harmonia*] is a form of mathematics, the [form] concerned with the ratios of differences between things that are heard, this [form] itself contributing to the good order that

---

ἐνδείξεως παρὰ τὴν τέχνην καὶ διὰ τῆς παρακολουθητικῆς ἐμπειρίας παρὰ τὸ ἔθος. τοῦτο δέ— ὅτι καὶ ὁ καθόλου λόγος εὑρίσκει μὲν τὸ εὖ—θεωρῶν, παρίστησι δὲ τὸ καταληφθὲν ἐνεργῶν, ἐξομοιοῖ δ' αὑτῷ τὸ ὑποκείμενον ἐθίζων, ὥστε εἰκότως καὶ τὴν κοινὴν τῶν παρὰ τὸν λόγων εἰδῶν ἐπιστήμην, ἰδίως δὲ καλουμένην μαθηματικήν, μὴ θεωρίας ἔχεσθαι τῶν καλῶν μόνης, ὥσπερ ἄν τινες ὑπολάβοιεν, ἀλλ' ἐνδείξεως ὁμοῦ καὶ μελέτης ἐξ αὐτῆς τῆς παρακολουθήσεως περιγινομένων.

17. Ibid. 3.5, D97: καὶ κοινὸν ἐν ἀμφοτέροις τοῖς γένεσι τό τε ἡρμοσμένον τῶν μερῶν ἐν τῷ κατὰ φύσιν ἑκατέρου καὶ τὸ ἀνάρμοστον ἐν τῷ παρὰ φύσιν.

arises out of the theory and understanding to people habituated [ἐθιζομένοις] in it . . ."[18] Hence, mathematicians produce and participate in *harmoniai* not only by playing music but also by bringing their souls into attunement. Through the science of harmonics they gain knowledge of harmonic ratios, they reconstitute the harmonic ratios among their souls' parts, and, through habituation, they exhibit the good order in their souls through good conduct. Theory, practice, and habituated experience are all integral to the ethical transformation.

The Beauty of Mathematical Objects

Again, *harmonia* produces good ends in the objects it moves, including good melody, good rhythm, good order, and good conduct. According to Ptolemy, harmonic ratios are not only good; they are also beautiful. After completing his survey of the harmonic ratios found in musical scale systems, Ptolemy states in *Harmonics* 3.3: "Since it is natural for a person who contemplates these [matters] immediately to be filled with wonder, if [he wonders] also at some other of the most beautiful things [καλλίστων], at the extreme rationality of the power of *harmonia* and at the way it finds and creates with perfect accuracy the differences between the forms that belong to it . . ."[19] Harmonic ratios are among the most beautiful of things, and all mathematical objects are to some degree beautiful. As quoted above, Ptolemy states, "it is to be expected that the general science that [encompasses] the species [of science] corresponding to [forms of] reason, which is properly called 'mathematics,' is not only a theoretical grasp of beautiful things [καλῶν] . . ."[20] Mathematics is more than the study of shape, number, size, place, time, and suchlike, as Ptolemy defines it in *Almagest* 1.1. It is the theoretical grasp, exhibition, and experience of beautiful things.

Before Ptolemy, both Plato and Aristotle characterized mathematical objects as beautiful. In *Metaphysics* M3, for instance, Aristotle distinguishes the good from the beautiful, and he asserts that mathematics treats both:

---

18. Ibid. 3.4, D94–95: καὶ ἡ θεωρητικὴ ταύτης ἐπιστήμη μαθηματικῆς ἐστιν εἶδος τὸ περὶ τοὺς λόγους τῶν ἀκουστῶν διαφορῶν, καὶ αὐτὸ διατεῖνον ἐπὶ τὴν ἐκ τῆς θεωρίας καὶ παρακολουθήσεως περιγινομένην τοῖς ἐθιζομένοις εὐταξίαν

19. Ibid. 3.3, D92: ἐπεὶ δ' ἀκόλουθον ἂν εἴη τῷ θεωρήσαντι ταῦτα τὸ τεθαυμακέναι μὲν εὐθύς, εἰ καί τι ἕτερον τῶν καλλίστων, τὴν ἁρμονικὴν δύναμιν ὡς λογικωτάτην καὶ μετὰ πάσης ἀκριβείας εὑρίσκουσάν τε καὶ ποιοῦσαν τὰς τῶν οἰκείων εἰδῶν διαφοράς

20. Ibid. 3.3, D93: ὥστε εἰκότως καὶ τὴν κοινὴν τῶν παρὰ τὸν λόγων εἰδῶν ἐπιστήμην, ἰδίως δὲ καλουμένην μαθηματικήν, μὴ θεωρίας ἔχεσθαι τῶν καλῶν μόνης

Since the good and the beautiful are different (for the [good] is always in an action, while the beautiful is also in immovable things), those who assert that the mathematical sciences say nothing about the beautiful or the good are mistaken. For they do say and show much [about them]; for if they do not name them in showing their effects and relations, it does not follow that they are not speaking about them. The chief forms of the beautiful are order, proportion, and definiteness, which are what the mathematical sciences show to the highest degree. Since these (I mean, for example, order and definiteness) evidently are causes of many things, clearly they must be speaking, too, about this sort of cause in some sense, i.e., the beautiful. But we will speak about these more intelligibly elsewhere.[21]

Unfortunately, Aristotle does not discuss this topic more intelligibly in any of his surviving texts, but it is evident that he takes mathematical objects, including order, proportion, and definiteness, to be beautiful. He calls them the greatest forms, or species, of beauty. According to Ptolemy, order and proportion are caused by reason—including harmonic reason—and, as mathematical, they are beautiful. Aristotle does not specify here what mathematics has to say about the good, but on this topic Ptolemy is clear. The objects of mathematics—which are not only contemplated but also exhibited and experienced—are good and beautiful.

In the *Timaeus*, Plato calls the entire cosmos beautiful. Timaeus explains that whether a created object is modeled after being or becoming determines whether or not it is beautiful:

So whenever the craftsman looks at what is always the same and, using a thing of that kind as a model, reproduces its form and power, then, of necessity, all that he so completes is beautiful [καλόν]; but were he to [look] at a thing that has come to be and use as a model something that has been begotten, [all that he so completes] would be not beautiful [οὐ καλόν].[22]

21. Aristotle, *Metaphysics* XIII, 1078a31–b6: ἐπεὶ δὲ τὸ ἀγαθὸν καὶ τὸ καλὸν ἕτερον (τὸ μὲν γὰρ ἀεὶ ἐν πράξει, τὸ δὲ καλὸν καὶ ἐν τοῖς ἀκινήτοις), οἱ φάσκοντες οὐδὲν λέγειν τὰς μαθηματικὰς ἐπιστήμας περὶ καλοῦ ἢ ἀγαθοῦ ψεύδονται. λέγουσι γὰρ καὶ δεικνύουσι μάλιστα· οὐ γὰρ εἰ μὴ ὀνομάζουσι τὰ δ' ἔργα καὶ τοὺς λόγους δεικνύουσιν, οὐ λέγουσι περὶ αὐτῶν. τοῦ δὲ καλοῦ μέγιστα εἴδη τάξις καὶ συμμετρία καὶ τὸ ὡρισμένον, ἃ μάλιστα δεικνύουσιν αἱ μαθηματικαὶ ἐπιστῆμαι. καὶ ἐπεί γε πολλῶν αἴτια φαίνεται ταῦτα (λέγω δ' οἷον ἡ τάξις καὶ τὸ ὡρισμένον), δῆλον ὅτι λέγοιεν ἂν καὶ τὴν τοιαύτην αἰτίαν τὴν ὡς τὸ καλὸν αἴτιον τρόπον τινά. μᾶλλον δὲ γνωρίμως ἐν ἄλλοις περὶ αὐτῶν ἐροῦμεν.

22. Plato, *Timaeus* 28a–b, translation modified from Zeyl, in Plato, *Complete Works*: ὅτου μὲν οὖν ἂν ὁ δημιουργὸς πρὸς τὸ κατὰ ταὐτὰ ἔχον βλέπων ἀεί, τοιούτῳ τινὶ προσχρώμενος

Because the Demiurge is good, he chooses to model the cosmos after what is eternal, and, having something eternally unchanging as its model, the cosmos is beautiful. Timaeus calls the cosmos "a perceptible god, image of the intelligible, its greatness, goodness, beauty [κάλλιστος], and perfection are unexcelled..."[23] The manner in which the Demiurge molds the cosmos is to give it order and proportion. He forms the four elements by amalgamating isosceles and scalene triangles, and, in this way, he makes the elements as perfect as possible. Even more, he forms the four elements in quantities proportionate to one another. After explaining why the cosmos consists of four elements, Timaeus declares, "This is the reason why the body of the cosmos was begotten out of these four such constituents, coordinating it by proportion, and they bestowed friendship upon it, so that, having come together into a unity with itself, it could not be undone by anyone but the one who had bound it together."[24] By imposing order and proportion on the cosmos, the Demiurge makes it beautiful. Ptolemy does not posit a divine craftsman, but, in a way analogous to the Demiurge, *harmonia*, which exists to some extent in all natural bodies, produces order and proportion in bodies' movements and configurations.

Moreover, Timaeus, like Ptolemy, describes the circular motion and harmony of souls. The Demiurge bestows on the cosmos one type of movement, rotation, which Timaeus associates with intelligence and understanding: "For, he imparted to it the movement suited to its body, that one of the seven [movements] which is especially associated with intellect and prudence..."[25] The world soul, rational and harmonious, moves circularly:

> the [soul] was woven [together with the body] from the center on out in every direction to the outermost limit of the heavens, and it covered [the limit of the heavens] all around on the outside, and, revolving within itself, it made a beginning of a divine principle of unceasing and intelligent life for all time. Now, the body of the heavens had come to be as a visible thing,

παραδείγματι, τὴν ἰδέαν καὶ δύναμιν αὐτοῦ ἀπεργάζηται, καλὸν ἐξ ἀνάγκης οὕτως ἀποτελεῖσθαι πᾶν· οὐ δ' ἂν εἰς γεγονός, γεννητῷ παραδείγματι προσχρώμενος, οὐ καλόν.

23. Ibid. 92c: εἰκὼν τοῦ νοητοῦ θεὸς αἰσθητός, μέγιστος καὶ ἄριστος κάλλιστός τε καὶ τελεώτατος

24. Ibid. 32b–c: καὶ διὰ ταῦτα ἔκ τε δὴ τούτων τοιούτων καὶ τὸν ἀριθμὸν τεττάρων τὸ τοῦ κόσμου σῶμα ἐγεννήθη δι' ἀναλογίας ὁμολογῆσαν, φιλίαν τε ἔσχεν ἐκ τούτων, ὥστε εἰς ταὐτὸν αὐτῷ συνελθὸν ἄλυτον ὑπὸ του ἄλλου πλὴν ὑπὸ τοῦ συνδήσαντος γενέσθαι.

25. Ibid. 34a: κίνησιν γὰρ ἀπένειμεν αὐτῷ τὴν τοῦ σώματος οἰκείαν, τῶν ἑπτὰ τὴν περὶ νοῦν καὶ φρόνησιν μάλιστα οὖσαν

but the [soul] is invisible, sharing in reason and harmony, and it came to be as the most excellent of all the things begotten by him who is himself most excellent of all that is intelligible and eternal.[26]

Human beings may mimic the reason and harmony of the world soul by modeling their own souls after it. Timaeus explains this affinity between the world soul and human souls accordingly:

> Now there is but one way to care for anything, and that is to provide the nourishment and movements that are proper to it. The movements that have an affinity to the divine [part] in us are the thoughts and revolutions of the universe; these, surely, each [of us] should follow, redirecting the revolutions in our heads that were thrown off course at birth by coming to learn the harmonies and revolutions of the universe, and so assimilate our faculty of understanding to that which is understood, in accordance with its original condition[27]

As we saw in the previous chapter, Ptolemy propounds a similar transformation of the soul by means of the study of the heavens, and here we see the significance of harmony for this transformation. In the *Timaeus*, the human soul learns the harmonies of the cosmos and assimilates the faculty of understanding to what is understood. Hence, according to Timaeus, the study of not only the heavens but also of harmony and rhythm effectuates the ordering of the soul's orbits.[28] For Ptolemy, it is because both human souls and heavenly bodies are moved by *harmonia* that human souls' understanding and love of astronomical objects alter and attune them.

Plato, like Ptolemy after him, does not limit beauty to harmonic objects. Timaeus further claims that everything beautiful is proportionate, harmonic or not: "Now all that is good is beautiful, and what is beautiful is not dispro-

---

26. Ibid. 36e–37a: ἡ δ' ἐκ μέσου πρὸς τὸν ἔσχατον οὐρανὸν πάντη διαπλακεῖσα κύκλῳ τε αὐτὸν ἔξωθεν περικαλύψασα, αὐτὴ ἐν αὑτῇ στρεφομένη, θείαν ἀρχὴν ἤρξατο ἀπαύστου καὶ ἔμφρονος βίου πρὸς τὸν σύμπαντα χρόνον. καὶ τὸ μὲν δὴ σῶμα ὁρατὸν οὐρανοῦ γέγονεν, αὐτὴ δὲ ἀόρατος μέν, λογισμοῦ δὲ μετέχουσα καὶ ἁρμονίας ψυχή, τῶν νοητῶν ἀεί τε ὄντων ὑπὸ τοῦ ἀρίστου ἀρίστη γενομένη τῶν γεννηθέντων.

27. Ibid. 90c–d: θεραπεία δὲ δὴ παντὶ παντὸς μία, τὰς οἰκείας ἑκάστῳ τροφὰς καὶ κινήσεις ἀποδιδόναι. τῷ δ' ἐν ἡμῖν θείῳ συγγενεῖς εἰσιν κινήσεις αἱ τοῦ παντὸς διανοήσεις καὶ περιφοραί· ταύταις δὴ συνεπόμενον ἕκαστον δεῖ, τὰς περὶ τὴν γένεσιν ἐν τῇ κεφαλῇ διεφθαρμένας ἡμῶν περιόδους ἐξορθοῦντα διὰ τὸ καταμανθάνειν τὰς τοῦ παντὸς ἁρμονίας τε καὶ περιφοράς, τῷ κατανοουμένῳ τὸ κατανοοῦν ἐξομοιῶσαι κατὰ τὴν ἀρχαίαν φύσιν

28. Ibid. 47c–e.

portionate; therefore, we must take it that if a living thing is to be in [good] condition, it will be proportionate."[29] Similarly, in Plato's *Philebus*, Socrates characterizes measure and proportion as beautiful,[30] and in the following passage, addressing Protarchus, he calls geometric objects, as well as certain colors and sounds, beautiful:

> Socrates: What I am saying may not be entirely clear immediately, but I must try to clarify. For, by the beauty of shapes, I am trying to express not what most people would suppose, such as of animals or of some pictures, but I mean, what the argument says, is something straight or round and plane and solid figures that are constructed from these by compasses, rulers, and squares, if you understand me. For I say these are not beautiful relatively, as others are, but they are by their nature always beautiful by themselves, and they provide some pleasures proper to them that are not at all similar to the [pleasures] of rubbing; and colors possess this type of beauty and pleasures, but do we now understand or how are we?
>
> Protarchus: I am trying, Socrates, but [I hope] you also will try to say it still more clearly.
>
> Socrates: What I am saying is that those sounds of notes that are smooth and bright and emit one clear tone are beautiful not in relation to something else but in and by themselves and following from these are pleasures that belong to them by nature.[31]

Socrates contends that mathematical objects are absolutely beautiful, whether they are straight lines, circles, two-dimensional objects more broadly, or three-dimensional objects. Moreover, the senses sight and hearing perceive beautiful things like colors and single, pure notes, respectively.

29. Ibid. 87c: πᾶν δὴ τὸ ἀγαθὸν καλόν, τὸ δὲ καλὸν οὐκ ἄμετρον· καὶ ζῷον οὖν τὸ τοιοῦτον ἐσόμενον σύμμετρον θετέον.

30. Plato, *Philebus* 64e.

31. Ibid. 51b–d: ΣΩ. Πάνυ μὲν οὖν οὐκ εὐθὺς δῆλά ἐστιν ἃ λέγω, πειρατέον μὴν δηλοῦν. σχημάτων τε γὰρ κάλλος οὐχ ὅπερ ἂν ὑπολάβοιεν οἱ πολλοὶ πειρῶμαι νῦν λέγειν, ἢ ζώων ἤ τινων ζωγραφημάτων, ἀλλ' εὐθύ τι λέγω, φησὶν ὁ λόγος, καὶ περιφερὲς καὶ ἀπὸ τούτων δὴ τά τε τοῖς τόρνοις γιγνόμενα ἐπίπεδά τε καὶ στερεὰ καὶ τὰ τοῖς κανόσι καὶ γωνίαις, εἴ μου μανθάνεις. ταῦτα γὰρ οὐκ εἶναι πρός τι καλὰ λέγω, καθάπερ ἄλλα, ἀλλ' ἀεὶ καλὰ καθ' αὑτὰ πεφυκέναι καί τινας ἡδονὰς οἰκείας ἔχειν, οὐδὲν ταῖς τῶν κνήσεων προσφερεῖς· καὶ χρώματα δὴ τοῦτον τὸν τύπον ἔχοντα [καλὰ καὶ ἡδονάς] ἀλλ' ἄρα μανθάνομεν, ἢ πῶς; ΠΡΩ. Πειρῶμαι μέν, ὦ Σώκρατες· πειράθητι δὲ καὶ σὺ σαφέστερον ἔτι λέγειν. ΣΩ. Λέγω δὴ ἤχὰς τῶν φθόγγων τὰς λείας καὶ λαμπράς, τὰς ἕν τι καθαρὸν ἱείσας μέλος, οὐ πρὸς ἕτερον καλὰς ἀλλ' αὑτὰς καθ' αὑτὰς εἶναι, καὶ τούτων συμφύτους ἡδονὰς ἑπομένας.

Following Plato, Ptolemy portrays mathematical objects as beautiful, and he maintains that sight and hearing are capable of perceiving beauty. As Socrates distinguishes between pleasure and beauty in the *Philebus*, so Ptolemy distinguishes the senses that perceive pleasure from the senses that also perceive beauty. While discussing reason, Ptolemy identifies sight and hearing as criteria of beauty:

> For this sort of power employs as its instruments and servants the highest and most marvelous of the senses, sight and hearing, which, of all [the senses], are most closely tied to the *hêgemonikon*, and which are the only [senses] that assess their objects not only with respect to pleasure but also, much more importantly, with respect to beauty. [...] But no one would classify the beautiful or the ugly as belonging to things touched or tasted or smelled, but only to things seen and things heard, such as shape and melody, or, again, the movements of the heavenly bodies and human actions; hence these, alone among the senses, give assistance with one another's apprehensions in many ways through the agency of the rational [part] of the soul, just as if they were truly sisters.[32]

Mathematical objects are beautiful, and sight and hearing—which of the senses are most closely affiliated with the *hêgemonikon*, the soul's ruling part—are the only senses capable of perceiving beauty. It is not that the other senses do not perceive mathematical objects—for touch perceives mathematical objects such as shapes—but sight and hearing are the only senses that perceive their beauty. Hearing, for instance, perceives the beauty of the harmonic ratios among musical pitches, and sight perceives the beauty of celestial bodies' movements and configurations.

Not only do sight and hearing individually perceive beauty, but they also work together in the apprehension of beautiful objects. Hearing relays what is seen by means of spoken expression, and sight displays what is heard by means of written illustration. When the two are combined—as when speech is accompanied by diagrams or letters, or when what is visible is described in poetic representation—the two senses reinforce one another such that the

---

32. Ptolemy, *Harmonics* 3.3, D93–94: Κέχρηται γὰρ ὀργάνοις ὥσπερ καὶ διακόνοις ἡ τοιαύτη δύναμις ταῖς ἀνωτάτω καὶ θαυμασιωτάταις τῶν αἰσθήσεων, ὄψει καὶ ἀκοῇ, τεταμέναις μὲν μάλιστα τῶν ἄλλων πρὸς τὸ ἡγεμονικόν, μόναις δὲ ἐκείνων οὐχ ἡδονῇ μόνῃ κρινούσαις τὰ ὑποκείμενα, πολὺ δὲ πρότερον τῷ καλῷ. [...] τὸ δὲ καλὸν ἢ αἰσχρὸν τῶν μὲν ἁπτῶν ἢ γευστῶν ἢ ὀσφραντῶν οὐδεὶς ἂν κατηγορήσαι, μόνων δὲ τῶν ὁρατῶν καὶ τῶν ἀκουστῶν, οἷον μορφῆς καὶ μέλους, ἢ πάλιν τῶν οὐρανίων κινήσεων καὶ τῶν ἀνθρωπίνων πράξεων, ὅθεν καὶ μόναι τῶν ἄλλων αἰσθήσεων τὰς ἀλλήλων καταλήψεις ἀντιδιακονοῦνται τῷ λογικῷ τῆς ψυχῆς πολλαχῇ, καθάπερ ὡς ἀληθῶς ἀδελφαὶ γινομένω.

soul can better learn about and contemplate the phenomena. Ptolemy concludes, "It is therefore not only by each one's apprehending its special object,
but also by their assisting one another in some way to learn about and consider
the things that are completed in accordance with the proper ratio, that these
[senses] and the most rational of the sciences that depend on them progres-
sively approach what is beautiful and what is useful . . ."[33] Ptolemy contrasts
the senses' perception of their special sensibles with their mutual cooperation
in the apprehension of phenomena. He does not go so far here as to identify
mathematical objects, which are beautiful, with common sensibles, despite
the fact that harmonic ratios are common sensibles, inasmuch as they are
both seen and heard among heavenly bodies and musical pitches, respectively.
Instead, the sense—sight or hearing—that does not perceive an object can
still represent it and in this way assist the soul's comprehension of it. Through
additional and compounding impressions, sight and hearing are able to make
beautiful objects more apprehensible to the rational part of the soul.

Harmonics

Although *harmonia* is a subspecies of moving principle that exists to some
degree in all natural bodies—and most especially in musical systems, human
souls, and the heavens—Ptolemy narrowly defines the science of *harmonia*, or
harmonics. He defines it in the very first lines of the *Harmonics*: "Harmonics is
a power that apprehends [Ἁρμονική ἐστι δύναμις καταληπτική] the distinctions
in sounds related to high and low [pitch] . . ."[34] The substantive *harmonikê*
modifies the implied noun *epistêmê*, signifying knowledge or science. Harmonics is the science that concerns the relations among musical pitches—i.e.,
harmonic ratios. Although these same ratios also characterize the relations
among the parts of the soul and heavenly bodies, the science of harmonics
is the study of harmonic ratios only among the phenomena that make these
ratios most apparent: musical pitches.

The way in which Ptolemy defines harmonics—identifying the science,
or art, with the power (δύναμις) by which one studies it, and signifying the
science by means of a substantive—was common in antiquity. This syntax
frequently arises in definitions of rhetoric. For instance, Aristotle defines rhetoric accordingly: "Now, let rhetoric be the power of observing in each case

33. Ibid. 3.3, D94: οὐ μόνον οὖν τῷ τὸ ἴδιον ἑκατέρᾳ καταλαμβάνειν, ἀλλὰ καὶ τῷ συναγωνίζεσθαί
πως ἀλλήλαις πρὸς τὸ μανθάνειν καὶ θεωρεῖν τὰ κατὰ τὸν οἰκεῖον συντελούμενα λόγον, ἐπὶ πλέον
τοῦ τε καλοῦ καὶ τοῦ χρησίμου διήκουσιν αὗταί τε καὶ τῶν κατ' αὐτὰς ἐπιστημῶν αἱ λογικώταται

34. Ibid. 1.1, D3: Ἁρμονική ἐστι δύναμις καταληπτικὴ τῶν ἐν τοῖς ψόφοις περὶ τὸ ὀξὺ καὶ τὸ
βαρὺ διαφορῶν

the possible means of persuasion. For this is a function of no other art . . ."[35] According to Aristotle, rhetoric (ῥητορική) is an art (τέχνη) and a power (δύναμις). The structure of this definition of rhetoric—a substantive identified with a power—became standard.[36] Dionysius Thrax, a second-century BCE grammarian, defines rhetoric with the following: "ῥητορική ἐστι δύναμις τεχνική [. . .]."[37] Dionysius Halicarnassensis, the first-century BCE historian and teacher of rhetoric at Rome, uses the same formulation in his *De imitatione*,[38] as do several later writers, including Troilus, the late fourth- to early fifth-century CE sophist, in his *Prolegomena in Hermogenis artem rhetoricam*;[39] Ammonius Hermeiou, the Platonic philosopher who established the tradition of Aristotelian commentary in Alexandria in the late fifth to early sixth century CE, in his *In Porphyrii isagogen sive quinque voces*;[40] Joannes Doxapatres, the eleventh-century Constantinopolitan rhetorician, in his *Prolegomena in Aphthonii progymnasmata*;[41] and the authors of the *Epitome artis rhetoricae*,[42] *Prolegomena in librum περὶ στάσεων*,[43] *Introductio in prolegomena Hermogenis artis rhetoricae*,[44] *Prolegomena in artem rhetoricam*,[45] and *Synopses artis rhetoricae*.[46] With this formulation, these writers indicate that rhetoric is an art (τέχνη), just as Aristotle does in the *Rhetoric*; however, unlike Aristotle, they do so by means of an adjectival construction. Rhetoric (a substantive) is a power (δύναμις) that operates as an art (τεχνική). Ptolemy uses the same syntax when defining harmonics. Harmonics (a substantive) is a power (δύναμις)

---

35. Aristotle, *Rhetoric* I, 1355b25–27: Ἔστω δὴ ἡ ῥητορικὴ δύναμις περὶ ἕκαστον τοῦ θεωρῆσαι τὸ ἐνδεχόμενον πιθανόν. τοῦτο γὰρ οὐδεμιᾶς ἑτέρας ἐστὶ τέχνης ἔργον

36. I must thank Alexander Jones for pointing me toward instances of this syntax. Although I here focus on the definition of rhetoric, other ancient authors used this formulation to define other fields of inquiry and art. For instance, in the first line of *Outlines of Pyrrhonism* 1.8, Sextus Empiricus defines skepticism: Ἔστι δὲ ἡ σκεπτικὴ δύναμις ἀντιθετικὴ [. . .]. Aristides Quintilianus, in the third century CE, defines the making of rhythm in *De musica* 1.19.13: Ῥυθμοποιία δέ ἐστι δύναμις ποιητική [. . .].

37. Dionysius Thrax, *Fragmenta*, fragment 53.2.

38. Dionysius Halicarnassensis, *De imitatione*, fragment 26.1.

39. Troilus, *Prolegomena in Hermogenis artem rhetoricam* 52.26–27.

40. Ammonius, *In Porphyrii isagogen sive quinque voces* 1.14.

41. Joannes Doxapatres, *Prolegomena in Aphthonii progymnasmata* 14.106.22.

42. Rhetorica Anonyma, *Epitome artis rhetoricae* 3.611.4–5; *Prolegomena in artem rhetoricam* 14.29.7–8, 14.30.12–13.

43. Anonymi in Hermogenem, *Prolegomena in librum περὶ στάσεων* 14.199.22–23.

44. Anonymi in Hermogenem, *Introductio in prolegomena Hermogenis artis rhetoricae* 14.283.14.

45. Anonymi in Hermogenem, *Prolegomena in artem rhetoricam* 14.349.8–9.

46. Anonymi in Hermogenem, *Synopses artis rhetoricae* 3.461.19.

that apprehends (καταληπτική). Just as the adjectival construction *tekhnikê* indicates that rhetoric is an art, *kataléptikê* indicates that harmonics is a type of apprehension, or *katalépsis*. In Ptolemy's corpus, the term *katalépsis* has lost its strong Stoic connotation of a firm grasp of some impression, and it instead denotes any type of apprehension. Here the type of apprehension is one of knowledge. Harmonics is a science.

The student of harmonics is the *harmonikos*. Ptolemy states the student's objective in *Harmonics* 1.2: "The aim of the student of harmonics (ἁρμονικοῦ) must be to preserve in all respects the rational hypotheses of the *kanôn* as never in any way conflicting with the perceptions that correspond to the estimation of most people . . ."[47] Ptolemy alludes here to his criterion of truth, the cooperation of reason and perception in the construction of knowledge. Indeed, he discusses the criterion in depth in the opening chapters of the *Harmonics* when establishing the proper method of harmonics. Two traditions developed in antiquity for harmonics' study, one purely rational and the other inordinately empirical. In *Harmonics* 1.2, Ptolemy labels these two camps the Pythagoreans and the Aristoxenians. According to Ptolemy, the former prioritize number theory and ignore the lack of correspondence between some of the ratios they posit (or reject) in harmonics and what ratios actually are perceived as harmonic. Historically, it would be more appropriate to label these theorists "Platonists" rather than "Pythagoreans," as the fifth- and fourth-century BCE Pythagoreans did attend to both reason and hearing when studying harmonics, and Plato criticized them for studying the numbers in perceptible harmonies rather than the numbers themselves.[48] Nevertheless, the Platonic manner of studying harmonics became identified with Pythagoreanism, and Ptolemy appropriates this identification. Ptolemy's account of the debate between Pythagoreans and Aristoxenians reads, as Andrew Barker has observed, as a paraphrase of Didymus, the music theorist who lived in the first century CE and whom Porphyry accuses Ptolemy of plagiarizing.[49] While Ptolemy criticizes the Pythagoreans for ignoring sense perception, he arraigns the Aristoxenians for the opposite reason. Rather than ignoring perception, they give too much weight to what they hear and they misuse reason, fitting numbers not to the ratios but to the intervals between sounds and even

47. Ptolemy, *Harmonics* 1.2, D5: ἁρμονικοῦ δ' ἂν εἴη πρόθεσις τὸ διασῶσαι πανταχῇ τὰς λογικὰς ὑποθέσεις τοῦ κανόνος μηδαμῇ μηδαμῶς ταῖς αἰσθήσεσι μαχομένας κατὰ τὴν τῶν πλείστων ὑπόληψιν

48. See Plato, *Republic* VII, 531b–c. Cf. *Timaeus* 35b–36b for rational scale building.

49. Barker, *Science of Harmonics*, 440, n. 11. See Porphyry, *Commentary on the Harmonics of Claudius Ptolemaeus* 28.23–26 and, for his comment on Ptolemy's alleged plagiarism, 5.11–15.

applying numbers to divisions that are inconsistent with what is perceived.[50] Ptolemy carves a middle path—the cooperation of perception and reason, the description of ratios that underlie what is perceived—and so, as Carl Huffman has acknowledged, he develops a methodology that is reminiscent of, but more sophisticated than, that of the early Pythagoreans.[51]

With his criterion of truth, Ptolemy maintains that the rational and perceptive parts of the human soul must cooperate. They have different capabilities, but each is essential in the construction of knowledge. In *Harmonics* 1.1, Ptolemy describes their capabilities as follows:

> rather, hearing is concerned with the matter and the modification, and reason is concerned with the form and the cause, since it is in general characteristic of the senses to discover what is approximate and to receive what is accurate, and of reason to receive what is approximate and to discover what is accurate. For since matter is defined and delimited only by form, and modifications only by the causes of movements, and since of these the former [i.e., the matter and the modification] belong to sense perception, the latter [i.e., the form and the cause] to reason, it suitably follows that the apprehensions of the senses are defined and delimited by the [apprehensions] of reason, first submitting to them the distinctions they have grasped more roughly, so far as regards those things that are conceived of by means of sense perception, and being guided by them to the [distinctions] that are accurate and agreed upon.[52]

The senses relay perceptions, which only approximate what is observed, to the rational part of the soul, which grasps what is accurate and, furthermore, directs the senses to make accurate, rather than only approximate, observations. Ptolemy illustrates this bidirectional cooperation of reason and perception by means of an example, that of drawing a circle: "Thus just as a circle constructed by sight alone often appears to be accurate, until the [circle]

50. Ptolemy gives more detailed criticisms of the Pythagoreans and Aristoxenians in *Harmonics* 1.5 and 1.9, especially.

51. Huffman, "Response to Barker," 425.

52. Ptolemy, *Harmonics* 1.1, D3: ἀλλ' ἡ μὲν ἀκοὴ παρὰ τὴν ὕλην καὶ τὸ πάθος, ὁ δὲ λόγος παρὰ τὸ εἶδος καὶ τὸ αἴτιον, ὅτι καὶ καθόλου τῶν μὲν αἰσθήσεων ἴδιόν ἐστι τὸ τοῦ μὲν σύνεγγυς εὑρετικόν, τοῦ δὲ ἀκριβοῦς παραδεκτικόν, τοῦ δὲ λόγου τὸ τοῦ μὲν σύνεγγυς παραδεκτικόν, τοῦ δ' ἀκριβοῦς εὑρετικόν. ἐπειδὴ γὰρ ὁρίζεται καὶ περαίνεται μόνως ἡ μὲν ὕλη τῷ εἴδει, τὰ δὲ πάθη τοῖς αἰτίοις τῶν κινήσεων, καὶ ἔστι τούτων τὰ μὲν αἰσθήσεως οἰκεῖα, τὰ δὲ λόγου, παρηκολούθησεν εἰκότως τὸ καὶ τὰς αἰσθητικὰς διαλήψεις ὁρίζεσθαι καὶ περαίνεσθαι ταῖς λογικαῖς, ὑποβαλλούσας μὲν πρώτας ἐκείναις τὰς ὁλοσχερέστερον λαμβανομένας διαφορὰς ἐπί γε τῶν δι' αἰσθήσεως νοητῶν, προσαγομένας δὲ ὑπ' ἐκείνων ἐπὶ τὰς ἀκριβεῖς καὶ ὁμολογουμένας.

produced by reason brings [sight] to a recognition of the one that is really accurate . . ."[53] If an individual draws a circle by hand, then he may judge what he drew to in fact be a circle, but he would be wrong. When he subsequently draws a circle with a compass, an instrument which instantiates rational principles, he observes and recognizes what is truly a circle. In this way, reason guides perception away from the approximate towards the precise and true, and the perceptive part of the soul perceives more accurately.

In the case of harmonics, reason and hearing cooperate in the same way. An individual may hear two notes and wrongly suppose that the relation they make is, for instance, a concord, but, when he hears the notes that truly make a concord, then he hears more accurately. Again, a rational instrument is required for accurate observation. Ptolemy explains as follows:

> For the ears, similarly, which with the eyes are most especially the servants of the theoretical and rational part of the soul, there is needed some method, derived from reason, for the things that they are not naturally capable of judging accurately, against which they will not bear witness but which they will agree is correct. The instrument of this kind of method, then, is called a harmonic *kanôn* [κανὼν ἁρμονικός], from the common predication and from its straightening [κανονίζειν] the things in sense perception that are inadequate to reveal the truth.[54]

The harmonic *kanôn* is a monochord, a single-stringed instrument with a movable bridge and graduated ruler, which one may use to measure lengths along the string. Ptolemy also uses instruments with many strings, either eight or fifteen, and he tunes the strings using the movable bridges. By constructing the *kanôn* in accordance with reason and playing it—moving the bridge to create the exact ratios that define musical relations—the sense of hearing shifts from observing only approximately to confirming what is accurate. Furthermore, reason guides the senses to observe distinctions that are not only accurate but also agreed upon. This reference to what is agreed upon recalls Ptolemy's account of knowledge (ἐπιστήμη) in *On the Kritêrion and Hêgemonikon*, where knowledge corresponds to a judgment that is "most clear and agreed upon"

<hr/>

53. Ibid. 1.1, D3–4: ὥσπερ οὖν ὁ μόνῃ τῇ ὄψει περιενεχθεὶς κύκλος ἀκριβῶς ἔχειν ἔδοξε πολλάκις, ἕως ἂν ὁ τῷ λόγῳ ποιηθεὶς εἰς ἐπίγνωσιν αὐτὴν μεταγάγοι τοῦ τῷ ὄντι ἀκριβοῦς

54. Ibid. 1.1–2, D5: τὸν αὐτὸν τρόπον καὶ ταῖς ἀκοαῖς διακόνοις οὔσαις μάλιστα μετὰ τῶν ὄψεων τοῦ θεωρητικοῦ καὶ λόγον ἔχοντος μέρους τῆς ψυχῆς, δεῖ τινος ἀπὸ τοῦ λόγου, πρὸς ἃ μὴ πεφύκασι κρίνειν ἀκριβῶς, ἐφόδου, πρὸς ἣν οὐκ ἀντιμαρτυρήσουσιν ἀλλ' ὁμολογήσουσιν οὕτως ἔχειν. Τὸ μὲν οὖν ὄργανον τῆς τοιαύτης ἐφόδου καλεῖται κανὼν ἁρμονικός, ἀπὸ τῆς κοινῆς κατηγορίας καὶ τοῦ κανονίζειν τὰ ταῖς αἰσθήσεσιν ἐνδέοντα πρὸς τὴν ἀλήθειαν παρειλημμένος.

(τρανωτάτη καὶ ὁμολογουμένη).[55] The interplay of reason and perception produces more precise observations, and the judgments the rational part of the soul makes, as a consequence of this interplay, are knowledge. In the science of harmonics, this knowledge consists in true and agreed-upon judgments of the ratios that characterize the relations among musical pitches.

Harmonic Ratios in the Human Soul

The final chapters of the *Harmonics* have received the least scholarly attention, but they are the most essential to Ptolemy's ethical theory. In them Ptolemy turns from the study of music to the instantiation of harmonic ratios in human souls and among heavenly bodies. *Harmonics* 3.5–7 treat the structure of the human soul, and *Harmonics* 3.8–16 (the last three chapters are no longer extant) examine harmonic relations in the heavens. Because musical systems, human souls, and the heavens share the same harmonic configuration, Ptolemy is able to map the ratios that describe the relations among musical pitches onto the formal structures within the human soul and among heavenly bodies.

In *Harmonics* 3.5, Ptolemy presents three alternative accounts of the human soul. In each, he divides the soul into the same number of parts and species, but what these parts and species are, and therefore what functions they enact in the soul, vary among the three accounts. I present an intellectual history of these accounts in chapter 7. Here I focus on the soul's harmonic structure. In describing the human soul as harmonic, Ptolemy follows both the Platonic and Stoic traditions.[56] His ascription of ratios, in particular, to the soul has a likely precedent in the work of Moderatus of Gades, a first-century CE Pythagorean. Moderatus's contributions remain only in fragments, but in his *De anima* Iamblichus ascribes to him the theory that the soul is a mathematical harmony that encompasses ratios (λόγους περιεχούσῃ).[57] For Ptolemy, too, the soul is composed of ratios. How the soul encompasses them is made apparent by correspondences between the parts of the soul and the homophones and concords in music.

Ptolemy divides the soul into three principal parts, which he first labels the intellectual part (νοερόν), the perceptive part (αἰσθητικόν), and the part that maintains a state (ἑκτικόν). As musical systems also consist of three principal

55. Ptolemy, *On the Kritērion*, La7. Cf. Ptolemy, *Geography* 2.1.2, where he contrasts that which is nearest the truth (ἐγγυτάτω τῆς ἀληθείας), and more or less agreed upon (ὡς ἐπίπαν ὁμολογούμενον), with that which is estimated more roughly (ὁλοσχερέστερον).

56. See, for instance, Plato, *Republic* IV, 443d–e, and *Timaeus* 47d. On the Stoic use of the harmonic metaphor, see Long, *Stoic Studies*, 202–23.

57. Iamblichus, *De anima* 364, from Stobaeus, *Eclogae* 1.49.32.

parts—the homophone of the octave, the concord of the fifth, and the concord of the fourth—it is possible to construct correspondences between them.
Ptolemy associates the homophone with the intellectual part of the soul, the
concord of the fifth with the soul's perceptive part, and the concord of the
fourth with the soul's part that maintains a state:

> Well then, there are three primary parts of the soul—the intellectual, the
> perceptive, and the [part] that maintains a state—and there are three pri-
> mary forms of homophones and concords—the homophone of the octave
> and the concords of the fifth and the fourth—and so the octave is attuned
> to the intellectual [part]—for in each there is the greatest simplicity, equal-
> ity, and homogeneity—the fifth to the perceptive [part], and the fourth to
> the [part] that maintains a state. For the fifth is closer to the octave than is
> the fourth, as it is more concordant on account of the fact that the [arith-
> metic] excess is closer to equality, and the perceptive [part] is closer to the
> intellectual [part] than is the [part] that maintains a state because it, too,
> has a share in some kind of apprehension.[58]

Ptolemy justifies the correspondences between the individual parts of the soul
and the principal forms of homophones and concords. First, the octave and
the intellectual part of the soul share the qualities of simplicity, equality, and
homogeneity. Ptolemy remarks in *Harmonics* 1.7 that homophones, despite
being two distinct notes, sound as if they are a single note,[59] and the intellectual part of the soul similarly is a unity. In *On the Kritêrion and Hêgemonikon*,
Ptolemy calls the faculty of thought undivided (ἀμέριστον) in substance and
he portrays the intellect as simple in nature.[60] Hence, in sharing the properties
of simplicity, equality, and homogeneity, the octave and the intellectual part
of the soul correspond to one another.

It then remains for Ptolemy to correlate each of the concords with a remaining part of the soul. He argues that it is appropriate to correlate the concord of
the fifth with the perceptive part because of how they each relate to the part
in their respective systems that is simplest and undivided. In the case of the

58. Ptolemy, *Harmonics* 3.5, D95–96: Ἔστι τοίνυν τὰ μὲν πρῶτα τῆς ψυχῆς μέρη τρία, νοερόν,
αἰσθητικόν, ἑκτικόν, τὰ δὲ πρῶτα τῶν ὁμοφώνων καὶ συμφώνων εἴδη τρία, τό τε διὰ πασῶν
ὁμόφωνον καὶ σύμφωνα τό τε διὰ πέντε καὶ διὰ τεσσάρων, ὥστε ἐφαρμόζεσθαι τὸ μὲν διὰ πασῶν
τῷ νοερῷ—πλεῖστον γὰρ ἐν ἑκατέρῳ τὸ ἁπλοῦν καὶ ἴσον καὶ ἀδιάφορον—τὸ δὲ διὰ πέντε τῷ
αἰσθητικῷ, τὸ δὲ διὰ τεσσάρων τῷ ἑκτικῷ. τοῦ τε γὰρ διὰ πασῶν ἐγγυτέρω τὸ διὰ πέντε παρὰ τὸ διὰ
τεσσάρων, ὡς συμφωνότερον διὰ τὸ τὴν ὑπεροχὴν πλησιαιτέραν ἔχειν τοῦ ἴσου, καὶ τοῦ νοεροῦ τὸ
αἰσθητικὸν ἐγγύτερον παρὰ τὸ ἑκτικὸν διὰ τὸ μετέχειν τινὸς καὶ αὐτὸ καταλήψεως.

59. Ibid. 1.7, D15.

60. Ptolemy, *On the Kritêrion*, La21 and La5.

concords, the terms of their ratios determine which is closer to equality. In general, one epimoric ratio is closer to equality than another when the difference between the two terms is a larger portion of the second term. The homophone of the octave exhibits equality, because the difference between the two terms in the octave's ratio (2:1) is identical to the second term. The ratio of the concord of the fifth (3:2) yields a difference of one half of the second term, and the ratio of the concord of the fourth (4:3) yields a difference of one third of the second term. Therefore, the "arithmetic excess" of the concord of the fifth is closer to equality than is the excess of the concord of the fourth, and so the concord of the fifth is more closely related to the octave than is the concord of the fourth.

Correspondingly, the perceptive part of the soul is closer to the intellectual part than is the part that maintains a state because, just as the intellectual part experiences types of apprehension, or *katalêpsis*, in a way so, too, does the perceptive part, but the part that maintains a state has no share in apprehension. Ptolemy's criterion of truth, the interplay of reason and perception, no doubt underlies this ascription of a type of apprehension to the perceptive part of the soul. After all, harmonic knowledge, which Ptolemy defines as a type of apprehension, depends on the cooperation of perception and reason. The perceptive part of the soul relays to the intellectual part the approximate, rough-and-ready distinctions it makes. These approximations partake of a kind of apprehension, which the intellectual part of the soul transforms into another kind of apprehension—either opinion or, when the distinctions are analyzed skillfully, knowledge. Hence, the perceptive part of the soul is closer to the intellectual part than is the part that maintains a state. Because the concord of the fifth is closer to the octave than is the concord of the fourth, and because the octave corresponds to the intellectual part, the concord of the fifth corresponds to the perceptive part of the soul and the concord of the fourth corresponds to the remaining part, the part of the soul that maintains a state.

Ptolemy further justifies these correspondences by presenting a hierarchy among the principal parts in the human soul and in music:

Now, just as things that have a state do not always have sense perception, and neither do things that have sense perception always have intellect, but, conversely, things that have sense perception always have a state, and things that have intellect always have both a state and sense perception, in the same way where there is a fourth there is not always a fifth, and neither where there is a fifth is there always an octave, but, conversely, where there is a fifth there is always a fourth, and where there is an octave there are always both a fifth and a fourth, because the former are made up of the less perfect melodic intervals and combinations, and the latter of the more perfect.[61]

Living things can have the part of the soul that maintains a state and not the perceptive part nor the intellectual part, but the other two parts of the soul cannot exist independently. The perceptive part requires the coexistence of the part that maintains a state, and the intellectual part requires the coexistence of both the perceptive part and the part that maintains a state. Ptolemy observes a similar hierarchy in musical systems. A scale system that contains a fourth does not necessarily contain a fifth or an octave; one that contains a fifth must contain a fourth but not necessarily an octave; one that contains an octave necessarily contains a fifth and a fourth. These hierarchical relationships lend themselves to the correspondences Ptolemy establishes between the three principal parts of the soul and the three principal forms of homophones and concords.

Thereafter, Ptolemy divides the parts of the soul into species, and the number of species each part has corresponds to the number of species in the homophone or concord analogous to it. In ancient Greek music theory, the species signify different possibilities for ordering the melodic intervals of a tetrachord, an attunement of four notes where the highest and lowest notes together mark a concord of the fourth. The tetrachord is the smallest possible scale system. It is taken to be prior to smaller, melodic intervals, and so whereas in other musical traditions the intervals of the notes are prior to the fourth, and the fourth is what it is because it is the sum of smaller intervals, in the ancient Greek schema the tetrachord comes first and the intermediate divisions between the base note and the fourth are movable. The sum of the ratios comprising the first two intervals can be smaller, intermediate, or greater than the third, remaining, interval, depending on the genus in which the tetrachord is set—enharmonic, chromatic, or diatonic—as well as how the harmonic theorist establishes the parameters of the three genera. The species are possibilities for how one arranges the melodic intervals of the tetrachord in whichever genus.

By way of illustration, the diagram in figure 5.2 contains two tetrachords, both in the diatonic genus. They are labeled "Tetrachord 1" and "Tetrachord 2,"

---

61. Ptolemy, *Harmonics* 3.5, D96: ἐπειδὴ ὥσπερ ἐν οἷς μὲν ἕξις, οὐ πάντως αἴσθησις, οὐδὲ ἐν οἷς αἴσθησις, καὶ νοῦς πάντως· ἀνάπαλιν δὲ ἐν οἷς αἴσθησις, καὶ ἕξις πάντως, καὶ ἐν οἷς νοῦς, καὶ ἕξις καὶ αἴσθησις πάντως, οὕτως ὅπου μὲν τὸ διὰ τεσσάρων, οὐ πάντως καὶ τὸ διὰ πέντε, οὐδ' ὅπου τὸ διὰ πέντε, καὶ τὸ διὰ πασῶν πάντως· ἀνάπαλιν δὲ ὅπου τὸ διὰ πέντε, καὶ τὸ διὰ τεσσάρων πάντως, καὶ ὅπου τὸ διὰ πασῶν, καὶ τὸ διὰ πέντε καὶ τὸ διὰ τεσσάρων πάντως, ὅτι τὰ μὲν τῶν ἀτελεστέρων ἐστὶν ἐμμελειῶν τε καὶ συγκρίσεων ἴδια, τὰ δὲ τῶν τελειοτέρων.

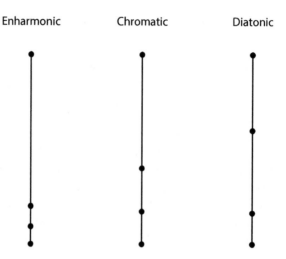

FIGURE 5.1. Tetrachords in the three genera.

and they are conjoined at note "D," such that the fourth note of Tetrachord 1 coincides with the first note of Tetrachord 2. The first species of the concord of the fourth ranges from the first note to the fourth note of either of the tetrachords, 1 or 2; the second species ranges from the second note of Tetrachord 1 to the second note of Tetrachord 2, and the third species ranges from the third note of Tetrachord 1 to the third note of Tetrachord 2. Hence, the concord of the fourth has three species. If we replicated this procedure, we would see that the concord of the fifth has four species and the octave has seven species. The parts of the soul also have species. In other words, they each have a certain number of possibilities for how one can order, or internally configure, them. The part that maintains a state, like the concord of the fourth, has three species; the perceptive part, like the concord of the fifth, has four species; the intellectual part, like the octave, has seven species.

In *Harmonics* 3.6, Ptolemy compares the musical genera—the enharmonic, chromatic, and diatonic—to the genera of the soul's species. These psychological genera are the three theoretical and three practical parts of philosophy. The former are, as they are in *Almagest* 1.1, physics, mathematics, and theology, and the latter are ethics, domestics, and politics. Ptolemy presents an analogy between the size of a tetrachord's first two intervals compared with the third interval in the various genera and the magnitude, or magnificence, of the various parts of philosophy, in which the soul participates. Because the first two intervals in the enharmonic genus are smallest, the enharmonic genus corresponds to physics and ethics, which are smaller in magnitude than their corresponding parts of philosophy. The diatonic genus, which has the largest

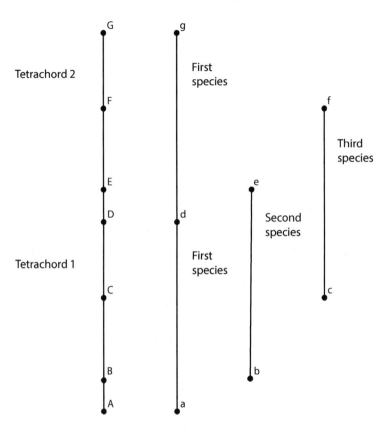

FIGURE 5.2. Species of the fourth.

sum of the first and second intervals, corresponds to theology and politics, because of the similarity in their order and magnitude. The chromatic genus, intermediate between the enharmonic and diatonic, corresponds to the inter-mediate sciences, mathematics and domestics. Ptolemy explains that mathe-matics is intermediate because it is involved, or engaged (ἀναστρέφεται), to a high degree in physics and theology. He may be alluding here to how, as he explains in *Almagest* 1.1, mathematical objects inhere in both physical bodies, which he characterizes as mainly sublunary, and aethereal bodies. Domestics, on the other hand, shares with ethics its privacy and subordination and with politics in being social and controlling.

Lastly, in *Harmonics* 3.7 Ptolemy compares modulations in character caused by changes in life circumstances to modulations of *tonoi*, where a *tonos* is a sequence of intervals that, located within a particular range of pitch, exhibits one of the seven species of the octave. Just as a modulation of *tonos* alters a melody, so a change in life circumstances furnishes a different way of life.

Peace, for example, makes souls more steady and fair; war makes them more bold and disdainful. Poverty makes them more moderate and thrifty; plenitude makes them more liberal and undisciplined. Listening to music also has the power to alter human souls. The activities of a melody can relax, invigorate, stupefy, etc. Ptolemy explains how this type of change occurs:

> Indeed, our souls are affected outright in sympathy with the very activities of a melody, recognizing the kinship, as it were, of the ratios belonging to its particular kind of constitution and being molded by any movements specific to the peculiar attributes of the melodies [...] as the melody itself modulates in different ways at different times and draws our souls to the conditions constituted from the likeness of the ratios.[62]

Because human souls are characterized by the same ratios as exist among musical pitches, their perception of a melody transforms them. The way it transforms them is by means of sympathy; souls are affected by what they have in common with a melody—i.e., harmonic ratios. Modulations of *tonoi* have different effects on the soul. A modulation to the higher *tonoi* is more stimulating; a modulation to the lower *tonoi* is more sedating. Hence, Ptolemy maintains that it is appropriate to compare the intermediate *tonoi*, those *tonoi* around the Dorian, to moderate and stable ways of life, the higher *tonoi*, like the Mixolydian, to more vigorous ways of life, and the lower *tonoi*, like the Hypodorian, to more sluggish ways of life. Ptolemy seems to prefer the intermediate, moderate and stable, *tonoi*, as he recounts a common story about Pythagoras, in which he advised individuals to start their mornings by listening to a soothing melody and so attune their souls, transforming them into a pure, settled, and concordant state.

Harmonic Ratios in the Heavens

After examining the human soul, Ptolemy turns to the third and final system of complete and rational objects: the heavens. He devotes the remainder of the *Harmonics* to delineating the harmonic relations that exist among the stars' movements and configurations. First, in *Harmonics* 3.8–9 he demonstrates how the geometry of the zodiacal circle encompasses the ratios of the homophones and concords. He relates the zodiac to the double octave, which he argues

---

62. Ibid. 3.7, D99–100: τοιγάρτοι καὶ ταῖς ἐνεργείαις αὐταῖς τῆς μελῳδίας συμπάσχουσιν ἡμῶν ἄντικρυς αἱ ψυχαί, τὴν συγγένειαν ὥσπερ ἐπιγινώσκουσαι τῶν τῆς ἰδίας συστάσεως λόγων καὶ τυπούμεναί τισι κινήμασιν οἰκείοις ταῖς τῶν μελῶν ἰδιοτροπίαις [. . .] ἄλλοτε ἄλλως τοῦ μέλους αὐτοῦ τε μεταβάλλοντος καὶ τὰς ψυχὰς ἐξάγοντος ἐπὶ τὰς ἐκ τῆς ὁμοιότητος τῶν λόγων συνισταμένας διαθέσεις.

in *Harmonics* 2.4 is the "complete *systêma*," the smallest scale system that includes homophones and concords as well as their species. Ptolemy maps the double octave onto the zodiacal circle. He does not indicate which zodiacal signs occur at which notes, but he places the two equinoxes at the beginning/end and the middle of the two-octave system. Accordingly, homophones lie in direct opposition, at opposite ends of the diameter of the zodiacal circle. In this way, homophones correspond to the configurations of stars that lie in direct opposition, and stars in opposition have the most active astrological relationship.

In *Harmonics* 3.9, the division of the zodiacal circle into four parts produces the complete set of celestial configurations that are harmonically concordant and, in their astrological power, active. Ptolemy divides the zodiacal circle accordingly:

> For, let us draw a circle, AB, and divide it, starting from some one point, such as A, into two equal [parts] by means of [line] AB, into three equal [parts] by means of [line] AC, into four equal [parts] by [line] AD, and into six equal [parts] by [line] CB. Then arc AB will make the configuration of diametrical opposition, AD that of the square, AC that of the triangle, and CB that of the hexagon. And the ratios of the arcs, [starting] from the same point, that is, from A, will include those of the homophones and the concords and that of the tone besides, as we can see if we suppose the circle to consist of twelve segments, since this is the first number to have a half, a third, and a fourth part.[63]

By dividing the circle in this way, Ptolemy is able to demonstrate the existence of the ratios of the homophones, concords, and tone in the zodiacal circle. If the circle consists of twelve equal segments, then the ratio of arc ABD to arc ABC manifests the 9:8 ratio of the tone. The 2:1 ratio of the octave exists in the relation of the entire circle to the semicircle, the eight segments of arc ABC to the four segments of arc AC, and the six segments of arc AB to the three segments of arc AD. The 3:2 ratio of the concord of the fifth exists in the relation of the entire circle to arc ABC, in the relation of arc ABD to arc AB, and the relation of arc AB to arc AC. The 4:3 ratio of the concord of the fourth exists in

---

63. Ibid. 3.9, D102: ἐὰν γὰρ ἐκθώμεθα κύκλον τὸν ΑΒ καὶ διέλωμεν αὐτὸν ἀπὸ τοῦ αὐτοῦ σημείου, οἷον τοῦ Α, εἰς μὲν δύο ἴσα τῇ ΑΒ, εἰς δὲ τρία ἴσα τῇ ΑΓ, εἰς δὲ τέσσαρα ἴσα τῇ ΑΔ, εἰς δὲ ἓξ ἴσα τῇ ΓΒ, ἡ μὲν ΑΒ περιφέρεια ποιήσει τὴν διάμετρον στάσιν, ἡ δὲ ΑΔ τὴν τετράγωνον, ἡ δὲ ΑΓ τὴν τρίγωνον, ἡ δὲ ΓΒ τὴν ἑξάγωνον. καὶ περιέξουσιν οἱ λόγοι τῶν ἀπὸ τοῦ αὐτοῦ σημείου, τουτέστι πάλιν τοῦ Α, λαμβανομένων περιφερειῶν, τούς τε τῶν ὁμοφώνων καὶ τοὺς τῶν συμφώνων καὶ ἔτι τὸν τονιαῖον, ὡς ἐξέσται σκοπεῖν ὑποθεμένοις τὸν κύκλον τμημάτων ιβ′, διὰ τὸ πρῶτον εἶναι τὸν ἀριθμὸν τοῦτον τῶν ἥμισυ καὶ τρίτον καὶ τέταρτον ἐχόντων μέρος.

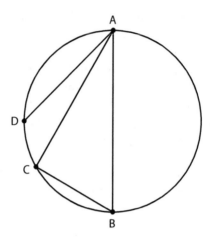

FIGURE 5.3. Four divisions of the zodiacal circle.

the relation of the whole circle to arc ABD, the relation of arc ABC to arc AB, and the relation of arc AC to arc AD. The circle also encompasses the 3:1 ratio of the octave and a fifth, the 4:1 ratio of the double octave, and the 8:3 ratio of the octave and a fourth. Attributing twelve segments to the zodiacal circle also justifies the traditional division of the zodiac into twelve signs.

The location of the ratios of the homophones and concords in the zodiacal circle explains why it is that certain relations among zodiacal signs are effective astrologically. It turns out that the ratios that characterize the octave, the fifth, and the fourth also describe the relations of opposition, trine, and quartile in astrology. Again, the signs across a diameter, which exhibit the 2:1 ratio of the octave, are in opposition. Because the relation of the entire circle to arc ABC, or arc AB to arc AC, manifests the ratio 3:2, the concord of the fifth underlies the astrological aspect of trine; because the relation of the entire circle to arc ABD manifests the ratio 4:3, the concord of the fourth underlies the astrological aspect of quartile. In other words, the mathematical relations, the configurations of the stars in these harmonic ratios, explain the stars' resultant physical effects when they are in these relationships.

As Noel Swerdlow has observed, Ptolemy also divides the zodiacal circle harmonically in *Tetrabiblos* 1.14:[64]

> We may learn from the following why only these intervals were accepted. For, the explanation of the [interval] in accordance with opposition is immediately manifest, inasmuch as it makes the signs [meet] on one

64. Swerdlow, "Ptolemy's *Harmonics*," 155.

straight line. But if we take the two parts and the two superparticulars most important also in concordance, and if the parts (the half and the third) [be applied] to the [interval] of opposition, [composed] of two right angles, the half makes the [interval] of the quartile and the third the sextile and the trine[65]

It is because opposition, trine, quartile, and sextile are characterized by the ratios of the octave and concords that zodiacal signs in these relations are astrologically active.

Ptolemy adds in *Tetrabiblos* 1.14 that in another sense the aspects trine and sextile are called concordant, and quartile and opposition discordant, because the former aspects are composed of signs that are of the same gender, entirely feminine or entirely masculine, whereas the latter are composed of signs of opposite genders. Opposition, for instance, links one masculine sign with one feminine. This discordance is not an actual, harmonic discordance, however. Harmonic ratios still underlie the aspects of quartile and opposition. Rather, the genders of the signs consist in another typology, where the aspects are called (καλοῦνται) concordant and discordant analogously. Moreover, the masculine and feminine signs do not cancel out each other's power. Instead, they shape the type of effect the aspect has on sublunary events. When joined in opposition, the most active relationship, the effect is especially powerful.

In *Harmonics* 3.9, Ptolemy explains that the relations of zodiacal signs that are one or five of twelve segments apart are not significant because the ratios that describe them are not harmonically concordant. The 12:1 and 12:11 ratios of adjacent signs are melodic but not concordant. The 12:5 and 12:7 ratios of signs that are five segments apart are neither melodic nor concordant. Correspondingly, Ptolemy explains in *Tetrabiblos* 1.17 that signs that are adjacent or five apart are disjunct:

> The segments which simply have not one relation of the aforementioned familiarities with one another are called "disjunct" and "alien." These are the ones that belong neither to the [familiarities] of commanding or obeying, nor of beholding or equal power, and furthermore they are found to be entirely without share in the four aspects set forth—opposition, trine, quartile, and sextile—and they verily are either one or five signs apart, inasmuch

---

65. Ptolemy, *Tetrabiblos* 1.14.2, H52: δι᾽ ἣν δὲ αἰτίαν αὗται μόναι τῶν διαστάσεων παρελήφθησαν, ἐκ τούτων ἂν μάθοιμεν. τῆς μὲν γὰρ κατὰ τὸ διάμετρον αὐτόθεν ἐστὶν ὁ λόγος φανερός, ἐπειδήπερ ἐπὶ μιᾶς εὐθείας ποιεῖται τὰς συναντήσεις. λαμβανομένων δὲ τῶν δύο μεγίστων καὶ διὰ συμφωνίας μορίων τε καὶ ἐπιμορίων, μορίων μὲν πρὸς τὴν τῶν β᾽ ὀρθῶν διάμετρον (τοῦ τε ἡμίσους καὶ τοῦ τρίτου), τὸ μὲν εἰς δύο τὴν τοῦ τετραγώνου πεποίηκε, τὸ δὲ εἰς τρία τὴν τοῦ ἑξαγώνου καὶ τὴν τοῦ τριγώνου

as those which are one [sign] apart are, as it were, turned away from one another and, though they are two, encompass the angle of one, and those that are five [signs] apart divide the whole circle into unequal [segments], while the other aspects make an equal division of the perimeter.[66]

Whether or not the relations between zodiacal signs are characterized by the ratios of the octave and concords determines whether or not they are in an astrologically active relationship—opposition, trine, quartile, or sextile—or in a relation that is disjunct or alien and, consequently, astrologically insignificant. Thus, harmonic ratios explain astrological phenomena.

In the remaining chapters of the *Harmonics*, Ptolemy leaves aside astrology and examines the existence of harmonic ratios among astronomical phenomena. In the astrological case, the geometry of the zodiacal circle—and the instantiation of harmonic ratios in the relations among the circle's arcs—determines the effective power of the zodiacal signs and the stars in them. In the following three astronomical cases, the stars' movements through the heavens resemble certain other relations in musical systems. In *Harmonics* 3.10–12, Ptolemy analyzes three principal movements of stars: movement in length (κατὰ μῆκος), movement in depth (κατὰ βάθος), and movement in breadth (κατὰ πλάτος). These translate to the diurnal rotation of the stars, or, conversely, their movement eastward along the ecliptic; their motion closer and further away from the earth; and their motion in declination north and south of the celestial equator.

With respect to movement in length, the stars' rising and setting correspond to the rising and falling of musical pitches. When a star is near the horizon, when rising or setting, its movement in the heavens corresponds to the absolutely lowest notes with respect to pitch. When at the zenith, a star's movement corresponds to the absolutely highest note. As a star rises, so too notes progress and pitch rises; as a star sets, notes descend and pitch falls. When a star is below the horizon, its movement is analogous to silence.

Second, movements in depth, toward and away from the earth, correspond to the genera in music: the enharmonic, chromatic, and diatonic. Again, what is relevant in the genera is the relative magnitude of the first two intervals

---

66. Ptolemy, *Tetrabiblos* 1.17.1, H56–57: Ἀσύνδετα δὲ καὶ ἀπηλλοτριωμένα καλεῖται τμήματα, ὅσα μηδένα λόγον ἁπλῶς ἔχει πρὸς ἄλληλα τῶν προκατειλεγμένων οἰκειώσεων. ταῦτα δέ ἐστιν ἃ μήτε τῶν προστασσόντων ἢ ὑπακουόντων τυγχάνει μήτε τῶν βλεπόντων ἢ ἰσοδυναμούντων, ἔτι τε καὶ τῶν ἐκκειμένων τεσσάρων συσχηματισμῶν (τοῦ τε διαμέτρου καὶ τοῦ τριγώνου καὶ τοῦ τετραγώνου καὶ τοῦ ἑξαγώνου) κατὰ τὸ παντελὲς ἀμέτοχα καταλαμβανόμενα καὶ ἤτοι δι᾽ ἑνὸς ἢ διὰ πέντε γινόμενα δωδεκατημορίων, ἐπειδήπερ τὰ μὲν δι᾽ ἑνὸς ἀπέστραπταί τε ὥσπερ ἀλλήλων καὶ δύο αὐτὰ ὄντα ἑνὸς περιέχει γωνίαν, τὰ δὲ διὰ πέντε εἰς ἄνισα διαιρεῖ τὸν ὅλον κύκλον, τῶν ἄλλων σχηματισμῶν εἰς ἴσα τὴν τῆς περιμέτρου διαίρεσιν ποιουμένων.

in the tetrachord compared with the third, remaining interval. In the case of stars' movements in depth, what is significant is the distance of the star—least, intermediate, or greatest—in its revolution about the earth. These relative distances are determined through a comparison of the star's velocity at different points in its revolution. Ptolemy assumes that a star's velocity around a circle is uniform and that a combination of circular movements causes a change in the star's velocity, as well as its distance from the earth. At an intermediate distance, a star's velocity is intermediate, and, because the divisions of the tetrachord in the chromatic genera are intermediate compared with the tetrachords in the other genera, intermediate movements in depth correspond to the chromatic genera. Stars' least movements, presumably least velocities, correspond to the enharmonic genera, since in enharmonic tetrachords the first two intervals are the smallest, and stars' greatest movements, again presumably greatest velocities, correspond to the diatonic genera, since the first two intervals in a diatonic tetrachord are never smaller than the remaining interval.

With respect to the least and greatest velocities, Ptolemy acknowledges that it is not simply the case that the least velocity corresponds to the least distance and that the greatest velocity corresponds to the greatest distance, as one might expect. Rather, the least or greatest velocity may occur either when a star is closest to or furthest from the earth. The particular relation of velocity to distance depends on the star. In Ptolemy's astronomical models, the sun appears to move fastest when it is closest to the earth and slowest when it is furthest away. This relationship of velocity to distance occurs in both the epicyclic and eccentric models, which Ptolemy describes in *Almagest* 3.3 when accounting for the apparent irregularity in the sun's revolution. If the earth were situated at the center of the sun's orbit—which Ptolemy, like Greek mathematicians and philosophers before him, took to be circular—then the four seasons would be of equal duration. In fact, the seasons are not of equal length, and so the center of the sun's orbit cannot coincide with the earth. Ptolemy presents two models to account for this apparent irregularity: the eccentric and the epicyclic. In the eccentric model, the center of the sun's orbit does not coincide with the earth, and in the epicyclic model the sun lies on a small circle, called the epicycle, whose center lies on another circle, called the deferent, whose center is the earth. The phenomena that result from these two models are identical. In the *Almagest*, Ptolemy chooses the eccentric model for the sun over the epicyclic model, but what is relevant in the *Harmonics* is the velocity of the sun at its least and greatest distances from the earth. In the eccentric model, the sun only appears to change its velocity from the perspective of the earth; its velocity remains constant. In the epicyclic model, the resultant motion of the sun is such that it moves fastest when it is closest to the earth and slowest when it is

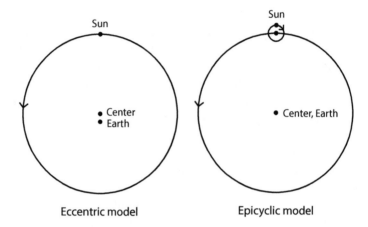

FIGURE 5.4. Solar models.

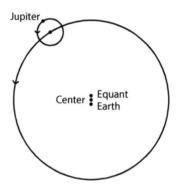

FIGURE 5.5. Jupiter's model. (For a discussion of the equant, see chapter 6.)

furthest away, because the sun's epicycle rotates in the opposite direction to the deferent. The combination of the circular motions produces a change in velocity. Because the sun has least, intermediate, and greatest velocities in the epicyclic model, that model—rather than the eccentric—is relevant to the *Harmonics'* account of movement in depth.

In the case of the five planets, their epicycles revolve in the same direction as their deferents, and therefore in the opposite direction to the sun's epicycle. Consequently, these stars move fastest when they are furthest from the earth, rather than when they are closest, and they create retrograde loops, which are visible from the earth. So, Ptolemy cannot say that the slowest or fastest velocity of a star corresponds to a distance closest or furthest from the earth, because the velocities of the various wandering stars differ in this regard.

The third type of movement stars experience, movement in breadth, corresponds to modulation in *tonoi*. Ptolemy justifies this correspondence by making reference to how modulation in *tonoi* and movement in breadth relate to musical genera and movement in depth, respectively. Just as a shift between *tonoi* does not produce a change in genus, so, too, movement in breadth does not affect movement in depth. Ptolemy maps the *tonoi* onto the northerly and southerly movements of the wandering stars. Because the Dorian *tonos* lies in the middle of the seven *tonoi*, it corresponds to movements at the celestial equator. The two *tonoi* at the extremes, the Mixolydian and the Hypodorian, correspond to the wandering stars' movements at the most northerly and southerly positions—i.e., at the tropics, the circles that are parallel to the celestial equator and coincide with the ecliptic's most northerly and southerly distances from the celestial equator. The remaining four *tonoi* correspond to movements at parallels intermediate between the tropics and the celestial equator, two above and two below the celestial equator. Each of the seven parallels—the celestial equator, the two tropics, and the four intermediate circles—intersect with the ecliptic at the points separating zodiacal signs. Thus, corresponding to the seven *tonoi*, movements in breadth mark off the boundaries of the twelve zodiacal signs.

One might suppose that in delineating these correspondences between musical relations and celestial phenomena Ptolemy portrays the heavens as making music. After all, much of Ptolemy's theory of *harmonia* derives from the Platonic tradition, and Plato describes music in the heavens in the *Republic*'s myth of Er. Socrates depicts a cosmic system of eight nested whorls, whose rims appear like circles and correspond to the wandering and fixed stars. Socrates proclaims, "Up above on each of its circles stood a Siren, who was carried around [by its rotation], emitting a single sound, one note, and from all

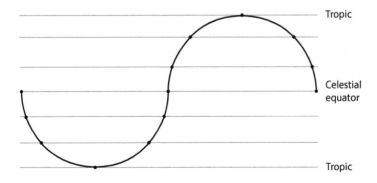

FIGURE 5.6. Movement in breadth.

eight in concordance came a single harmony."[67] Aristotle argues against this Pythagorean doctrine, that the stars produce an audible harmony, in *De caelo* II, 290b12–291a26. He notes that neither do we hear this music nor do we experience the force that such loud sounds, proportionate in intensity to the size of the bodies moving, should produce in us. For Ptolemy, too—perhaps influenced by empirical arguments like Aristotle's—there are no sirens sitting atop the heavenly spheres, and none of the stars make music. Rather, the same ratios that describe the relations in musical systems exist in the movements and configurations of heavenly bodies. Swerdlow has observed that because the wandering stars' periods for movement in depth are different from their periods for movement in breadth, all of the heavenly bodies' movements corresponding to the genera will manifest in conjunction with every one of the movements corresponding to the seven *tonoi*, with a complete cycle taking approximately fifty-nine years for Saturn, seventy-one years for Jupiter, seventy-nine years for Mars, eight years for Venus, and forty-six years for Mercury.[68] Combining the movements of every one of the stars, wandering and fixed, the complexity of this harmonic system is astounding. Heavenly *harmonia* is eternally prolific.

Conclusion

Because the heavens are immutable, *harmonia* produces harmonic ratios in the heavens for all eternity. Human souls, on the other hand, are mortal and corruptible. When in their proper attunement, human souls are configured according to harmonic ratios, but they can and do fall out of attunement. Of all of the complete and rational entities moved by *harmonia*—musical systems, human souls, and heavenly bodies—only the human soul requires an external exemplar, an instantiation of harmonic ratios in a body physically unlike itself on which to model its formal configuration. In *Almagest* 1.1, Ptolemy explains that this exemplar is astronomical. The contemplation of astronomical objects transforms the mathematician's soul to a similar state of constancy, good order, commensurability, and calm. In the *Harmonics*, he implies that harmonics, and not just astronomy, can serve as the catalyst for this psychological transformation. While musical pitches are not eternal, the relations within mu-

---

67. Plato, *Republic* X, 617b: ἐπὶ δὲ τῶν κύκλων αὐτοῦ ἄνωθεν ἐφ᾿ ἑκάστου βεβηκέναι Σειρῆνα συμπεριφερομένην, φωνὴν μίαν ἱεῖσαν, ἕνα τόνον· ἐκ πασῶν δὲ ὀκτὼ οὐσῶν μίαν ἁρμονίαν συμφωνεῖν.

68. Swerdlow uses the periods Ptolemy gives in *Almagest* 9.3 for when a planet nearly completes integral numbers of synodic and zodiacal revolutions: Swerdlow, "Ptolemy's *Harmonics*," 161.

sical systems are constant. If the student of harmonics fails to create and exhibit harmonic ratios by means of a monochord, the failure is not in the musical relations but in the actualization of the student's harmonic faculty, or power of *harmonia*. It is because human beings can exercise their harmonic faculty through the repeated production of harmonic ratios among musical pitches—however fleeting individual musical pitches may be—that the theoretical study, exhibition, and experience of harmonics, in addition to astronomy, attunes the soul. Thus evoking *Timaeus* 47c–d, in which Plato's character Timaeus maintains that the study of not only the heavens but also harmony facilitates the ordering of the soul's orbits, Ptolemy indicates that two mathematical sciences—harmonics and astronomy—provide paths by which a human being can achieve the ultimate objective, the harmonic attunement of the soul, its transformation into a state resembling the movements and configurations of the stars.

6

Harmonics and Astronomy

FOR PTOLEMY, harmonics and astronomy are more than independent means of attaining good order in the human soul. They have a special, figuratively familial relationship. Each is attendant on either sight or hearing, the two senses perceptive of beauty, and they each examine one set of complete and rational objects. Harmonics is the mathematical science of the relations among musical pitches, the complete and rational objects perceptible only by hearing. Astronomy is the mathematical science of the movements and configurations of heavenly bodies, the complete and rational objects perceptible only by sight. Moreover, harmonics and astronomy are complementary inasmuch as they each utilize arithmetic or geometry, which Ptolemy calls the "indisputable methods" or "instruments" of mathematics. This chapter examines in what ways and to what degree harmonics and astronomy are complementary, particularly with respect to their epistemic efficacy. In *Almagest* 1.1, Ptolemy claims that mathematics provides sure and incontrovertible knowledge. In his study of harmonics and astronomy, does he maintain this position? Does Ptolemy consider his harmonic and astronomical judgments to be sure and incontrovertible knowledge? I argue that Ptolemy's *Harmonics* and *Almagest* indicate that harmonics does attain this epistemic success, but astronomy does so only in certain respects. How astronomy falters epistemologically has consequences for Ptolemy's practice of astronomy as well as the relevance of the various parts of his astronomy to his ethical system.

Cousin Sciences

In the *Harmonics*, Ptolemy associates harmonics and astronomy on account of their relationship to hearing and seeing. He proclaims, "related to sight and to the movements in place of the things that are only seen—that is, the heavenly bodies—is astronomy [ἀστρονομία], and related to hearing and to the movements in place, once again, of the things that are only heard—that is, sounds—is harmonics [ἁρμονική] . . ."[1] Three sets of objects are the most complete and rational in the cosmos: musical scale systems, human souls, and the heavens. As I discuss in the next chapter, the human soul is, according to Ptolemy, imperceptible. The effects it has in and through the body are perceptible, but the soul itself is imperceptible. Of the complete and rational objects in the cosmos, then, only the relations among musical pitches and the movements and configurations of heavenly bodies are perceptible, and they are perceptible by different senses. The relations among musical pitches are perceptible only by hearing; heavenly bodies' movements and configurations are perceptible only by sight. Although harmonic ratios are common sensibles—as they are perceptible by both hearing and seeing—music and the heavens, in which harmonic ratios inhere, are each perceptible by only one sense. It is significant that hearing and seeing, in particular, perceive these phenomena, as they are the only senses that perceive objects inasmuch as they are beautiful and not merely pleasurable. When sight and hearing perceive astronomical and harmonic objects, they can and do perceive them as beautiful.

Harmonics and astronomy are also complementary in that each is dependent on either arithmetic or geometry. In *Almagest* 1.1, Ptolemy calls arithmetic and geometry "indisputable methods." He says "that only the mathematical [genus of the theoretical part of philosophy] can provide sure and incontrovertible knowledge to its practitioners, if one approaches it rigorously, for its demonstration proceeds by indisputable methods, both arithmetic and geometry . . ."[2] In *Harmonics* 3.3, Ptolemy relates these indisputable methods to harmonics and astronomy. He says of harmonics and astronomy that "they employ both arithmetic and geometry as indisputable instruments to [study] the quantity and quality of the primary movements, and they are, as it were, cousins, born of the sisters sight and hearing, and they have been reared by both arithmetic and geometry like [children] most closely related in their stock [γένους]."[3] According to Ptolemy, arithmetic and geometry are

1. Ptolemy, *Harmonics* 3.3, D94: παρὰ μὲν τὴν ὄψιν καὶ τὰς κατὰ τόπον κινήσεις τῶν μόνως ὁρατῶν, τουτέστι τῶν οὐρανίων, ἀστρονομία, παρὰ δὲ τὴν ἀκοὴν καὶ τὰς κατὰ τόπον πάλιν κινήσεις τῶν μόνως ἀκουστῶν, τουτέστι τῶν ψόφων, ἁρμονική

2. Ptolemy, *Almagest* 1.1, H6: μόνον δὲ τὸ μαθηματικόν, εἴ τις ἐξεταστικῶς αὐτῷ προσέρχοιτο, βεβαίαν καὶ ἀμετάπιστον τοῖς μεταχειριζομένοις τὴν εἴδησιν παράσχοι ὡς ἂν τῆς ἀποδείξεως δι' ἀναμφισβητήτων ὁδῶν γιγνομένης, ἀριθμητικῆς τε καὶ γεωμετρίας

3. Ptolemy, *Harmonics* 3.3, D94: χρῶμεναι μὲν ὀργάνοις ἀναμφισβητήτοις ἀριθμητικῇ τε καὶ γεωμετρίᾳ πρός τε τὸ ποσὸν καὶ τὸ ποιὸν τῶν πρώτων κινήσεων, ἀνεψιαὶ δ' ὥσπερ καὶ αὐταί, γενόμεναι μὲν ἐξ ἀδελφῶν ὄψεως καὶ ἀκοῆς, τεθραμμέναι δὲ ὡς ἐγγύτατω πρὸς γένους ὑπ' ἀριθμητικῆς τε καὶ γεωμετρίας.

not mathematical sciences. They are the mathematical sciences' methods or instruments. Mathematical sciences study mathematical objects—e.g., shape, number, size, place, time, and suchlike, as Ptolemy lists them in *Almagest* 1.1— in specific sets of bodies. Astronomy, for instance, studies not all movements and configurations but, rather, the movements and configurations of only heavenly bodies. Harmonics studies the distinctions related to high and low pitch among sounds, where a sound, according to *Harmonics* 1.1, is a modification of air that has been struck. Even though harmonic ratios exist not only among musical pitches but also in human souls and the heavens, Ptolemy defines harmonics narrowly, as the study of these ratios among only musical pitches. Hence, a mathematical science's scope encompasses one discrete set of phenomena. Arithmetic and geometry are not mathematical sciences, as they do not study mathematical objects in one discrete set of bodies. Instead, they are the paths or instruments by means of which mathematical sciences progress. A mathematical science employs arithmetic or geometry to study the mathematical objects in a particular domain. Harmonics appropriates the number theory of arithmetic to describe the ratios among musical pitches; astronomy appropriates the points, lines, circles, and spheres of geometry to describe the movements and configurations of heavenly bodies.

In addition, Ptolemy applies a familial metaphor to harmonics and astronomy's relationship. They are cousins, descended from the sisters sight and hearing. Ptolemy's association of harmonics and astronomy has a precedent in earlier Greek philosophy, but it was not a given. In the *Metaphysics*, for instance, Aristotle compares harmonics not to astronomy but to optics. When arguing against the separate, transcendent existence of mathematical objects, Aristotle juxtaposes the objects of harmonics and optics:

> For besides the perceptible things there similarly will be the things with which astronomy is concerned and the things with which geometry is concerned; but how is it possible that there be a heaven and its parts, or anything else whatsoever with movement [besides the perceptible]? Also the optical things and the harmonic things will be similarly [separate]; for there will be sound and sight besides the perceptible and individual [sounds and sights][4]

As the studies of sights and sounds, respectively, optics and harmonics are complementary. Aristotle again mentions harmonics and optics in concert

---

4. Aristotle, *Metaphysics* XIII, 1077a1–6: περὶ ἃ γὰρ ἡ ἀστρολογία ἐστίν, ὁμοίως ἔσται παρὰ τὰ αἰσθητὰ καὶ περὶ ἃ ἡ γεωμετρία· εἶναι δ' οὐρανὸν καὶ τὰ μόρια αὐτοῦ πῶς δυνατόν, ἢ ἄλλο ὁτιοῦν ἔχον κίνησιν; ὁμοίως δὲ καὶ τὰ ὀπτικὰ καὶ τὰ ἁρμονικά· ἔσται γὰρ φωνή τε καὶ ὄψις παρὰ τὰ αἰσθητὰ καὶ τὰ καθ' ἕκαστα

when explaining how each studies its objects: "The same account concerns harmonics and optics; for neither considers [its objects] *qua* seeing or *qua* sound, but *qua* lines and numbers (indeed, the latter are attributes proper to the former), and mechanics likewise . . ."[5] Again, the counterpart of harmonics is optics. For Aristotle, these two sciences are the studies of all sounds and sights, respectively, and not just particular sets of audible and visible objects.

Aristotle juxtaposes harmonics and optics in the *Posterior Analytics* as a consequence of their relative status in scientific hierarchies. Optics is subordinate to geometry, just as harmonics is subordinate to arithmetic. He explains, "nor [can one prove] by another science what pertains to a different [science], except such as relate to one another in such a way that one falls under the other, as the optical [relate] to geometry and the harmonic to arithmetic."[6] Moreover, Aristotle describes astronomy as comparable not to optics and harmonics but, rather, to arithmetic and geometry. Discussing the claims that arithmetic and geometry make about their objects, he states the following:

> Proper too are the things which are assumed to exist, concerning which the science theorizes what belongs to them in themselves—as arithmetic [theorizes] units, and geometry [theorizes] points and lines. For they assume that these things exist and exist such-and-such. As for the attributes of these in themselves, they assume what each means, as arithmetic [assumes] what odd or even or square or cube [means], and geometry [assumes] what the irrational or inflection or verging [means], and they show that they are by means of the common things and from what have been demonstrated. Astronomy proceeds in the same way.[7]

Astronomy, like arithmetic and geometry, assumes that certain objects exist and that these objects have certain properties. In this way, astronomy is on the same order of the scientific hierarchy as arithmetic and geometry.

5. Ibid., 1078a14–17: ὁ δ᾽ αὐτὸς λόγος καὶ περὶ ἁρμονικῆς καὶ ὀπτικῆς· οὐδετέρα γὰρ ᾗ ὄψις ἢ ᾗ φωνὴ θεωρεῖ, ἀλλ᾽ ᾗ γραμμαὶ καὶ ἀριθμοί (οἰκεῖα μέντοι ταῦτα πάθη ἐκείνων), καὶ ἡ μηχανικὴ δὲ ὡσαύτως

6. Aristotle, *Posterior Analytics* I, 75b14–17: οὐδ᾽ ἄλλη ἐπιστήμη τὸ ἑτέρας, ἀλλ᾽ ἢ ὅσα οὕτως ἔχει πρὸς ἄλληλα ὥστ᾽ εἶναι θάτερον ὑπὸ θάτερον, οἷον τὰ ὀπτικὰ πρὸς γεωμετρίαν καὶ τὰ ἁρμονικὰ πρὸς ἀριθμητικήν. Cf. ibid., 76a22–25.

7. Ibid., 76b3–11: Ἔστι δ᾽ ἴδια μὲν καὶ ἃ λαμβάνεται εἶναι, περὶ ἃ ἡ ἐπιστήμη θεωρεῖ τὰ ὑπάρχοντα καθ᾽ αὑτά, οἷον μονάδας ἡ ἀριθμητική, ἡ δὲ γεωμετρία σημεῖα καὶ γραμμάς. ταῦτα γὰρ λαμβάνουσι τὸ εἶναι καὶ τοδὶ εἶναι. τὰ δὲ τούτων πάθη καθ᾽ αὑτά, τί μὲν σημαίνει ἕκαστον, λαμβάνουσιν, οἷον ἡ μὲν ἀριθμητικὴ τί περιττὸν ἢ ἄρτιον ἢ τετράγωνον ἢ κύβος, ἡ δὲ γεωμετρία τί τὸ ἄλογον ἢ τὸ κεκλάσθαι ἢ νεύειν, ὅτι δ᾽ ἔστι, δεικνύουσι διά τε τῶν κοινῶν καὶ ἐκ τῶν ἀποδεδειγμένων. καὶ ἡ ἀστρολογία ὡσαύτως.

At the same time, Aristotle implicitly contrasts astronomy with optics and harmonics. As in the *Metaphysics*, he describes optics as subordinate to geometry and harmonics as subordinate to arithmetic, but astronomy remains on the same hierarchic level as arithmetic and geometry. It is a superior science. Aristotle proclaims, "Such are those that relate to one another in such a way that one falls under the other, as the optical [relate] to geometry, the mechanical to stereometry, the harmonic to arithmetic, and the phenomena to astronomy."[8] Astronomy is a science that, like geometry and arithmetic, has a science subordinate to it. In the same way as geometry and arithmetic explain optical and harmonic phenomena, astronomy explains celestial appearances, the phenomena (τὰ φαινόμενα). Moreover, optics, harmonics, and astronomy are, according to *Physics* II, 194a7–8, the "more physical of the mathematical sciences" (τὰ φυσικώτερα τῶν μαθημάτων) as, for instance, geometry studies the lines in physical bodies *qua* mathematical but optics studies them *qua* physical. If one were to incorporate this characterization of the more physical of the mathematical sciences into the hierarchical description of the sciences in the *Posterior Analytics*, then the mathematical sciences that study their objects *qua* physical would be subordinate to the sciences that study them *qua* mathematical. Nevertheless, Aristotle repeatedly portrays astronomy as one of the superior sciences, on the order of arithmetic and geometry, whereas harmonics and optics are subordinate sciences. Ptolemy appropriates Aristotle's systematization of the sciences, as superior or subordinate, but he then applies it to the complementarity of harmonics and astronomy. Again, arithmetic and geometry are not mathematical sciences, according to Ptolemy, but rather methods or instruments, and just as the mathematical science of harmonics employs the mathematical instrument of arithmetic, so the mathematical science of astronomy employs the mathematical instrument of geometry.

Ptolemy appropriates the familial metaphor of harmonics and astronomy's relationship from the Platonic tradition. When outlining the education of the philosopher-king in *Republic* VII, Plato includes fields of inquiry that turn the soul from becoming to being, from perceptible phenomena to intelligible objects, and among these fields he includes astronomy and harmonics. After describing the proper way to study astronomy, Socrates continues with Glaucon:

> Socrates: Besides the one we [have discussed], there is its counterpart.
> Glaucon: What is that?

---

8. Ibid., 78b35–39: τοιαῦτα δ' ἐστὶν ὅσα οὕτως ἔχει πρὸς ἄλληλα ὥστ' εἶναι θάτερον ὑπὸ θάτερον, οἷον τὰ ὀπτικὰ πρὸς γεωμετρίαν καὶ τὰ μηχανικὰ πρὸς στερεομετρίαν καὶ τὰ ἁρμονικὰ πρὸς ἀριθμητικὴν καὶ τὰ φαινόμενα πρὸς ἀστρολογικήν.

Socrates: It seems likely that, as the eyes have fixed upon astronomical [motion], so the ears fix upon harmonic motion, and these sciences [viz. astronomy and harmonics] are like sisters of one another, as the Pythagoreans say, and we agree, Glaucon. Do we not?

Glaucon: We do.[9]

In their relationships to sight and hearing, astronomy and harmonics are like sisters, a simile of kinship Socrates ascribes to the Pythagoreans. For Ptolemy, the senses sight and hearing—rather than the sciences—are like sisters and, as children of sight and hearing, astronomy and harmonics are like cousins, reared by the mathematical methods geometry and arithmetic. In this way, Ptolemy blends the familial, complementary relationship of harmonics and astronomy in the Platonic tradition with an Aristotelian hierarchy of superior and subordinate sciences.

Sure and Incontrovertible Knowledge?

Harmonics and astronomy are counterparts, but do they have equal claims to knowledge? Do they both, as Ptolemy implies in *Almagest* 1.1, produce sure and incontrovertible knowledge? In the *Harmonics*, Ptolemy implies that harmonics does generate knowledge, but he reveals in the *Almagest* that astronomy does so only to a limited degree. Some of astronomy's conclusions are sure and incontrovertible, but others are neither sure nor incontrovertible. To begin with, the objective of the student of harmonics is to determine which ratios describe the relations among musical pitches and thereafter confirm that creating these ratios on a monochord, the harmonic *kanôn*, or multistringed *kanones* generates pitches that sound like the homophones, concords, or melodic relations that he is aiming to produce. The sense of hearing alone cannot determine or confirm these ratios. Reason must interact with hearing and guide it toward making more precise perceptions. The harmonic *kanôn* embodies rational principles and, as a result, it "straightens" perception. Ptolemy explains, "Then, the instrument of this kind of method is called a harmonic *kanôn*, from the common predication and from its straightening [κανονίζειν] the things in sense perception that are inadequate to reveal the

9. Plato, *Republic* VII, 530d: Πρὸς τούτῳ, ἦν δ᾽ ἐγώ, ἀντίστροφον αὐτοῦ.

Τὸ ποῖον;

Κινδυνεύει, ἔφην, ὡς πρὸς ἀστρονομίαν ὄμματα πέπηγεν, ὡς πρὸς ἐναρμόνιον φορὰν ὦτα παγῆναι, καὶ αὗται ἀλλήλων ἀδελφαί τινες αἱ ἐπιστῆμαι εἶναι, ὡς οἵ τε Πυθαγόρειοί φασι καὶ ἡμεῖς, ὦ Γλαύκων, συγχωροῦμεν. ἢ πῶς ποιοῦμεν;

Οὕτως, ἔφη.

truth."[10] Not just any set of sounds lends itself to harmonic study. The *kanôn* renders audible phenomena, specifically the distinctions between higher and lower pitches, more apprehensible.[11] Ptolemy asserts, "But the string stretched over what is called the *kanôn* will show us the ratios of the concords more accurately and more readily . . ." (ἀκριβέστερόν τε καὶ προχειρότερον).[12] The *kanôn* engenders some practical difficulties, but it accomplishes this straight-ening better than other instruments, such as *auloi* or *syringes*.[13] By imposing the ratios on the string, the student of harmonics creates pitches that are, for instance, audibly concordant: "When something of this kind has been grasped and the measuring rod has been divided in the ratios of the concords that have been set out, by the shifting of the bridge to each point of division we shall find that the differences of the appropriate notes agree most accurately [ἀκριβέστατον] with what are heard."[14] Thus, the student of harmonics con-firms that the ratios he imposes on the *kanôn* are the true ratios of harmonic phenomena.

As the quotes in the previous paragraph indicate, the concept of accuracy (ἀκριβεία) is central to Ptolemy's epistemology. The *kanôn* is the best har-monic instrument because it more accurately produces harmonic phenomena. Students of harmonics can confirm that their divisions of the *kanôn* agree most accurately with what are perceived to be concords. I argue that Ptolemy lends two meanings to the term "accuracy." In one sense, it denotes precision. After all, Ptolemy maintains in *Harmonics* 1.1 that, without the aid of instru-ments, the senses perceive objects only approximately, whereas reason grasps them precisely, "since it is in general characteristic of the senses to discover what is approximate [σύνεγγυς] and to receive what is accurate [ἀκριβοῦς], and of reason to receive what is approximate [σύνεγγυς] and to discover what

10. Ptolemy, *Harmonics* 1.2, D5: Τὸ μὲν οὖν ὄργανον τῆς τοιαύτης ἐφόδου καλεῖται κανὼν ἁρμονικός, ἀπὸ τῆς κοινῆς κατηγορίας καὶ τοῦ κανονίζειν τὰ ταῖς αἰσθήσεσιν ἐνδέοντα πρὸς τὴν ἀλήθειαν παρειλημμένος.

11. For an intellectual history of the term *kanôn*, particularly in the Epicurean tradition and as related to the criterion of truth, see Striker, *Essays on Hellenistic Epistemology*, 31–33. For a comprehensive study of the monochord, see Creese, *Monochord*.

12. Ptolemy, *Harmonics* 1.8, D17: ἡ δὲ ἐπὶ τοῦ καλουμένου κανόνος διατεινομένη χορδὴ δείξει μὲν ἡμῖν τοὺς λόγους τῶν συμφωνιῶν ἀκριβέστερόν τε καὶ προχειρότερον

13. See Ptolemy, *Harmonics* 2.12, D66.

14. Ptolemy, *Harmonics* 1.8, D18–19: ἔπειτα τοῦ τοιούτου καταληφθέντος καὶ καταδιαιρεθέντος τοῦ κανονίου τοῖς ἐκκειμένοις τῶν συμφωνιῶν λόγοις, εὑρήσομεν ἐκ τῆς ἐφ᾽ ἕκαστον τμῆμα τοῦ μαγαδίου παραγωγῆς ὁμολογουμένας ταῖς ἀκοαῖς ἐπὶ τὸ ἀκριβέστατον τὰς τῶν οἰκείων φθόγγων διαφοράς.

is accurate [ἀκριβοῦς]."[15] Accuracy is the opposite of approximation, because the former is more precise than the latter.

In another sense, Ptolemy suggests that an account of phenomena that is accurate is true. In *Almagest* 3.1, he associates what is not true (μὴ ἀληθῶς) with what is not accurate (μὴ ἀκριβῶς),[16] and in *Harmonics* 1.1 he explains that reason determines not only what is accurate but also what is accepted:

> For since matter is defined and delimited only by form, and modifications only by the causes of movements, and since of these the former [i.e., the matter and the modification] belong to sense perception, the latter [i.e., the form and the cause] to reason, it suitably follows that the apprehensions of the senses are defined and delimited by the [apprehensions] of reason, first submitting to them the distinctions they have grasped more roughly, so far as regard those things that are conceived of by means of sense perception, and being guided by them to the [distinctions] that are accurate [ἀκριβεῖς] and agreed upon.[17]

Reason guides sense perception and makes it capable of more precise observation, but this improvement is more than a matter of increased precision. Ptolemy's reference to what are agreed upon alludes to his definition of knowledge in *On the Kritêrion and Hêgemonikon*, where knowledge corresponds to a judgment that is "most clear and agreed upon" (τρανωτάτη καὶ ὁμολογουμένη).[18] When the senses perceive distinctions that are agreed upon, they perceive the true distinctions. Consequently, the senses are able to confirm that certain judgments are true.

Correspondingly, judgments that are only close to the truth, rather than absolutely true, are only more or less agreed upon. Ptolemy makes this distinction in *Geography* 2.1.2 when discussing the coordinates in longitude and latitude of localities on earth. He contrasts the near certainty of the coordinates

---

15. Ibid. 1.1, D3: ὅτι καὶ καθόλου τῶν μὲν αἰσθήσεων ἴδιόν ἐστι τὸ τοῦ μὲν σύνεγγυς εὑρετικόν, τοῦ δὲ ἀκριβοῦς παραδεκτικόν, τοῦ δὲ λόγου τὸ τοῦ μὲν σύνεγγυς παραδεκτικόν, τοῦ δ' ἀκριβοῦς εὑρετικόν.

16. Ptolemy, *Almagest* 3.1, H200.

17. Ptolemy, *Harmonics* 1.1, D3: ἐπειδὴ γὰρ ὁρίζεται καὶ περαίνεται μόνως ἡ μὲν ὕλη τῷ εἴδει, τὰ δὲ πάθη τοῖς αἰτίοις τῶν κινήσεων, καὶ ἔστι τούτων τὰ μὲν αἰσθήσεως οἰκεῖα, τὰ δὲ λόγου, παρηκολούθησεν εἰκότως τὸ καὶ τὰς αἰσθητικὰς διαλήψεις ὁρίζεσθαι καὶ περαίνεσθαι ταῖς λογικαῖς, ὑποβαλλούσας μὲν πρώτας ἐκείναις τὰς ὁλοσχερέστερον λαμβανομένας διαφορὰς ἐπί γε τῶν δι' αἰσθήσεως νοητῶν, προσαγομένας δὲ ὑπ' ἐκείνων ἐπὶ τὰς ἀκριβεῖς καὶ ὁμολογουμένας.

18. Ptolemy, *On the Kritêrion*, La7.

of well-trodden places with the less certain, and more roughly determined, coordinates of less-traveled places:

> Here we shall begin the detailed guide, but we first make the following observation, that the numbers of degrees in longitude and latitude of well-trodden places are to be considered as nearest to the truth [ἐγγυτάτω τῆς ἀληθείας] because more or less agreed upon [ὡς ἐπίπαν ὁμολογούμενον] accounts of them have been passed down without interruption; but [the coordinates] of the [places] that have not been so traveled, because of the sparseness and lack of confirmation of the research, have been estimated more roughly [ὁλοσχερέστερον] according to their proximity to the more trustworthily determined positions or relative configurations[19]

The geographical coordinates of the most visited localities are generally agreed upon and, accordingly, should be considered as nearest to the truth.

In *Harmonics* 1.1, Ptolemy illustrates the interplay of reason and perception that produces the perception and confirmation of true judgments. He presents the example of drawing a circle. Here, what are more and less accurate correspond to what is true and what only appears to be, but is not, true:

> Thus just as a circle constructed by sight alone often appears to be accurate [ἀκριβῶς], until the [circle] produced by reason brings [sight] to a recognition of the one that is really accurate [τοῦ τῷ ὄντι ἀκριβοῦς], so if some determined difference between sounds is found by hearing alone, it sometimes will seem at first neither to fall short of nor to exceed what is proper, but when there is tuned against it the one that is picked out according to its proper ratio, it often will be exposed as not being so, when the hearing, through the comparison, recognizes the more accurate [ἀκριβεστέραν] as something legitimate [γνησίαν], as it were, beside the bastardy [νόθον] of the other.[20]

19. Ptolemy, *Geography* 2.1.2, translation modified from Berggren and Jones, *Ptolemy's "Geography"*: Ἀρξώμεθα δ' ἐντεῦθεν τῆς κατὰ μέρος ὑφηγήσεως ἐκεῖνο προλαβόντες, ὅτι τὰς μὲν τῶν τετριμμένων τόπων μοιρογραφίας μήκους τε καὶ πλάτους ἐγγυτάτω τῆς ἀληθείας ἔχειν νομιστέον διὰ τὸ συνεχὲς καὶ ὡς ἐπίπαν ὁμολογούμενον τῶν παραδόσεων· τὰς δὲ τῶν μὴ τοῦτον τὸν τρόπον ἐφοδευθέντων, ἕνεκεν τοῦ σπανίου καὶ ἀδιαβεβαιώτου τῆς ἱστορίας ὁλοσχερέστερον ἐπιλελογίσθαι κατὰ συνεγγισμὸν τῶν πρὸς τὸ ἀξιοπιστότερον εἰλημμένων θέσεων ἢ σχηματισμῶν

20. Ptolemy, *Harmonics* 1.1, D3–4: ὥσπερ οὖν ὁ μόνῃ τῇ ὄψει περιενεχθεὶς κύκλος ἀκριβῶς ἔχειν ἔδοξε πολλάκις, ἕως ἂν ὁ τῷ λόγῳ ποιηθεὶς εἰς ἐπίγνωσιν αὐτὴν μεταγάγοι τοῦ τῷ ὄντι ἀκριβοῦς, οὕτω κἂν μόνῃ τῇ ἀκοῇ ληφθῇ τις ὡρισμένη διαφορὰ ψόφων, δόξει μὲν εὐθὺς ἐνίοτε μήτε ἐνδεῖν τοῦ μετρίου, μήτε ὑπερβάλλειν, ἐφαρμοσθείσης δὲ τῆς κατὰ τὸν οἰκεῖον λόγον ἐκλαμβανομένης ἀπελεγχθήσεται πολλάκις οὐχ οὕτως ἔχουσα, τῆς ἀκοῆς ἐπιγινωσκούσης τῇ παραθέσει τὴν ἀκριβεστέραν ὡσανεὶ γνησίαν τινὰ παρ' ἐκείνην νόθον.

When one draws a circle by hand, what is drawn appears to be circular, but when one draws a circle by means of a compass, then sight correctly perceives the second as a circle and the first no longer appears to be a circle. The same goes for pitches created with and without the harmonic *kanôn*.

Accurate circles and pitches are, according to Ptolemy, legitimate. What he means by legitimate becomes clear soon after the above passage, when he discusses the accumulation of observational error:

> and therefore this sort of deficiency of sense perceptions does not miss the truth [ἀληθείας] by much when it is simply a matter of discovering whether there is or is not a difference between them, nor again does it in perceiving the amounts of excess between differing things, so long as the [amounts] consist in larger parts of the things to which they belong. But in the case of comparisons concerned with lesser parts more [deficiency] accumulates and in these [comparisons] it is then evident, the more so as the things [compared] have finer parts. The cause is that the deviation from the truth [τὸ παρὰ τὴν ἀλήθειαν], being smallest when it is [taken] just once, cannot yet make the accumulation of the small [amount] perceptible when only a few comparisons have been made, but when [the comparisons are] taken more times together it is then remarkable and altogether easy to observe.[21]

The deficiency of sense perception yields a deviation from, or missing of, the truth. Reason, however, has the ability to direct sense perception to hit the truth. Accurate observations, those observations that are guided by reason, recognize the legitimate and confirm the true. Inaccurate observations, bereft of reason's guidance, embrace the "bastard" and miss the truth.

Ptolemy's opposition of legitimacy with bastardy suggests an atomist influence on his epistemology. Democritus makes a similar distinction in a fragment Sextus Empiricus attributes to him:

> but in his *Canons* he says that there are two [kinds of] knowledge [γνώσεις], the one by means of sense perceptions, and the other by means of thought, of which he calls the one by means of thought "legitimate" [γνησίην], bearing witness to its trustworthiness in judging truth [ἀληθείας], while the one by means of sense perceptions he names "dark" [σκοτίην], depriving it of

---

21. Ibid., D4: καὶ τοίνυν ἡ τοιαύτη τῶν αἰσθήσεων ἔνδεια πρὸς μὲν τὸ γνωρίσαι τὸ διάφορον ἁπλῶς ἢ τὸ μὴ πρὸς αὐτάς, οὐ παραπολὺ ἂν διαμαρτάνοι τῆς ἀληθείας, οὐδ' αὖ πρὸς τὸ θεωρῆσαι τὰς τῶν διαφερόντων ὑπεροχὰς τάς γοῦν ἐν μείζοσι μέρεσιν ὧν εἰσι λαμβανομένας. ἐπὶ δὲ τῶν κατὰ ἐλάττονα μόρια παραβολῶν πλείων ἂν συνάγοιτο καὶ ἤδη κατάφορος αὐταῖς καὶ μᾶλλον ἐπὶ τῶν μᾶλλον λεπτομερεστέρων. αἴτιον δὲ ὅτι τὸ παρὰ τὴν ἀλήθειαν καθάπαξ βραχύτατον ὂν ἐν μὲν ταῖς ὀλιγάκις γινομέναις παραβολαῖς οὐδέπω τὴν ἐπισυναγωγὴν τοῦ βραχέος αἰσθητὴν δύναται ποιεῖν, ἐν δὲ ταῖς πλεονάκις ἀξιόλογον ἤδη καὶ παντάπασιν εὐκατανόητον.

fixity in the discernment of what is true. He says word for word: "Of means of knowing there are two forms, the one legitimate [γνησίη], and the other dark [σκοτίη]; and of the dark [form] these are all of them together: sight, hearing, smell, taste, and touch; the other [form] is legitimate, and separated from this [viz. the dark]." Then, preferring the legitimate one to the dark one, he adds these words: "When the dark one [σκοτίη] is no longer able to see in the direction of further smallness nor to hear nor to smell nor to taste nor to sense by touch other things in the direction of further fineness [...]." Therefore, also according to him, reason is a criterion, which he calls "legitimate knowing" [γνησίην γνώμην].[22]

According to Democritus, and similarly for Ptolemy, the legitimate is the true, and its converse, whether "dark" or "bastard," strays from the truth. In addition, Democritus maintains that the senses are limited in their ability to perceive fine distinctions, just as Ptolemy proposes that the senses on their own perceive only approximately. A circle drawn by sight alone appears to be accurate but is not. Reason guides sense perception to the recognition of the circle that really is accurate (τοῦ τῷ ὄντι ἀκριβοῦς). This opposition of the image with the reality is markedly Platonic, and it is on this point that Ptolemy's account diverges from Democritus's and, indeed, Plato's epistemology. For Democritus, the senses are limited absolutely in their ability to observe what is true, and in Plato's *Republic* the intelligible realm is inaccessible to the senses. For Ptolemy, reason transfigures the senses. They gain the ability to distinguish the accurate from the inaccurate, to observe and confirm what is true. In the case of harmonics, the student of harmonics uses the criterion of truth—the interplay of reason and perception—to determine what ratios describe the relations among musical pitches, he imposes the ratios on the *kanôn*, and, upon hearing the pitches that are played, he confirms that these exact ratios indeed produce what are observably and truly melodious. Thus, harmonics yields certain knowledge.

22. Sextus Empiricus, *Adversus mathematicos* 7.138–39 (DK 68B11): ἐν δὲ τοῖς Κανόσι δύο φησὶν εἶναι γνώσεις, τὴν μὲν διὰ τῶν αἰσθήσεων τὴν δὲ διὰ τῆς διανοίας, ὧν τὴν μὲν διὰ τῆς διανοίας γνησίην καλεῖ, προσμαρτυρῶν αὐτῇ τὸ πιστὸν εἰς ἀληθείας κρίσιν, τὴν δὲ διὰ τῶν αἰσθήσεων σκοτίην ὀνομάζει, ἀφαιρούμενος αὐτῆς τὸ πρὸς διάγνωσιν τοῦ ἀληθοῦς ἀπλανές. λέγει δὲ κατὰ λέξιν· "γνώμης δὲ δύο εἰσὶν ἰδέαι, ἡ μὲν γνησίη, ἡ δὲ σκοτίη· καὶ σκοτίης μὲν τάδε σύμπαντα, ὄψις ἀκοὴ ὀδμὴ γεῦσις ψαῦσις· ἡ δὲ γνησίη, ἀποκεκριμένη δὲ ταύτης." εἶτα προκρίνων τῆς σκοτίης τὴν γνησίην ἐπιφέρει λέγων· "ὅταν ἡ σκοτίη μηκέτι δύνηται μήτε ὁρῆν ἐπ' ἔλαττον μήτε ἀκούειν μήτε ὀδμᾶσθαι μήτε γεύεσθαι μήτε ἐν τῇ ψαύσει αἰσθάνεσθαι, ἀλλ' ἐπὶ λεπτότερον <...>." οὐκοῦν καὶ κατὰ τοῦτον ὁ λόγος ἐστὶ κριτήριον, ὃν γνησίην γνώμην καλεῖ.

The case is different with astronomy. Ptolemy takes some but not all aspects of his astronomical models to be certain. First of all, Ptolemy maintains, like so many philosophers and mathematicians before him, that the heavens are spherical. He supports this hypothesis with several arguments, mathematical and physical, which he presents in *Almagest* 1.3 and I paraphrase here:

1. The size and mutual distances of the stars do not appear to change.[23]
2. The spherical shape of the heavens is the only hypothesis that accounts for sundial constructions' agreement with the phenomena.
3. The motion of the heavenly bodies is unhindered and the most easily moved, or free, of all motions. The freest motion belongs, among solid shapes, to the sphere.
4. The heavens are greater in size than all other bodies and, therefore, they should have the shape with the greatest volume: the sphere.
5. The aether consists of parts that are more fine and homogeneous than the parts that compose all other bodies. Bodies with parts like each other have surfaces that are like each other. The only solid with surfaces like each other is the sphere. Because the heavens consist of aether, the heavens must be spherical.
6. Similarly, because the aether's constituent parts are spherical, the heavens are spherical.

Ptolemy first presents arguments concerning mathematical phenomena—sizes, distances, motions, and constructions—and then supplements them with arguments appealing to the heavens' component element: aether. He uses the most assertive language in the second, mathematical argument. He maintains "the impossibility according to any other hypothesis than only this one that constructions of sundials accord [with the phenomena] . . ."[24] If the heavens had any other shape than a sphere, then sundial constructions would not accord with the appearances, but they do. The hypothesis of the heavens' spherical shape, and no other, accounts for why sundial constructions agree with the phenomena.

In *Almagest* 1.8, Ptolemy proclaims that the hypotheses he has discussed thus far, including the heavens' spherical shape, will be confirmed in subsequent

23. Ptolemy notes one exception, that stars appear larger at the horizon. He gives a physical and optical explanation for this phenomenon. Moisture in the atmosphere makes the stars close to the horizon appear greater in size, just as objects placed in water appear bigger than they are. *Optics* 3.59 provides a psychological explanation.

24. Ptolemy, *Almagest* 1.3, H13: τό τε μὴ δύνασθαι κατ' ἄλλην ὑπόθεσιν τὰς τῶν ὡροσκοπίων κατασκευὰς συμφωνεῖν ἢ μόνην ταύτην

chapters: "It will be enough, then, that these hypotheses—which of necessity are taken as an introduction to the particular expositions and what are consequent upon them—have been outlined summarily just so far, since they both will be confirmed and completely borne witness [ἐπιμαρτυρηθησομένας] by the very accordance with the phenomena of what we shall demonstrate consequently and one after another."[25] The participle *epimarturêthêsomenas* is related to the noun *epimarturêsis*, which is a technical term in Epicurean epistemology that was used across school boundaries in Roman Empire philosophy.[26] It means "witnessing," or the verification of a proposition by a phenomenon. Its opposite is *antimarturêsis*, counter-witnessing, or witnessing against, where a proposition is refuted by a phenomenon.[27] Sextus Empiricus defines these terms in *Adversus mathematicos*: "Of opinions, then, according to Epicurus some are true and others are false, the true being those which both bear witness to [ἐπιμαρτυρούμεναι] and do not bear witness against [οὐκ ἀντιμαρτυρούμεναι] the clear [perception], and the false being those which both bear witness against [ἀντιμαρτυρούμεναι] and do not bear witness to [οὐκ ἐπιμαρτυρούμεναι] the clear [perception]."[28] Ptolemy imbues the term *epimarturêsis* with an Epicurean connotation. The hypotheses presented in *Almagest* 1.3–7 are confirmed and borne witness by the phenomena, what are perceived.

The term *epimarturêsis* appears again in Ptolemy's corpus only in *Tetrabiblos* 4.9 alongside the verb *epimartureô*, where it operates as a technical term denoting the casting of an aspect, but Ptolemy also uses the verb *antimartureô*.[29] It appears in both the *Almagest* and the *Harmonics*, and it conveys the Epicurean meaning. In *Harmonics* 1.1, Ptolemy argues that a method is needed to produce results that the sense of hearing would not contradict:

> In the same way also for the ears, which with the eyes are most especially servants of the theoretical and rational part of the soul, there is needed some method, derived from reason, to deal with the things that they are

25. Ibid. 1.8, H26: Ταύτας μὲν δὴ τὰς ὑποθέσεις ἀναγκαίως προλαμβανομένας εἰς τὰς κατὰ μέρος παραδόσεις καὶ τὰς ταύταις ἀκολουθούσας ἀρκέσει καὶ μέχρι τῶν τοσούτων ὡς ἐν κεφαλαίοις ὑποτετυπῶσθαι βεβαιωθησομένας τε καὶ ἐπιμαρτυρηθησομένας τέλεον ἐξ αὐτῆς τῆς τῶν ἀκολούθως καὶ ἐφεξῆς ἀποδειχθησομένων πρὸς τὰ φαινόμενα συμφωνίας.

26. See Long, "Ptolemy on the Criterion," 155.

27. See Asmis, *Epicurus' Scientific Method*, 351–52.

28. Sextus Empiricus, *Adversus mathematicos* 7.211: οὐκοῦν τῶν δοξῶν κατὰ τὸν Ἐπίκουρον αἱ μὲν ἀληθεῖς εἰσιν αἱ δὲ ψευδεῖς, ἀληθεῖς μὲν αἵ τε ἐπιμαρτυρούμεναι καὶ οὐκ ἀντιμαρτυρούμεναι πρὸς τῆς ἐναργείας, ψευδεῖς δὲ αἵ τε ἀντιμαρτυρούμεναι καὶ οὐκ ἐπιμαρτυρούμεναι πρὸς τῆς ἐναργείας.

29. See Ptolemy, *Tetrabiblos* 4.9.9, H338, and 4.9.10, H339.

not naturally capable of judging accurately, [a method] against which they will not bear witness [οὐκ ἀντιμαρτυρήσουσιν] but which they will agree is correct.[30]

Ptolemy seeks a method that relies on reason but will not conflict with sense perception, a method not contradicted by or borne witness against the phenomena. He likewise discusses counter-witnessing in *Almagest* 1.7 when contesting the supposition that the earth rotates:

> Now, some people, more plausibly so they suppose, agree with the above, since they do not have anything to say against these [arguments], but they think that nothing will bear witness against [ἀντιμαρτυρήσειν] them, if, for instance, they supposed the heavens to be immovable and the earth to rotate from west to east about the same axis [as the heavens] approximately one revolution a day[31]

The verb *antimartureô* denotes the observation of a phenomenon that, in this case, would bear witness against the theory that the earth rotates.

Some historians have characterized Ptolemy's argument for the earth's immobility as a physical argument, because it concerns the absence of centrifugal effects on sublunary bodies.[32] If this argument were physical, then it would place Ptolemy's astronomy on epistemically dubious foundations because physics is, according to Ptolemy, conjectural. Nevertheless, he makes a point of setting aside physical considerations. For the sake of the argument, he concedes against nature (παρὰ φύσιν) that the most fine and light bodies either move not at all or no than bodies of the opposite nature and vice versa, that the most thick and heavy bodies move quickly and uniformly.[33] The ensuing argument consists in an analysis of movements, which are, according to Ptolemy, the subject matter of mathematics. If the earth moved quickly, he argues, then certain movements would result: objects not standing on the earth would appear to have an identical motion in the opposite direction to the earth's rotation; no objects in the air would ever be seen

---

30. Ptolemy, *Harmonics* 1.1, D5: τὸν αὐτὸν τρόπον καὶ ταῖς ἀκοαῖς διακόνοις οὔσαις μάλιστα μετὰ τῶν ὄψεων τοῦ θεωρητικοῦ καὶ λόγον ἔχοντος μέρους τῆς ψυχῆς, δεῖ τινος ἀπὸ τοῦ λόγου, πρὸς ἃ μὴ πεφύκασι κρίνειν ἀκριβῶς, ἐφόδου, πρὸς ἣν οὐκ ἀντιμαρτυρήσουσιν ἀλλ᾽ ὁμολογήσουσιν οὕτως ἔχειν.

31. Ptolemy, *Almagest* 1.7, H24: ἤδη δέ τινες, ὡς γ᾽ οἴονται, πιθανώτερον, τούτοις μὲν οὐκ ἔχοντες, ὅ, τι ἀντείποιεν, συγκατατίθενται, δοκοῦσι δὲ οὐδὲν αὐτοῖς ἀντιμαρτυρήσειν, εἰ τὸν μὲν οὐρανὸν ἀκίνητον ὑποστήσαιντο λόγου χάριν, τὴν δὲ γῆν περὶ τὸν αὐτὸν ἄξονα στρεφομένην ἀπὸ δυσμῶν ἐπ᾽ ἀνατολὰς ἑκάστης ἡμέρας μίαν ἔγγιστα περιστροφήν

32. See, for instance, Lloyd, "Saving the Appearances," 216.

33. Ptolemy, *Almagest* 1.7, H24.

moving eastward, because the earth's rotation would overtake them. Even if the air were carried around in the same direction and at the same speed as the earth, and even if compound objects in the air, too, were carried around, as if they were fused to the air, then these objects would never appear to have any motion whatsoever. None of these (lack of) movements, however, are observed. Objects are seen moving through the air in all directions. Therefore, the observed movements, in contradiction to the opinions of "some people," bear witness against the hypothesis of the earth's rotation, and so Ptolemy is certain that the earth is immobile. Similarly, in *Almagest* 1.3 he argues that the heavens must be spherical because the phenomena bear witness against all other hypotheses: "so in the beginning they apprehended the aforementioned concept through [considerations] only such as these, but thereafter, in accordance with successive contemplation, they understood that everything else accorded with these, since absolutely all phenomena bear witness against [ἀντιμαρτυρούντων] the alternative concepts."[34] The phenomena bear witness against all other hypotheses of the heavens' shape. Therefore, the heavens must be spherical.

As certain as Ptolemy is that the heavens are spherical, he is certain that the heavens consist of eccentric and epicyclic spheres. To reconcile the heavens' uniform circular motion with celestial bodies' seemingly irregular movements, Ptolemy accepts, like mathematicians before him, that some heavenly spheres are eccentric and/or epicyclic. He describes these hypotheses in *Almagest* 3.3 when accounting for the sun's apparent irregularity:

> It is possible for what is responsible for the appearance of irregularity to be produced in accordance with two hypotheses, which are the most primary and simple. For when their movement is considered with respect to the circle that is conceived of in the plane through the middle of the zodiac and homocentric to the cosmos, so its center is not different from our point of view, then we can suppose that their regular movements are produced either upon circles that are not homocentric with the cosmos or upon [circles] that are homocentric, but not simply on these [circles], but rather on different [circles] that are carried by the first [circles] and are called "epicycles." For, it will be shown that, in accordance with either of these hypotheses, the appearance to our eyes of the [stars'] traversal in equal

---

34. Ibid. 1.3, H11: ὡς τὴν μὲν ἀρχὴν διὰ μόνα τὰ τοιαῦτα τὴν προειρημένην ἔννοιαν αὐτοὺς λαβεῖν, ἤδη δὲ κατὰ τὴν ἐφεξῆς θεωρίαν καὶ τὰ λοιπὰ τούτοις ἀκόλουθα κατανοῆσαι πάντων ἁπλῶς τῶν φαινομένων ταῖς ἑτεροδόξοις ἐννοίαις ἀντιμαρτυρούντων.

times of unequal arcs of the circle through the middle of the zodiac and homocentric to the cosmos is possible.[35]

Stars' apparently irregular movements are not in fact irregular. The eccentric and epicyclic hypotheses maintain the regular circular motion of heavenly bodies while at the same time accounting for their irregular appearances. Motion along a circle eccentric to the earth will give the appearance of traversing unequal arcs in equal times even though the motion is in fact uniform. Correspondingly, the combination of the two circles' movements in the epicyclic hypothesis gives the appearance of irregular motion but the movements along the individual circles are regular. Ptolemy considers the existence of eccentric and epicyclic spheres in the heavens to be necessarily true, because these hypotheses reconcile the heavens' regular circular motion with the stars' apparently irregular movements.

In the case of the sun's model, Ptolemy was confronted with a choice. While the other wandering stars require a combination of the eccentric and epicyclic hypotheses, the sun has a single anomaly accountable by either one. Each hypothesis is consistent with the phenomena. Choosing between the two hypotheses, discerning which truly describes the sun's system, Ptolemy could not appeal to observation, and so he introduces a rational principle as the criterion. In *Almagest* 3.4, having explained that either the eccentric or epicyclic hypothesis could produce the sun's anomaly, he concludes "but it would be more reasonable [εὐλογώτερον] to fit it to the eccentric hypothesis, since it is more simple and is accomplished by one [movement], and not by two movements."[36] Ptolemy appeals to mathematical simplicity. Because one movement, rather than two, brings about the sun's anomaly in the eccentric hypothesis, it is more reasonable to associate it, rather than the epicyclic hypothesis, with the sun's system. Appealing to reason, Ptolemy leans on one of the key components of his dually rational and empirical criterion of truth.

35. Ibid. 3.3, H216–17: τὸ δ' αἴτιον τῆς ἀνωμάλου φαντασίας κατὰ δύο μάλιστα τὰς πρώτας καὶ ἁπλᾶς ὑποθέσεις ἐνδέχεται γίνεσθαι. τῆς γὰρ κινήσεως αὐτῶν θεωρουμένης πρὸς τὸν ὁμόκεντρόν τε τῷ κόσμῳ καὶ ἐν τῷ ἐπιπέδῳ τοῦ διὰ μέσου τῶν ζῳδίων νοούμενον κύκλον, ὡς ἀδιαφορεῖν πρὸς τὸ κέντρον αὐτοῦ τὴν ἡμετέραν ὄψιν, αὐτοὺς ἤτοι κατὰ μὴ ὁμοκέντρων τῷ κόσμῳ κύκλων ὁμαλὰς ὑποληπτέον ποιεῖσθαι τὰς κινήσεις ἢ κατὰ ὁμοκέντρων μέν, οὐχ ἁπλῶς δὲ ἐπ' αὐτῶν, ἀλλ' ἐπὶ ἑτέρων ὑπ' ἐκείνων φερομένων, καλουμένων δὲ ἐπικύκλων. καθ' ἑκατέραν γὰρ τούτων τῶν ὑποθέσεων ἐνδεχόμενον φανήσεται τὸ ἐν ἴσοις αὐτοὺς χρόνοις ἀνίσους φαίνεσθαι ταῖς ὄψεσιν ἡμῶν διερχομένους τοῦ διὰ μέσων τῶν ζῳδίων κύκλου ὁμοκέντρου τῷ κόσμῳ περιφερείας.

36. Ibid. 3.4, H232: εὐλογώτερον δ' ἂν εἴη περιαφθῆναι τῇ κατ' ἐκκεντρότητα ὑποθέσει ἁπλουστέρᾳ οὔσῃ καὶ ὑπὸ μιᾶς, οὐχὶ δὲ ὑπὸ δύο κινήσεων, συντελουμένῃ.

Ptolemy's choice between the eccentric and epicyclic hypotheses in *Almagest* 3.4 served as one of the crucial examples adduced by historians when arguing for Ptolemy's instrumentalism. For most of the twentieth century the instrumentalist, as opposed to the realist, interpretation of ancient Greek astronomy took hold, owing in part to Pierre Duhem's influential *To Save the Phenomena*.[37] In brief, the position of scientific realism is that (some aspects of) scientific theories correspond (to some degree) to what truly exists. An instrumentalist, on the other hand, takes scientific theories to be mere fictions, devices that serve some purpose, such as the prediction of phenomena. In Duhem's instrumentalist interpretation, Ptolemy did not believe that his astronomical models corresponded to what exists in the heavens. When he was deciding between the eccentric and epicyclic hypotheses for his solar model, he was not determining which one was true; he was deciding between what he took to be mere contrivances.[38] Regarding *Almagest* 13.2, Duhem proclaims, "Certainly, Ptolemy means to indicate in this passage that the many motions he compounds in the *Syntaxis* to determine the trajectory of a planet have no physical reality; only the resultant motion is actually produced in the heavens."[39] The instrumentalist interpretation echoed through the twentieth century in such prominent texts as Dreyer's *History of the Planetary Systems from Thales to Kepler* and Sambursky's *The Physical World of Late Antiquity*. G.E.R. Lloyd, however, dealt a decisive blow to the instrumentalist interpretation with his "Saving the Appearances," establishing that a close and accurate analysis of the Greek texts shows that all of the ancient Greek astronomers about whom we have sufficient evidence to render judgment were realists. In support of Ptolemy's realism, Lloyd points to the *Almagest*'s framework of physical assumptions, which appears in his discussions of aether in *Almagest* 1.3 and element theory in *Almagest* 1.7.[40] Indeed, historians now make quick work of arguing for Ptolemy's realism with the simple observation that he wrote the *Planetary Hypotheses*, which renders his astronomical models physical.[41] I would note, however, that Ptolemy's astronomical and, in general, mathematical realism is not dependent on his physical realism.[42] It is not

---

37. Duhem, "ΣΩΖΕΙΝ ΤΑ ΦΑΙΝΟΜΕΝΑ." The same year, this series of articles was reprinted by A. Hermann as a book under the same title.

38. Duhem, *To Save the Phenomena*, 18.

39. Ibid., 17.

40. Lloyd, "Saving the Appearances," 215–16.

41. For an analysis of Ptolemy's aethereal physics, see chapter 8.

42. G.E.R. Lloyd discusses mathematical realism—although not Ptolemy's in particular—in the introduction to "Saving the Appearances" in *Methods and Problems*, 250–51.

only because Ptolemy had a physical, cosmological theory that we know that he believed in the reality of the astronomical objects he describes. According to his Aristotelian metaphysical framework, mathematical objects simply are real—they exist in the universe—and Ptolemy is certain that eccentric and epicyclic spheres, in particular, exist in the heavens because of his criterion of truth, the exercise of reason and perception, which in his epistemological system generates knowledge.

Ptolemy does not display the same confidence, however, in other aspects of his astronomical models. For instance, he rejects the possibility that mathematicians can know the exact periods of heavenly phenomena. In *Almagest* 3.1, he examines the limits to observation that prevent human beings from accurately determining the tropical year, the annual return of the sun to the same point on the ecliptic. Ptolemy maintains that the tropical year is constant even though a potential error of one quarter of a day exists in his observations:

> There has been, then, in these observations not any discrepancy worthy of mention, even though it is possible with these [observations] for there to be an error of up to a quarter of a day not only in solstice observations but even in equinoctial [observations]; for even if the positioning or gradua- tion of the instruments deviates from exactness [παραλλάξῃ τῆς ἀκριβείας] by only 1/3600th of the circle through the poles of the equinoctial circle, the sun corrects a concession of such size in declination when it is near its intersections with the equator by moving ¼° in longitude on the ecliptic, so that the discrepancy differs up to approximately one quarter of a day.[43]

Although a possible error of one quarter of a day is small enough for Ptolemy to discount when affirming the constancy of the tropical year, the existence of any possible error, whatever its magnitude, is sufficient to prevent him from claiming that he or anyone else has or ever will determine the exact length of the tropical year.

Ptolemy recognizes that if one increases the length of time between obser- vations, then the accuracy of the determination of the tropical year, or any other periodic return, increases; however, because the length of time between

---

43. Ptolemy, *Almagest* 3.1, H196–97: οὐδ' ἐν ταύταις ἄρα ταῖς τηρήσεσιν γέγονέ τις ἀξιόλογος διαφορὰ καίτοι δυνατοῦ ὄντος οὐ μόνον περὶ τὰς τροπικὰς τηρήσεις, ἀλλὰ καὶ περὶ τὰς ἰσημερινάς, γίγνεσθαί τι παρ' αὐτὰς διαμάρτημα καὶ μέχρι δ' μιᾶς ἡμέρας· κἂν γὰρ τῷ τρισχιλιοστῷ καὶ ἑξακοσιοστῷ μόνῳ μέρει τοῦ διὰ τῶν πόλων τοῦ ἰσημερινοῦ κύκλου παραλλάξῃ τῆς ἀκριβείας ἡ θέσις ἢ καὶ διαίρεσις τῶν ὀργάνων, τὴν τοσαύτην κατὰ πλάτος παραχώρησιν ὁ ἥλιος διορθοῦται πρὸς τοῖς ἰσημερινοῖς τμήμασιν τέταρτον μιᾶς μοίρας κατὰ μῆκος ἐπὶ τοῦ λοξοῦ κύκλου κινηθείς, ὥστε καὶ τὴν διαφωνίαν μέχρι δ' μιᾶς ἡμέρας ἔγγιστα διενεγκεῖν.

observations is necessarily limited, the true period cannot be known. He explains as follows:

> but one can apprehend nearly accurately [ἔγγιστα ἀκριβῶς] such a period of revolution the more time is found between the observations compared. And such holds not only in this case, but also for all periodic revolutions; for the error which is from the deficiency of the observations themselves, even when they are performed accurately, is, in reference to their perception side by side, small and approximately the same whether they appear after a long or a small interval of time, but the distribution over a smaller number of years makes the yearly error, as well as the [error] accumulated out of this [yearly error] in the course of the longer interval of time, greater, but [the distribution] over a greater number [of years makes the error] smaller. Hence, it is proper to consider it sufficient if we endeavor to take into account as great an increase in accuracy [ἐγγύτητι] of the periodic hypotheses as is possible to lay claim to between the time of our observations and those [observations] we have which are both ancient and accurate [ἀκριβῶν], and we must not purposely be neglectful of the proper examination [of the observations], but we must hold assertions about the whole of eternity, or even a period of time many times in length that which is in accordance with the observations, to be foreign to both a love of learning and a love of truth [φιλαληθείας].[44]

Observation is limited in scope and precision. A possible error of at least one quarter of a day is built into the instruments Ptolemy and his predecessors used to make observations, and the length of time between their observations is contingent upon and, therefore, limited by the collective lifetime of astronomers and their records. Ptolemy adds in *Almagest* 9.2 that, because the available observational records of the five planets are relatively recent,

44. Ibid. 3.1, H202–3: λαμβάνοιτο δ' ἂν ἔγγιστα ἀκριβῶς ἡ τοιαύτη ἀποκατάστασις, ὅσῳ ἂν ὁ μεταξὺ τῶν συγκρινομένων τηρήσεων χρόνος πλείων εὑρίσκηται. καὶ οὐ μόνον ἐπὶ ταύτης τὸ τοιοῦτον συμβέβηκεν, ἀλλὰ καὶ ἐπὶ πασῶν τῶν περιοδικῶν ἀποκαταστάσεων· τὸ γὰρ παρὰ τὴν αὐτῶν τῶν τηρήσεων ἀσθένειαν, κἂν ἀκριβῶς μεθοδεύωνται, γινόμενον διάψευσμα βραχὺ καὶ τὸ αὐτὸ ἔγγιστα ὑπάρχον ὡς πρὸς τὴν παρ' αὐτὰ αἴσθησιν ἐπί τε τῶν διὰ μακροῦ καὶ ἐπὶ τῶν δι' ὀλίγου χρόνου φαινομένων εἰς ἐλάττονα μὲν ἐπιμεριζόμενον ἔτη μεῖζον ποιεῖ τὸ ἐνιαύσιον ἁμάρτημα καὶ τὸ ἐκ τούτου κατὰ τὸν μακρότερον χρόνον ἐπισυναγόμενον, εἰς πλείονα δὲ ἔλασσον. ὅθεν αὔταρκες προσήκει νομίζειν, ἐάν, ὅσον ὁ μεταξὺ χρόνος ἡμῶν τε καὶ ὧν γε ἔχομεν παλαιῶν ἅμα καὶ ἀκριβῶν τηρήσεων δύναται προσποιῆσαι τῇ τῶν περιοδικῶν ὑποθέσεων ἐγγύτητι, τοσοῦτον καὶ αὐτοὶ πειραθῶμεν συνεισενεγκεῖν καὶ μὴ ἑκόντες ἀμελήσωμεν τῆς προσηκούσης ἐξετάσεως, τὰς δὲ περὶ ὅλου τοῦ αἰῶνος ἢ καὶ τοῦ μακρῷ τινι πολλαπλασίου τοῦ κατὰ τὰς τηρήσεις χρόνου διαβεβαιώσεις ἀλλοτρίας φιλομαθείας τε καὶ φιλαληθείας ἡγώμεθα.

predictions of planetary movements are unsure (ἀβέβαιον).[45] He concludes that mathematicians cannot know and should not claim that their values for any periodic revolution are true, that they are valid for all eternity. The best they can do is offer an approximation, such as Ptolemy's approximation of the tropical year, which in the sexagesimal system of base sixty is 365;14,48 days. Ptolemy calls this value "for us the closest approximation as any that can be derived from what are available."[46]

Ptolemy's attention to observational error corresponds with his esteem for Hipparchus, the second-century BCE mathematician whom Ptolemy treats as his principal predecessor in astronomy. In *Almagest* 3.1, Ptolemy ascribes to Hipparchus a "love of truth" (φιλαλήθεια).[47] At the same time as he corrects Hipparchus's ascription of a second anomaly to the sun, and an inconstant tropical year, Ptolemy commends him for prioritizing observation and even rationalizes Hipparchus's mistake. Ptolemy asserts, "However, I believe that Hipparchus himself also had recognized that there is nothing sufficient in such [arguments] for attributing a second irregularity to the sun and only had wished, owing to his love of truth [φιλαληθείας], not to suppress anything that in any way could lead to suspicion of some [irregularity]."[48] Similarly, Ptolemy quotes Hipparchus's *On the Displacement of the Solsticial and Equinoctial Points*, where he addresses observational error in his calculation of the tropical year. Ptolemy introduces the fragment as follows:

> For, in his "On the displacement of the solsticial and equinoctial points," first setting out both the summer and winter solstices he supposes to have been observed accurately and in succession, even he concedes that there is not such inconsistency in them to proclaim, by means of them, that there is an inequality in the length of the year; for he comments on them as follows: "From these observations, then, it is clear that the differences in the lengths of the year are very small indeed. However, in the case of the solstices, I confess that both we and Archimedes erred up to a quarter of a day in the observation and calculation [of the lengths]. But it is possible to perceive accurately the irregularity in the lengths of the year from the [equinoxes] observed on the bronze ring set up in Alexandria in what is called the 'Square Stoa,' which is supposed to indicate the equinoctial day,

---

45. Ibid. 9.2, HII208.

46. Ibid. 3.1, H208: εἴη ἂν ἔγγιστα ἡμῖν ὡς ἔνι μάλιστα ἐκ τῶν παρόντων εἰλημμένον.

47. Cf. Ptolemy, *On the Kritêrion*, La5, for Ptolemy's use of the term φιλαληθέστατος.

48. Ptolemy, *Almagest* 3.1, H200: ἀλλ' οἶμαι καὶ τὸν Ἵππαρχον συνεγνωκέναι μὲν καὶ αὐτόν, ὅτι μηδὲν ἐν τοῖς τοιούτοις ἔνεστιν ἀξιόπιστον πρὸς τὸ δευτέραν τινὰ τῷ ἡλίῳ προσάπτειν ἀνωμαλίαν, βεβουλῆσθαι δὲ μόνον ὑπὸ φιλαληθείας μὴ σιωπῆσαί τι τῶν ἐνίους εἰς ὑποψίαν ὁπωσδήποτε, δυναμένων ἐνεγκεῖν.

during which from one of its parts [to the other] the concave surface is first illuminated."[49]

In the fragment, Hipparchus, like Ptolemy after him, recognizes degrees of precision in observation and takes potential error into account. Ptolemy praises Hipparchus for his attention to observation, and in *Almagest* 9.2 he commends him for not even attempting to construct hypotheses of the five planets, since he faced a considerable dearth of observational records.[50] Conversely, Ptolemy condemns those mathematicians who do not pay sufficient attention to the limits of observation and observational records when calculating the periods of heavenly phenomena.

Different Methods, Different Claims to Truth

Why is it that Ptolemy considers his harmonic hypotheses to be accurate, precise and true, but not all aspects of his astronomical hypotheses? I argue that this epistemological disparity follows not from some difference in the objects that harmonics and astronomy study nor human beings' access to the two sets of phenomena, but rather from the sciences' methods. Whether a science depends entirely upon Ptolemy's criterion of truth—the interplay of perception and reason—or extends beyond it determines whether the science's claims are known surely and incontrovertibly. Furthermore, this differential reliance on the criterion of truth correlates with the sciences' dependence on or independence of the indisputable methods, or instruments, of mathematics: arithmetic and geometry. Even though Hero of Alexandria distinguishes only geometrical demonstration as indisputable, it turns out that it is crucial for Ptolemy's epistemology of mathematics that both arithmetic and geometry lend indisputability to the sciences that utilize them.

Ptolemy's harmonics exemplifies the proper utilization of the criterion of truth. The student of harmonics relies on sense perception but compensates

---

49. Ibid. 3.1, H194–95: ἐκθέμενος γὰρ τὸ πρῶτον ἐν τῷ Περὶ τῆς μεταπτώσεως τῶν τροπικῶν καὶ ἰσημερινῶν σημείων τὰς δοκούσας αὐτῷ ἀκριβῶς καὶ ἐφεξῆς τετηρῆσθαι θερινάς τε καὶ χειμερινὰς τροπὰς ὁμολογεῖ καὶ αὐτὸς μὴ τοσοῦτον ἐν αὐταῖς εἶναι τὸ διάφωνον, ὥστε δι' αὐτὰς ἀνισότητα καταγνῶναι τοῦ ἐνιαυσίου χρόνου· ἐπιλέγει γὰρ αὐταῖς οὕτως· "ἐκ μὲν οὖν τούτων τῶν τηρήσεων δῆλον, ὅτι μικραὶ παντάπασιν γεγόνασιν αἱ τῶν ἐνιαυτῶν διαφοραί. ἀλλ' ἐπὶ μὲν τῶν τροπῶν οὐκ ἀπελπίζω καὶ ἡμᾶς καὶ τὸν Ἀρχιμήδη καὶ ἐν τῇ τηρήσει καὶ ἐν τῷ συλλογισμῷ διαμαρτάνειν καὶ ἕως τετάρτου μέρους ἡμέρας. ἀκριβῶς δὲ δύναται κατανοεῖσθαι ἡ ἀνωμαλία τῶν ἐνιαυσίων χρόνων ἐκ τῶν τετηρημένων ἐπὶ τοῦ ἐν Ἀλεξανδρείᾳ κειμένου χαλκοῦ κρίκου ἐν τῇ τετραγώνῳ καλουμένη στοᾷ, ὃς δοκεῖ διασημαίνειν τὴν ἰσημερινὴν ἡμέραν, ἐν ᾗ ἂν ἐκ τοῦ ἑτέρου μέρους ἄρχηται τὴν κοίλην ἐπιφάνειαν φωτίζεσθαι."

50. Ibid. 9.2, HII210.

for its inherent weakness by bringing reason to bear on it. Reason guides the sense of hearing to perceive and confirm the true relations among musical pitches. Part of why the student of harmonics is able to grasp these true relations is that he is able to appropriate the whole-number ratios of arithmetic number theory. He adopts from arithmetic those ratios that correspond to the relations among musical pitches. The process by which the student of harmonics determines which ratios explain harmonic phenomena involves several stages—determining the homophones and concords, the tetrachords, and the melodic relations of scale systems—but all the stages are alike in kind. At every stage, he determines and demonstrates the arithmetic ratios that characterize the relations among musical pitches. In this way, the student of harmonics employs Ptolemy's criterion of truth and transfers the indisputability of arithmetic to harmonics.

While harmonics adopts the ratios of arithmetic, astronomy appropriates the objects of geometry: points, lines, circles, and spheres. In other words, when Ptolemy describes the heavens as composed of spheres, he employs geometry. He claims that certain geometric objects—namely, spheres—exist, and that they exist in eccentric and epicyclic configurations in the heavens. So far the method of astronomy resembles harmonics. Just as the student of harmonics adopts the whole-number ratios of arithmetic and determines which exist among harmonic phenomena, so, too, the student of astronomy appropriates the objects of geometry and determines which exist in the heavens. In *Almagest* 1.3, Ptolemy argues that the heavens are spherical, and in *Almagest* 3.3 he concludes that the heavens contain eccentric and epicyclic spheres.

Thereafter, Ptolemy approximates the periods of celestial phenomena, he ascertains the number and arrangement of the spheres that exist in stars' systems, he calculates the models' parameters, and he determines the absolute sizes of the lunar and solar systems. For each star—except the moon, whose model he revisits in Book 5—Ptolemy progresses through two fundamentally different stages. The first involves the construction of an abstract model of the star's system of spheres; the second consists in the quantification of the model. Having made observations and analyzed observational records, Ptolemy determines in the first stage how many irregularities a star appears to experience when moving through the heavens. The number of irregularities reveals how many spheres exist in the star's system as well as their relative arrangement. In this initial phase of the construction of an astronomical model, Ptolemy takes an approach similar to the method ancient Greek mathematicians employed when studying geometry. Just as Euclid, for instance, does not quantify the geometric figures he constructs in the *Elements*, so, too, Ptolemy determines the number and arrangement of the spheres in a star's system but does not yet calculate the parameters.

The second stage, the quantification of the model, is different in kind from the first stage. Geometry provides the objects of the first stage, but the mathematician cannot derive from geometry nor determine through reason the quantitative aspects of the models. The student of harmonics derives every aspect of his theory from arithmetic by discovering the whole-number ratios underlying the phenomena, but the student of astronomy must go beyond geometry. He affirms that the heavens contain eccentric and epicyclic spheres—reason and perception have confirmed these hypotheses—but in order to determine the dimensions and periods of the spheres he must rely on observation alone. Only by extrapolating from a series of observations, which are sufficiently accurate and as far apart in time as possible, can the mathematician attempt to calculate the quantitative parameters of his astronomical models. Yet, as Ptolemy recognizes, observation is plagued by potential error that independently of reason cannot be corrected, and the scope of astronomical observations is limited by the collective lifetime of human beings and their records. Hence, the mathematician cannot know the exact periods of celestial phenomena nor the exact dimensions of the heavenly spheres.

This lack of certainty appears time and again in the *Almagest* in Ptolemy's use of approximations. For instance, he begins *Almagest* 4.3, on the individual mean motions of the moon, with the following:

> If, then, we multiply the mean daily motion of the sun which has been demonstrated, approximately [ἔγγιστα] 0;59,8,17,13,12,31°/d, by the days of one [mean synodic] month, 29;31,50,8,20d, and we add to the result the 360° of one circle, we will get the mean motion of the moon in longitude during one synodic month as approximately [ἔγγιστα] 389;6,23,1,24,2,30,57°. Dividing this by the proposed number of days in a month, we get the mean daily motion of the moon in longitude as approximately [ἔγγιστα] 13;10,34,58,33,30,30°.[51]

Ptolemy's values for periodic phenomena consist of long chains of sexagesimal fractions, and, because they do not even approximate simple ratios, there is no way for reason to resolve them. In fact, it is not even clear whether they represent rational or irrational numbers. Ptolemy carries the values out to six or so places merely because he takes this level of precision to be sufficient for his

---

51. Ibid. 4.3, H278, translation modified from Toomer, in Ptolemy, *Almagest*: Ἐὰν τοίνυν τὸ ἀποδεδειγμένον μέσον τοῦ ἡλίου κίνημα ἡμερήσιον ⊽ ν̅θ̅ η̅ ι̅ζ̅ ι̅γ̅ ι̅β̅ λ̅α̅ ἔγγιστα πολλαπλασιάσωμεν ἐπὶ τὰς τοῦ ἑνὸς μηνὸς ἡμέρας κ̅θ̅ λ̅α̅ ν̅ η̅ κ̅ καὶ τοῖς γενομένοις προσθῶμεν ἑνὸς κύκλου μοίρας τ̅ξ̅, ἕξομεν, ἃς ἐν τῷ ἑνὶ μηνὶ μέσως ἡ σελήνη κινεῖται κατὰ μῆκος μοίρας τ̅π̅θ̅ ϛ̅ κ̅γ̅ α̅ κ̅δ̅ β̅ λ̅ ν̅ζ̅ ἔγγιστα. ταύτας ἐπιμερίσαντες εἰς τὰς προκειμένας τοῦ μηνὸς ἡμέρας ἕξομεν ἡμερήσιον μέσον κίνημα μήκους μοίρας ι̅γ̅ ι λ̅δ̅ ν̅η̅ λ̅γ̅ λ̅ λ̅ ἔγγιστα.

calculations. He presents this practical viewpoint when calculating the sun's mean daily motion: "For, since it has been demonstrated by us that one revolution is 365;14,48 days, if we divide the latter into the 360° of the circle, we get the mean daily motion of the sun as approximately [ἔγγιστα] 0;59,8,17,13,12,31°. For, it will be sufficient to carry out divisions to this number [i.e., six] of sexagesimal places."[52] Whether the values for astronomical phenomena are commensurable or incommensurable does not seem to concern Ptolemy. What is troublesome is their inelegance and cumbersomeness on a practical level. These values are not the simple ratios of harmonics. They are the results of extensive calculations from values that themselves are only approximations.

As discussed in the previous chapter, harmonic ratios do exist in the heavens. They describe the relations among zodiacal signs as well as the heavenly bodies' movements in length, depth, and breadth, but they do not describe all the quantities in the heavens. The heavenly spheres' sizes and periods are not harmonic. Thus, harmonics adopts the ratios of arithmetic number theory but it does not adopt the objects of geometry; conversely, astronomy appropriates the objects of geometry but not number theory. Harmonics and astronomy each utilize only one of the indisputable mathematical instruments—arithmetic or geometry—and the certainty of their propositions extends only as far as they rely on an indisputable instrument and the criterion of truth. Harmonics entirely depends on the interplay of reason and perception, and it adopts all of its quantities from number theory. Astronomy, on the other hand, utilizes reason and perception to establish that spheres, eccentric and epicyclic, exist in the heavens, but the determination of these spheres' sizes and periods depends on observation alone. The certainty of astronomy extends only as far as the existence of the spheres, the objects it adopts from geometry, but astronomy stretches beyond geometry in the quantification of the models. This latter stage in Ptolemy's astronomy proceeds by neither of the indisputable mathematical methods, and so the quantitative aspects of his astronomical models are not indisputable. The periods and sizes of the heavenly spheres are only approximations.

Of course, the student of harmonics does not merely determine what ratios describe harmonic phenomena; he also demonstrates that the ratios are correct. He imposes them on the *kanôn* and confirms that they produce pitches that are audibly homophonic, concordant, or melodic. The student of astronomy seems to have no instrument by which to confirm any aspect

<hr />

52. Ibid. 3.1, H209: τῆς γὰρ μιᾶς ἀποκαταστάσεως ἀποδεδειγμένης ἡμερῶν τ̅ξ̅ε̅ ι̅δ̅ μ̅η̅, ἐὰν ἐπιμερίσωμεν εἰς ταύτας τὰς τοῦ ἑνὸς κύκλου μοίρας τ̅ξ̅, ἕξομεν τὸ ἡμερήσιον μέσον κίνημα τοῦ ἡλίου μοιρῶν ∘ ν̅θ̅ η̅ ι̅ζ̅ ι̅γ̅ ι̅β̅ λ̅α̅ ἔγγιστα· ἀρκέσει γὰρ μέχρι τοσούτων ἑξηκοστῶν τοὺς μερισμοὺς τούτων ποιεῖσθαι.

of his hypotheses. The sundial, for instance, is critical in the establishment of the heavens' sphericity, but the student of astronomy cannot reproduce the heavens in the sundial like the student of harmonics reproduces ratios in the *kanôn*. According to Ptolemy, astronomy's analogue to the harmonic *kanôn* is the heavens themselves. In *Harmonics* 1.2, he juxtaposes them: "The aim of the student of harmonics must be to preserve in all respects the rational hypotheses of the *kanôn* as never in any way conflicting with the perceptions that correspond to the estimation of most people, just as the [aim] of the astronomer is to preserve the hypotheses of the heavenly movements as concordant with the observed courses . . ."[53] Even if Ptolemy were to construct a model of the heavens, like the didactic tool he describes in Book 1 of the *Planetary Hypotheses*, it would not demonstrate the veracity of his astronomical hypotheses. Only the perception of the actual phenomena, whether heard from a *kanôn* or seen in the heavens, can confirm that the hypotheses concerning them are true.

The procedures by which one makes observations of the *kanôn* and the heavens are obviously different. A human being can impose ratios on the *kanôn* but must observe the heavens from afar. Nevertheless, I argue that, inasmuch as the *kanôn* and the heavens are domains that furnish human beings with observations of mathematical objects, they are different only in degree. Although mathematicians cannot impose geometric shapes and motions onto heavenly aether, they do witness certain astronomical phenomena that are, like musical relations played on the *kanôn*, indisputable upon observation. A lunar eclipse, for instance, appears the same from anywhere on earth. Deliberating on which observations to take into account when constructing a lunar model, Ptolemy mandates the following:

> but for the general apprehensions in these [topics] especially to take heed of the demonstrations which not only [cover] a long period of time but also deal with the very observations corresponding with lunar eclipses; for by means of only these [observations] can the positions of the moon accurately be discovered; all others, which are observed either by means of the passages [of the moon] near fixed stars or by means of instruments or by means of solar eclipses, can err greatly on account of lunar parallax[54]

53. Ptolemy, *Harmonics* 1.2, D5: ἁρμονικοῦ δ' ἂν εἴη πρόθεσις τὸ διασῶσαι πανταχῇ τὰς λογικὰς ὑποθέσεις τοῦ κανόνος μηδαμῇ μηδαμῶς ταῖς αἰσθήσεσι μαχομένας κατὰ τὴν τῶν πλείστων ὑπόληψιν, ὡς ἀστρολόγου τὸ διασῶσαι τὰς τῶν οὐρανίων κινήσεων ὑποθέσεις συμφώνους ταῖς τηρουμέναις παρόδοις

54. Ptolemy, *Almagest* 4.1, H265: ἀλλὰ πρὸς μὲν τὰς καθόλου καταλήψεις ἐκείναις μάλιστα προσέχειν τῶν ἀποδείξεων, ὅσαι μὴ μόνον ἐκ τοῦ πλείονος χρόνου, ἀλλὰ καὶ ἀπ' αὐτῶν τῶν κατὰ τὰς σεληνιακὰς ἐκλείψεις τηρήσεων λαμβάνονται· διὰ μόνων γὰρ τούτων ἀκριβῶς ἂν οἱ τόποι τῆς σελήνης εὑρίσκοιντο τῶν ἄλλων, ὅσαι ἤτοι διὰ τῶν πρὸς τοὺς ἀπλανεῖς ἀστέρας παρόδων ἢ διὰ

The moon's location is especially difficult to observe because of its parallax. It is close enough to the earth that it appears to be in a different location, with respect to the backdrop of the fixed stars, when observed from different positions on earth. Lunar eclipses obviate this difficulty by revealing the relative locations of the moon, earth, and sun independently of the observer.

A mathematician failing to observe a lunar eclipse—or any other celestial phenomenon—cannot make the eclipse repeat itself, but with patience and time that mathematician or another human being can observe another lunar eclipse, because all astronomical phenomena are periodic. Similarly, the student of harmonics cannot perceive that the ratio he imposed on the *kanôn* sometime in the past created a certain musical relation, but he can impose the ratio on the instrument again and present to the senses another case of the musical relation. The periodic nature of heavenly phenomena entail that observations of the *kanôn* and of the heavens are different only in degree. The student of harmonics can observe phenomena repeatedly and within a short (or long) period of time. The student of astronomy must wait a long period of time to observe a phenomenon again. Indeed, the wait in some cases is so long that it extends beyond a human lifetime, and so the student may consult the records of astronomers from previous centuries, as Ptolemy does in the *Almagest*.

The physical distance of a mathematician from his objects of study proves not to be a factor in Ptolemy's assessment of the sciences' individual claims to truth. After all, Ptolemy is certain that the heavens contain eccentric and epicyclic spheres, despite their distance. He is certain even though he cannot manipulate the heavens as he does the *kanôn*. Ptolemy does not introduce this thought experiment but, even if he were able to rise up to the aether and observe the heavens close at hand, even then he would not be able to determine the spheres' periods and sizes accurately rather than approximately. The proximity would no doubt increase the precision of his observations, but according to Ptolemy sense perception is essentially limited. Reason supplies the precision that guides sense perception toward making accurate observations, but reason is not a criterion in the determination of the heavenly spheres' sizes and periods. These calculations employ observation as their only criterion, and therefore the values Ptolemy presents for their sizes and periods are necessarily approximate. The only aspects of his mathematical models of which Ptolemy is certain are those that derive from both his criterion of truth and geometry, an indisputable method, or instrument, of mathematics.

τῶν ὀργάνων ἢ διὰ τῶν τοῦ ἡλίου ἐκλείψεων θεωροῦνται, πολὺ διαψευσθῆναι δυναμένων διὰ τὰς παραλλάξεις τῆς σελήνης

Deeming arithmetic and geometry indisputable, and harmonics and astronomy indisputable only to the degree that they appropriate the objects of arithmetic or geometry, Ptolemy echoes an epistemological sentiment that Aristotle relates in *Metaphysics* M3. Aristotle argues that the simpler an object is, the more accurate human beings' apprehension of it is:

> Indeed, the more [that our knowledge] concerns what are prior in definition and simpler, the greater the accuracy (that is, simplicity) [our knowledge] has, so that there is more [accuracy] where there is no magnitude than where there is magnitude, and most of all where there is no movement, though if there is movement [the accuracy] is greatest if it is the primary [movement]; for this is the simplest, and uniform [movement the simplest] of that.[55]

The more simple an object is, the more accurate human comprehension of it is; with more complexity comes less accuracy. Similarly, in the *Posterior Analytics* Aristotle remarks that one science is more accurate than another depending on how many items it posits:

> A science is more accurate than and prior to another science if the same one is both of the fact and of the reason why—but not [a science] of the fact separately from [the science] of the reason why—and if it does not concern what is underlying and the other does concern what is underlying, like arithmetic and harmonics, and if it [proceeds] from fewer things and the other [proceeds] from something additional, like arithmetic and geometry.[56]

Aristotle analyzes the relative accuracy of sciences in a relationship of subordination. Leaving aside the relationship between arithmetic and geometry, which Ptolemy seems to consider equally indisputable, Aristotle asserts that subordinate sciences are less accurate than their superior sciences. Arithmetic is more accurate than harmonics, and geometry is more accurate than optics.

Ptolemy's epistemology implies a similar hierarchy. Because arithmetic and geometry are not sciences—they do not study mathematical objects in discrete

---

55. Aristotle, *Metaphysics* XIII, 1078a9–13: καὶ ὅσῳ δὴ ἂν περὶ προτέρων τῷ λόγῳ καὶ ἁπλουστέρων, τοσούτῳ μᾶλλον ἔχει τὸ ἀκριβές (τοῦτο δὲ τὸ ἁπλοῦν ἐστίν), ὥστε ἄνευ τε μεγέθους μᾶλλον ἢ μετὰ μεγέθους, καὶ μάλιστα ἄνευ κινήσεως, ἐὰν δὲ κίνησιν, μάλιστα τὴν πρώτην· ἁπλουστάτη γάρ, καὶ ταύτης ἡ ὁμαλή.

56. Aristotle, *Posterior Analytics* I, 87a31–35: Ἀκριβεστέρα δ᾽ ἐπιστήμη ἐπιστήμης καὶ προτέρα ἥ τε τοῦ ὅτι καὶ διότι ἡ αὐτή, ἀλλὰ μὴ χωρὶς τοῦ ὅτι τῆς τοῦ διότι, καὶ ἡ μὴ καθ᾽ ὑποκειμένου τῆς καθ᾽ ὑποκειμένου, οἷον ἀριθμητικὴ ἁρμονικῆς, καὶ ἡ ἐξ ἐλαττόνων τῆς ἐκ προσθέσεως, οἷον γεωμετρίας ἀριθμητική.

sets of bodies in the cosmos—they are simple. Therefore, they are indisputable and can be employed as instruments by the mathematical sciences. Indeed, harmonics employs arithmetic and astronomy employs geometry; harmonics appropriates the objects of arithmetic, and astronomy appropriates the objects of geometry. Because harmonics simply utilizes arithmetic ratios to describe the relations among musical pitches, it produces sure and incontrovertible knowledge. Appropriating the objects of geometry, astronomy's hypotheses of eccentric and epicyclic spheres are sure and incontrovertible, but the calculations of the spheres' parameters and periods are not certain because the student of astronomy has extended his study beyond the bounds of geometry's indisputability. The mathematical sciences only produce knowledge insofar as they adhere to arithmetic or geometry. Once they exceed these boundaries, their propositions are neither sure nor incontrovertible. They only approximate the truth. Harmonics stays within these boundaries, but astronomy stretches beyond them.

Consequences

Although Ptolemy was certain of the truth of the nonquantified features of his astronomical models, centuries of astronomers debated their merits. Especially controversial was Ptolemy's equant point, the center of uniform motion, distinct from the deferent circle's own center, in the models of the five planets (see figure 5.5). That a circle can move uniformly around a point that is not its own center is not intuitive, and even Ptolemy may have conceded how difficult this hypothesis is to grasp in *Almagest* 9.2, where he mentions how he was compelled to use procedures that are not strictly rational (παρὰ τὸν λόγον), including proposing hypotheses based not on an evident first principle but as a consequence of sustained trial and application.[57] Ptolemy implies that this process must have involved some method, even if it is difficult to formulate what the method is, and in this way he justifies using apparently ad hoc procedures, provided they are consistent with the phenomena. This brief discussion may or may not allude to the discovery of the equant but, regardless, Ptolemy's astronomical models, and especially the equant, remained disputable in the eyes of his critics.

Controversial in another way are Ptolemy's observation reports. The uncertainty Ptolemy discovered in the quantitative aspects of his astronomical models may have been a contributing factor to the notorious fabrication of some of his observations. Denounced by Delambre, derided by

---

57. Ptolemy, *Almagest* 9.2, HII212. For an analysis of how Ptolemy may have discovered the equant, see Evans, "On the Function."

Robert Newton, and deemed by Alexander Jones to be the most irrefutable cases of falsification are the dates and times of equinox and solstice observations on which Ptolemy bases his solar model and, by extension, the rest of his astronomical theory, as his model of the sun serves as the foundation for every other model he presents in the *Almagest*.[58] Ptolemy claims in *Almagest* 3.1 that he made these observations by himself, and in *Almagest* 3.4 he calls his observations of the equinoxes and summer solstice "most accurate" (ἀκριβέστατα).[59] Nevertheless, he reports the times to the precision of an hour, which he would have known was impossibly precise given the inherent error of his instruments, and each of the observations is in agreement, within a fraction of an hour, with the date and time one can derive simply from adding the relevant number of years of 365;14,48 days from the observations of equinoxes by Hipparchus and the summer solstice by Meton and Euktemon, which Ptolemy pairs with his own observations to determine the length of the tropical year. What ultimately condemns Ptolemy's observations is the fact that they are all one day late; the date and time reported for each observation is approximately one day later than the actual event it describes. The error of one day far exceeds the potential error of Ptolemy's instruments, and no one has found a plausible reason for why a systematic error of one day should exist in these observations. Therefore, it is likely that Ptolemy fabricated or, at the very least, finessed these observation reports. I suggest that the inherent lack of certainty in the determination of heavenly bodies' periods and sizes motivated Ptolemy, at least in part, to cut these corners.

What effect does the lack of certainty in some aspects of Ptolemy's astronomy have for his ethical system? It may seem that harmonics would be a better path to the good life than astronomy. After all, harmonics yields sure and incontrovertible knowledge, while only some aspects of astronomy are certain; other aspects are at best approximations of the truth. Nevertheless, human beings use harmonic and astronomical objects as exemplars not only because they facilitate the proper attunement of their souls but also because, in their proper attunement, human souls resemble the divine. We must remember that Ptolemy's ethics constitutes a mathematical adaptation of the Platonic *telos* of *homoiôsis theôi*, becoming like a god. It is this quest to be like divine, astronomical objects that motivates Ptolemy's entire scientific program. Is it possible for the study of astronomy to provide the ethical transformation of the human soul if not every one of its judgments is true, surely and incontrovertibly? I

---

58. Delambre, *Astronomie ancienne*, 1, xxvi; 2, 107–14; *Astronomie du moyen âge*, lxvii–lxix. Newton, *Crime of Claudius Ptolemy*, 87–94. A. Jones, "Ptolemy," 175. See also A. Jones, "In Order That," 21; Britton, *Models and Precision*, 24–36; Evans, *History and Practice*, 209.

59. Ptolemy, *Almagest* 3.1, H203; 3.4, H233.

suggest that yes, it is. What is crucial for Ptolemy's ethics is that astronomy reveals the properties of celestial bodies' movements and configurations that human souls acquire in their virtuous transformation. These properties are the ones Ptolemy lists in *Almagest* 1.1: constancy, good order, and commensurability. Even though a mathematician can only approximate the heavenly spheres' sizes and periods, the very fact that the stars' movements and configurations are eternally periodic entails that they are constant, well ordered, and commensurable. It is these qualities that the mathematician emulates and loves; it is these qualities that ignite his ethical transformation. The mathematician can only approximate the heavenly spheres' quantitative features, but he is certain of their constancy, good order, and overall commensurability.

Should the mathematician study astronomy only to the extent that it has pronounced ethical benefits? In other words, having discovered the spheres' constancy, good order, and commensurability, should mathematicians end their study there, and contemplate only these qualities of astronomical objects, or should they continue further to calculate the quantitative features of the individual astronomical models? In the Hellenistic period, it was a commonly held view, endorsed especially by Stoic and Epicurean philosophers, that one should pursue sciences only inasmuch as they furnish the good life, and so they set limits on their study. Ptolemy, however, does not limit the sciences in this way. Even if the ethical transformation is possible without calculating the parameters and periods of every one of the astronomical models—for the sun, moon, fixed stars, and each of the five planets—Ptolemy indicates that astronomy, like all mathematical sciences, should be studied to its end.

7

Mathematizing the Human Soul[1]

FOR PTOLEMY, the study of the human soul is a physical science. What permits its classification as such is the soul's nature. According to Ptolemy, it is material. It consists of fine particles, which are smaller than the particles constituting the body. Because these particles are so small, the soul is imperceptible. Moreover, the soul is mortal; the particles constituting it disaggregate upon death and scatter. The combination of these qualities, mortality and imperceptibility, makes the soul unique in Ptolemy's cosmology. We saw in chapter 2 that in *Almagest* 1.1 Ptolemy distinguishes the Prime Mover principally by its imperceptibility. Indeed, in this text it seems as if the Prime Mover is the only object in Ptolemy's cosmos that is imperceptible, but Ptolemy's accounts of the soul reveal that it, too, is imperceptible. While the Prime Mover is imperceptible because it is immaterial, the human soul is imperceptible because its particles are too small to perceive. That Ptolemy takes the soul to permit any rigorous study, despite its imperceptibility, is at first surprising, for in *Almagest* 1.1 he indicates that the imperceptibility of the Prime Mover entails its complete ungraspability. In *On the Kritêrion and Hêgemonikon*, however, Ptolemy explains that one can study the soul by observing the effects it has in and through the body. In this way, its study remains an observational science.

Ptolemy investigates the soul's nature and structure in three texts: *On the Kritêrion and Hêgemonikon*, the *Harmonics*, and the *Tetrabiblos*. He mentions the soul's parts only in passing in the *Tetrabiblos*, and so I leave its discussion for the following chapter. Here I investigate the development of Ptolemy's psychological theory from *On the Kritêrion* to the *Harmonics*. I say "development," because I argue that Ptolemy composed the former before the latter and that,

---

1. The greater part of this chapter is modified from Jacqueline Feke, "Mathematizing the Soul: The Development of Ptolemy's Psychological Theory from *On the Kritêrion and Hêgemonikon* to the *Harmonics*," *Studies in History and Philosophy of Science* 43, no. 4 (December 2012): 585–94. Copyright © 2012 by Elsevier. https://www.sciencedirect.com/science/article/pii/S0039368112000428.

furthermore, comparison of the two texts reveals Ptolemy's move from a less to a more restrictive model of the soul. I argue that this move follows from Ptolemy's effort to mathematize the soul, to analyze its parts and species in concordance with mathematical, specifically harmonic, ratios. *On the Kritêrion* is Ptolemy's only text devoid of mathematics and, accordingly, it is his only study of a physical science that does not pursue the scientific method he propounds in *Almagest* 1.1: the persistent and necessary contribution of mathematics to physics. One could infer that the variance in the texts' methodologies—mathematical or not—results simply from Ptolemy's distinct motivations in writing them. Nevertheless, merely noting a difference in motivation does not explain all of the inconsistencies between the texts; what would seem to be avoidable incongruities in the psychological accounts would remain unexplained. A chronological argument, on the other hand, though it must remain speculative, explains the methodological discrepancy and accounts for the contradictory features of the two texts' psychological accounts. I argue that Ptolemy articulated his scientific method after composing *On the Kritêrion* and before completing the *Harmonics*. When Ptolemy mathematized the soul he sought to improve his account of the soul's structure, to make it epistemically sound. This endeavor to improve and mathematize his psychology resulted in discernible incongruities between *On the Kritêrion* and the *Harmonics*. Thus, we may parse the development of Ptolemy's scientific method into two stages. When he composed *On the Kritêrion*, he already had committed himself to his criterion of truth, the collaboration of perception and reason in the effort to generate knowledge. Thereafter, he developed his mature scientific method. In addition to the criterion, he committed himself entirely to the pursuit and application of mathematics.

The Nature and Structure of the Soul in
On the Kritêrion and Hêgemonikon

Ptolemy's most thorough account of the soul's nature resides in *On the Kritêrion*, but he labels the soul "mortal" only in *Harmonics* 3.4. This term, "mortal" (θνητός), arises in only one other instance in Ptolemy's corpus: in *Almagest* 1.1.[2] As we saw in chapter 2, Ptolemy argues that, of the three theoretical sciences, mathematics is intermediate between physics and theology for two reasons. First, while the object studied by theology—or, as the text implies, the Prime Mover—is imperceptible and physical objects are

---

2. The term "mortal" (θνατός) also appears in Epigram *Anthologia Palatina* 9.577, included in two of the three main branches of the *Almagest*'s manuscript tradition. Its authorship is unknown, and it is doubtful that it is by Ptolemy. See Tolsa, "The 'Ptolemy' Epigram."

perceptible, mathematical objects can be conceived of both with and without the aid of sense perception. Second, Ptolemy explains that the subject matter of mathematics "is an attribute of all beings without exception, both mortal and immortal, for those that are perpetually changing in their inseparable form, it changes with them, while for the eternal, which have an aethereal nature, it keeps their immovable form unmoved."[3] In other words, Ptolemy argues that mathematics is intermediate between physics and theology because mathematical objects exist in bodies that are sublunary and superlunary, mortal and immortal (θνητοῖς καὶ ἀθανάτοις), physical and theological. I discussed in chapter 2 how these two arguments for mathematics' intermediate status are inconsistent with respect to which objects are theological—the Prime Mover or aethereal bodies—but what is relevant to this chapter is that the category of physical objects remains constant. Mortality is a trait that distinguishes sublunary physical bodies from superlunary and theological objects. As mortal, the human soul is a sublunary physical body and, as such, it is the subject matter of physics.

The human soul, however, is unlike other physical bodies in that it is not perceptible and, in this way, it differs from the body. Ptolemy contrasts soul and body in *On the Kritērion*: "The parts in us being grasped according to the most general differentiae are body and soul. By 'body' we mean the [part] composed of bones, flesh, and similar perceptible things, and by 'soul' the [part] which is the cause of the movements occurring in or through these and which we can only grasp through its powers ..."[4] Despite its imperceptibility, the soul is apprehensible by means of the effects it causes in and through the body. Ptolemy refers to the soul's powers, or faculties (δυνάμεις), the term which Aristotle uses in the *De anima* to signify the soul's capacities.[5] According to Ptolemy, the soul is the cause of a human being's thinking (διανοεῖσθαι),

3. Ptolemy, *Almagest* 1.1, H6: ἀλλὰ καὶ τῷ πᾶσιν ἁπλῶς τοῖς οὖσι συμβεβηκέναι καὶ θνητοῖς καὶ ἀθανάτοις τοῖς μὲν αἰεὶ μεταβάλλουσι κατὰ τὸ εἶδος τὸ ἀχώριστον συμμεταβαλλομένην, τοῖς δὲ ἀιδίοις καὶ τῆς αἰθερώδους φύσεως συντηροῦσαν ἀκίνητον τὸ τοῦ εἴδους ἀμετάβλητον.

4. Ptolemy, *On the Kritērion*, La11, translation modified from Liverpool-Manchester Seminar, in Ptolemaeus, "On the Kriterion": τῶν καθ' ὅλας διαφορὰς λαμβανομένων ἐν ἡμῖν μερῶν, τὸ μέν ἐστι σῶμα, τὸ δὲ ψυχή. καὶ σῶμα μὲν καλοῦμεν τὸ ἐξ ὀστέων καὶ σαρκῶν καὶ τῶν τοιούτων αἰσθητῶν, ψυχὴν δὲ τὸ τῶν ἐν τούτοις ἢ διὰ τούτων κινήσεως αἴτιον, οὗ τῶν δυνάμεων μόνων ἀντιλαμβανόμεθα

5. *De anima* II, 417a9–418a6, illuminates what Aristotle means by "faculty" (δύναμις). "Faculties," or "capacities" as it also is translated, refer to the potentialities of living beings. A human being, for example, has the capacity for sense perception, which, like the other faculties, may be either potential or actual. When not sensing an object, a human being's faculty of sense perception lies dormant; it is potential. When exercising the faculty, when a human being senses an object, it is actualized.

as well as "both sensory and all other movements."[6] By observing the effects the soul's faculties have on the body, or the movements the soul causes in and through the body, one apprehends the soul's nature.

Although the soul and body have different capacities, in their material consistency they differ merely in degree. Examining the effects of death, Ptolemy explains that the soul consists of finer particles than the body:

the soul is so constituted as to scatter immediately to its proper elements, like water or breath released from a container, because of the preponderance of finer particles . . . the body, on the other hand, although it stays in the same state for a considerable time because of the thicker consistency of its matter, nevertheless is not seen to have any sensation or to make any movement of the kind it did previously.[7]

Recalling Epicurean materialism in particular, Ptolemy portrays the soul as consisting of fine particles that scatter upon release from the body; the body has a thicker consistency and is more material (ὑλικώτερον) than the soul.[8] Hence, the particles composing the soul are finer than the constituents of the body, and these fine particles are so small that they are imperceptible.

Ptolemy affirms that the soul consists of matter, either different kinds or a single sort, in the following passage:

Further, if soul is composed not of one and the same but of different kinds of material, it will be the individual characteristics of these different materials that shape the parts of the body that surround each of them to suit the properties of their own substance and so make them able to cooperate with the faculties of the soul; if, on the other hand, the underlying nature of the whole [soul] is one and the same, the variety of the [soul's] faculties will be produced by the differences in the surrounding parts of the body[9]

6. Ptolemy, *On the Kritêrion*, La11: τάς τε αἰσθητικὰς καὶ τὰς ἄλλας πάσας κινήσεις

7. Ibid., La12, translation modified from Liverpool-Manchester Seminar, in Ptolemaeus, "On the Kriterion": καθ' ἣν ἡ μὲν ψυχὴ διὰ τὴν ὑπερβολὴν τῆς λεπτομερείας καθάπερ ὕδωρ ἢ πνεῦμα τοῦ συνέχοντος ἀνεθὲν εὐθὺς εἰς τὰ οἰκεῖα στοιχεῖα πέφυκε χωρεῖν [. . .] τὸ δὲ σῶμα διὰ τὸ τῆς ὕλης παχυμερὲς ἐπιδιαμένον συχνὸν χρόνον ἐν ταῖς αὐταῖς καταστάσεσιν, ὅμως οὐδεμίαν φαίνεται ποιούμενον οὔτε αἴσθησιν οὔτε ὅλως κίνησίν τινα τῶν προτέρων.

8. Ibid., La19. Cf. Ptolemy, *Tetrabiblos* 3.12.1, H224. Boll observes that Ptolemy's portrayal of the soul as a substance finer and more movable than the body is evidence of his materialism: Boll, "Studien," 88. Epicurus uses the terms λεπτομέρεια and παχυμερές for finer and thicker consistencies of matter in his letters to Herodotus and Pythocles, and he defines the soul as a body consisting of finer particles in Diogenes Laertius, *Lives* 10.63.

9. Ptolemy, *On the Kritêrion*, La12, translation modified from Liverpool-Manchester Seminar, in Ptolemaeus, "On the Kriterion": καὶ εἰ μὲν οὐκ ἀπὸ μιᾶς καὶ τῆς αὐτῆς ὕλης ἀλλ' ἀπὸ

Ptolemy ultimately argues that the soul consists of different types of matter. Appropriating Aristotle's five-element theory, he claims that the human soul is composed of three of the five elements: air, fire, and aether. He justifies which elements he assigns to body or soul by appealing to the elements' activity or passivity: "Among the compounds, too, we apply the term 'body' properly to what is more material and less active and 'soul' to what moves both itself and [body]. It is therefore reasonable that the body should be classed in accordance with the elements of earth and water and the soul in accordance with the elements of fire, air, and aether . . ."[10] Ptolemy states here that the soul merely should be classed (τετάχθαι) in accordance with fire, air, and aether, but thereafter he makes a stronger claim and asserts that the soul is composed of these elements: "It will also be a consequence of this that the substance itself of the soul has a distinct nature akin to the elements of which it is composed, so that its nature will be both passive and active in its proper movements in proportion to the air and fire, but active only [in proportion to] the aether."[11] Thus, Ptolemy maintains that the soul is composed of the elements air, fire, and aether.

According to the members of the Liverpool-Manchester Seminar on ancient Greek philosophy, Ptolemy's particular ascription of passivity and activity (πάσχειν καὶ ποιεῖν) to the five elements is neither Peripatetic nor Stoic.[12] Following Boll, I would argue that Ptolemy joins these two traditions by amalgamating Aristotle's five-element theory with a Stoic conception of the passivity and activity of the elements.[13] Like the Stoics, Ptolemy portrays air and fire, the constituents of the Stoics' *pneuma*, as active in comparison with earth and water, which both the Stoics and Ptolemy depict as passive. Unlike the Stoics, Ptolemy appropriates Aristotle's fifth element, labels the

---

διαφόρων συνέστηκεν, ὑπὸ τῆς τούτων ἰδιότητος τὰ περιέχοντα ἑκάστην σώματα πρὸς τὸ οἰκεῖον τῆς οὐσίας διαμορφούμενα συνεργὰ γίνεται πρὸς τὰς δυνάμεις τῆς ψυχῆς, εἰ δὲ μία καὶ ἡ αὐτὴ πάσης ἐστὶν ὑπόστασις, τὸ ποικίλον τῶν δυνάμεων αὐτῆς ὑπὸ τῆς διαφορᾶς τῶν περιεχόντων σωμάτων ἀποτελεῖται

10. Ibid., L.a19: ἐπεὶ δὲ καὶ ἐν τοῖς συγκρίμασι σῶμα μὲν ἰδίως καλοῦμεν τὸ ὑλικώτερον καὶ ἀνενέργητον, ψυχὴν δὲ τὸ κινητικὸν κἀκείνου καὶ ἑαυτοῦ, τὸ μὲν σῶμα εὔλογον τετάχθαι κατὰ τὰ γῆς καὶ ὕδατος στοιχεῖα, τὴν δὲ ψυχὴν κατὰ τὰ πυρὸς καὶ ἀέρος καὶ αἰθέρος

11. Ibid., L.a20: ἀκόλουθον ἂν εἴη τούτοις καὶ τῆς ψυχῆς τὴν οὐσίαν αὐτὴν ἔχειν τινὰ διαφορὰν ἐφαρμόζουσαν τοῖς ποιοῦσιν αὐτὴν στοιχείοις, ὥστε ὅσῳ μέν ἐστιν ἀέρος καὶ πυρός, τούτην καὶ πάσχειν καὶ ποιεῖν ἐν ταῖς οἰκείαις κινήσεσιν, ὅσῳ δ' αἰθέρος, ταύτην ἐνεργεῖν μόνον.

12. Liverpool-Manchester Seminar, in Ptolemaeus, "On the Kriterion," 226, n. 13.4. On the Stoics' ascription of activity and passivity to the elements, see Long, *Stoic Studies*, 229. Cf. White, "Stoic Natural Philosophy," 135–36.

13. See Boll, "Studien," 89.

aether active, and defines air and fire as both passive and active. While Ptolemy's portrayal of the soul as consisting of air and fire stems from the Stoic conception of *pneuma*, his inclusion of aether as an elemental component of the soul proceeds from the Peripatetic tradition. In *Generation of Animals* II, 736b29–737a1, Aristotle describes a faculty of the soul that has to do with a body that is more divine than the elements, and he describes semen as containing within it a substance that is analogous to the element of the stars (τῷ τῶν ἄστρων στοιχείῳ). Providing a Hellenistic reference to this tradition, Cicero relates in *Academica* 1.7.26, "Aristotle deemed that there existed a certain fifth sort of element, in a class by itself and unlike the four that I have mentioned above, which was the source of the stars and of thinking minds."[14] Ptolemy appropriates this association of aether with the human soul, and he adapts it to a Stoic schema. Accordingly, he portrays aether as active, fire and air as active and passive, and these three elements as constituting the human soul.

Ptolemy presents in *On the Kritêrion* a different account of the elements' qualities than he offers in the *Almagest* and *Tetrabiblos*. In *Almagest* 1.1, as we saw in chapter 2, he advances a dichotomy between the sublunary elements that move toward the center of the cosmos and the sublunary elements that move away from the center. The elements that move toward the center, presumably earth and water, are heavy and passive (παθητικόν); the elements that move away from the center, presumably air and fire, are light and active (ποιητικόν). Note that here the elements are entirely passive or active. In *Tetrabiblos* 1.5, Ptolemy describes the four qualities, rather than the elements, as passive or active. In this regard, he follows Aristotle's *On Generation and Corruption*, but, in his ascription of activity and passivity to the qualities, he diverges from Aristotle. At *On Generation and Corruption* II, 329b23–31, Aristotle portrays both hot and cold as active principles, but Ptolemy, in reference to the four humors, characterizes hot and moist as active and cold and dry as passive.

To account for the inconsistency of Ptolemy's portrayals of activity and passivity, one could argue that Ptolemy merely contrasts the four sublunary elements, by virtue of their qualities, as wholly passive or active—as opposed to passive and/or active—when he examines them in isolation from the fifth element, as he does in the *Tetrabiblos*. This argument, however, is contradicted by *Almagest* 1.1, wherein Ptolemy also asserts that the mathematician distinguishes the corruptible elements from the incorruptible, or the sublunary elements from the aether, based on whether they move rectilinearly or circularly. Consequently, one is left with the distinct impression that Ptolemy simply did

---

14. Cicero, *Academica* 1.7.26, trans. Rackham: Quintum genus, e quo essent astra mentesque, singulare eorumque quattuor quae supra dixi dissimile Aristoteles quoddam esse rebatur.

not adhere to a single schema for the activity and passivity of the elements and their qualities.

In *On the Kritêrion*, Ptolemy proceeds to assign all five elements, including earth and water, to the soul's faculties. The soul consists of three faculties: the faculty of thought (διανοητικόν), the faculty of sense perception (αἰσθητικόν), and the faculty of impulse (ὁρμητικόν). The faculty of sense perception is around (περί) the passive elements, earth and water; the faculty of impulse is around the elements that are both passive and active, air and fire; the faculty of thought is around the element that is only active, aether. Ptolemy further divides the faculty of impulse into two parts: the appetitive (ὀρεκτικόν) and spirited (θυμικόν). The former has more air in its composition (ἀεροειδέστερον) and the latter has more fire (πυροειδέστερον). In general, the soul exists in greater proportion in the more hot and moist areas of the body, but each faculty has its own, distinct location(s).

The term Ptolemy uses for the faculty of impulse, *hormêtikon*, derives from the Stoic tradition, but it had become common intellectual property by the second century. For instance, in *Didaskalikos* 25.7 Alcinous portrays the faculty of impulse as a faculty of gods' souls that transforms into the spirited faculty (θυμοειδές) of human souls upon embodiment. In his *De anima*, Alexander of Aphrodisias consistently links the faculty of impulse with the appetitive faculty (ὀρεκτικόν), Aristotle's term for the appetitive faculty in the *De anima*, *Nicomachean Ethics*, and *De motu animalium*. Ptolemy appropriates the association of the faculty of impulse with the appetitive faculty for his own tripartite model of the human soul.

Ptolemy's spirited and appetitive parts evoke Plato's *Timaeus*. Plato situates the appetitive part (ἐπιθυμητικόν) lowest in the body. It is "in the area between the midriff and the boundary toward the navel," where the liver resides.[15] For Ptolemy, the appetitive part (ὀρεκτικόν) is situated slightly lower in the body. It is around the stomach and abdomen (περὶ τὴν γαστέρα καὶ τὸ ἦτρον), and its motions occur in the area below the "inward parts" (σπλάγχνα).[16] Plato's spirited part, or "the part that exhibits courage and spirit, which is ambitious," is located between the neck and midriff, and the heart resides in the guardhouse of this area.[17] Similarly, for Ptolemy the spirited part (θυμικόν) is located around the heart, and it is the *hêgemonikon*, or the soul's commanding faculty, with respect to living.

15. Plato, *Timaeus* 70e, trans. Zeyl, in Plato, *Complete Works*: εἰς τὸ μεταξὺ τῶν τε φρενῶν καὶ τοῦ πρὸς τὸν ὀμφαλὸν ὅρου

16. Ptolemy, *On the Kritêrion*, La21.

17. Plato, *Timaeus* 70a: τὸ μετέχον οὖν τῆς ψυχῆς ἀνδρείας καὶ θυμοῦ, φιλόνικον ὄν

Ptolemy's description of the spirited faculty as the *hêgemonikon* of living derives from a common tradition of depicting the heart as the source of a vital substance.[18] When examining the spirited part of the soul in *Timaeus* 70a–b, for example, Plato describes the heart as the source from which blood flows through the body. In *Generation of Animals* II, 740a1–23, Aristotle describes the heart as the first principle (ἀρχή) of an animal, because it is the first part of an embryo to develop and, once developed, it provides nourishment to the growing animal in the form of blood.[19] Post-Hellenistic philosophers adapted these earlier arguments to characterize the heart as the seat of the *hêgemonikon*. In *De anima* 39.21–40.3, for instance, Alexander of Aphrodisias argues that the *hêgemonikon* resides in the heart because, as the container of blood, it is the source of nutriment for the body. Ptolemy joins this tradition by designating the faculty located around the heart as the *hêgemonikon* with respect to living.

Like the faculty of impulse, Ptolemy's faculty of sense perception (αἰσθητικόν)—which Aristotle names as a faculty of the soul in the *De anima*— is multiple in location and capacity. It governs the contact of the sense organs with perceptible bodies and *phantasia*, the transmission of sense impression to the intellect (νοῦς). Concerning the senses, Ptolemy identifies five, each with its own location in the body. Touch, more material (ὑλικωτέραν) than the other senses, extends through the body's flesh and blood.[20] The other four senses exist in the parts of the body that are more easily penetrated and moist. Taste and smell are located lower in the body than sight and hearing, and, accordingly, they are closely related to the faculty of impulse. Residing physically higher in the body are sight and hearing. They are the more easily activated and valuable (τὰς μὲν μᾶλλον εὐκινητοτέρας καὶ τιμιωτέρας) of the senses, and they are closely connected with the faculty of thought, which Ptolemy deems the *hêgemonikon* with regard to both living and living well.[21] While the spirited part of the soul is the *hêgemonikon* with respect to only living, the senses sight and hearing are secondary *hêgemonika* of living well.

---

18. Boll traces Ptolemy's portrayal of the heart as the location of the *hêgemonikon* of living to the Stoic placement of the *hêgemonikon* in the heart, and he claims that Ptolemy's adoption of two *hêgemonika*, one in the heart and one in the brain, stems from the Pythagorean tradition: Boll, "Studien," 92. In support of this claim, he cites Aëtius, *De placitis reliquiae* 4.5, D391: περὶ τοῦ ἡγεμονικοῦ· Πυθαγόρας τὸ μὲν ζωτικὸν περὶ τὴν καρδίαν, τὸ δὲ λογικὸν καὶ νοερὸν περὶ τὴν κεφαλήν.

19. Cf. Aristotle, *Parts of Animals* III, 666a18–36.

20. Ptolemy, *On the Kritêrion*, La20.

21. Ibid.

The faculty of thought (διανοητικόν) is the most valuable and divine, in both capacity and substance, of the soul's faculties. It is undivided, and Ptolemy appropriates its location from the *Timaeus*. Ptolemy maintains that the faculty of thought is in the head, around the brain, and in *Timaeus* 44d the immortal part of the soul resides in the head.[22] Although Ptolemy's faculty of thought is single in location, like the other faculties it is multiple in capacity. He states, "it exhibits a capacity for forming opinions in accordance with its connection to the senses and a capacity for knowledge in accordance with its independent reexamination of external objects."[23] In addition, Ptolemy describes a rational faculty (λογικόν), which encompasses thought (διάνοια), the activity of the faculty of thought (διανοητικόν), and speech (διάλεκτος): "Of the rational faculty, by which the special property of human beings is defined, on the one hand, thought is the *logos* that is an internal analysis and repetition and differentiation of what has been remembered; on the other hand, speech is the vocal symbols through which what were thought are revealed to other people."[24] The agent of judgment, the intellect (νοῦς), uses thought, or internal *logos*, as the means by which it judges. Speech, on the other hand, makes no contribution to the process of judgment, because it is secondary to thought as an image is to an original.

The dichotomy between internal and external discourse was standard in ancient Greek philosophy. Plato and Aristotle distinguish between the two in *Sophist* 263e, *Philebus* 38e–39a, and *Posterior Analytics* I, 76b24–27, respectively, and the Stoics developed the distinction between internal and external *logos*, specifically.[25] In portraying uttered logos as an image (εἰκών) of internal *logos*, Ptolemy utilizes a distinctly Platonic metaphor. Internal *logos*, for Ptolemy, takes two forms: (1) opinion and supposition (δόξα καὶ οἴησις), and (2) knowledge and understanding (ἐπιστήμη καὶ γνῶσις).[26] Furthermore, the intellect makes judgments within two fields of inquiry, the theoretical and the practical, and the faculty of thought has a capacity for each of these fields. Thus, Ptolemy portrays internal thought as the activity of the faculty of thought. Thought

---

22. Ptolemy, *On the Kritêrion*, La21. Cf. Plato, *Timaeus* 90a.

23. Ptolemy, *On the Kritêrion*, La21, translation modified from Liverpool-Manchester Seminar, in Ptolemaeus, "On the Kriterion": τῇ μὲν δοξαστικῇ κατὰ τὴν πρὸς τὰς αἰσθήσεις συναφήν, τῇ δ' ἐπιστημονικῇ κατὰ τὴν ἐφ' αὑτῆς τῶν πραγμάτων ἀναπόλησιν.

24. Ibid., La6: τοῦ δὲ λογικοῦ, καθὸ τὸ τῶν ἀνθρώπων ἴδιον ὥρισται, διάνοια μέν ἐστιν ὁ λόγος ὁ ἐνδιάθετος διέξοδός τις οὖσα καὶ ἀναπόλησις καὶ διάκρισις τῶν μνημονευθέντων, διάλεκτος δὲ τὰ τῆς φωνῆς σύμβολα, δι' ὧν προφέρεται τοῖς πλησίον τὰ διανοηθέντα.

25. Cf. Plato, *Theaetetus* 189e. See Sextus Empiricus *Adversus mathematicos* 8.275–76.

26. Ptolemy, *On the Kritêrion*, La6.

takes the form of either opinion or knowledge, and the intellect has the capacities to make judgments in theoretical and practical philosophy.

Although Ptolemy appropriates the terms of his psychology from the Aristotelian, the Platonic, and, to a lesser extent, the Stoic traditions, he contends that observation would corroborate his account:

> Even if we do not apply what is reasonable and appropriate to the natures of things, we also could learn that these [natures] exist in this way from the movements of each faculty of the soul, provided we are willing to investigate in a way that is loving of truth, for we shall observe that the exertions that accompany these [movements], whether they are passive or active, always take place in the parts of the body recounted. The sensory [exertions] take place in each of the corresponding sense organs, the appetitive in the area below the liver, the spirited in the area around the heart—these last include cases of pleasure, pain, fear, and anger—and only the cognitive occur when the other parts of the body are at rest but the head is being filled, like exertions [caused] by the permanent internal movement.[27]

Ptolemy claims that both reason and perception substantiate his account. He employs reason when introducing the account by considering what is reasonable and appropriate to the soul's nature, and he asserts that observation of the movements the soul causes in and through the body would corroborate his exposition.

Ptolemy's claim that observation would corroborate his account may allude to the ancient medical traditions' attempts to discover the location(s) of the soul. In the second century, for instance, Galen devoted Books 1–6 of *De placitis Hippocratis et Platonis* to a demonstration (ἀπόδειξις) of the soul's parts and locations.[28] Galen supports his argument with empirical evidence derived from the dissection of animals. The fact that the rational soul is situated in the brain is evident from the observation that the brain is the source of the nerves; that the spirited soul is situated in the heart is evident from the observation

27. Ibid., Laz1: ὅτι δὲ ταῦτα οὕτως ἔχει, κἂν μὴ τὸ εὔλογον καὶ οἰκεῖον τῶν φύσεων ἐφαρμόζωμεν, μάθοιμεν ἂν καὶ ἐπ᾽ αὐτῶν τῶν καθ᾽ ἑκάστην δύναμιν τῆς ψυχῆς κινήσεων, εἰ φιλαληθῶς ἐθέλοιμεν σκοπεῖν· τὰς γὰρ ἐπ᾽ αὐταῖς γινομένας διατάσεις ἐάν τε πάσχωσιν ἐάν τε ποιῶσιν ἐν τοῖς κατειλεγμένοις μέρεσι τοῦ σώματος ἀεὶ συμβαινούσας κατανοήσομεν, τὰς μὲν αἰσθητικὰς ἐν ἑκάστῳ τῶν οἰκείων αἰσθητηρίων, τὰς δ᾽ ὀρεκτικὰς ἐν τοῖς ὑπὸ τὰ σπλάγχνα, τὰς δὲ θυμικὰς ἐν τοῖς περὶ τὴν καρδίαν—ἐν τούτοις γὰρ ἡδοναὶ καὶ λῦπαι καὶ φόβοι καὶ ὀργαί—μόνας δὲ τὰς διανοητικὰς τῶν μὲν ἄλλων ἠρεμούντων, τῆς δὲ κορυφῆς πληρουμένης, ὥσπερ καὶ διαδόσεις ὑπὸ τῆς ἐντὸς ἀεὶ κινήσεως.

28. Cf. Galen, *De propriis placitis* 6.1–6. See Hankinson, "Galen," 229.

that the heart is the source of the arteries; that the appetitive soul is situated in the liver is evident from the observation that the liver is the source of the veins. While Ptolemy simply claims that empirical evidence would support his psychological account, Galen actually provides empirical evidence, which substantiates his own account of the soul's parts and locations.

By claiming that his psychological account is founded on both reason and perception, Ptolemy grounds his psychology in the criterion of truth he outlines in the first portion of *On the Kritêrion and Hêgemonikon*. Investigating how it is that a human being judges objects for the sake of knowing the truth, he lists several elements involved in this judgment:

1. That being judged, or what is (τὸ ὄν)
2. That through which it is judged, or sense perception (αἴσθησις)
3. That which judges, or intellect (νοῦς)
4. That by which it is judged, or reason (λόγος)
5. That for the sake of which it is judged, or truth (ἀλήθεια)

The criterion of truth emerged as a fundamental concern in Hellenistic philosophy, and thereafter—arguably from the late Hellenistic period—some philosophers divided the criterion into parts. According to Diogenes Laertius, Potamo of Alexandria, the eclectic philosopher active near the end of the first century BCE, advanced a bipartite criterion of truth: the agent and that through which an object is judged.[29] Notably, Potamo's and Ptolemy's second parts are identical. In *Didaskalikos* 4.1, Alcinous identifies three components of his criterion, two of which are identical to Ptolemy's: that which judges, that being judged, and the process of judgment.[30] Refuting the criterion in *Adversus mathematicos* 7.35, Sextus Empiricus likewise identifies three components similar to Ptolemy's: the agent, that through which an object is judged, and the application.[31] Listing several components in his criterion of truth, Ptolemy follows a contemporary trend.

29. Diogenes Laertius, *Lives* 1.21: ἀρέσκει δ' αὐτῷ, καθά φησιν ἐν τῇ στοιχειώσει, κριτήρια τῆς ἀληθείας εἶναι· τὸ μὲν ὡς ὑφ' οὗ γίνεται ἡ κρίσις, τουτέστι τὸ ἡγεμονικόν· τὸ δὲ ὡς δι' οὗ, οἷον τὴν ἀκριβεστάτην φαντασίαν. For a study of Potamo's criterion of truth, see Hatzimichali, *Potamo of Alexandria*, 82–103.

30. Alcinous, *Didaskalikos* 4.1: Ἐπεὶ οὖν ἔστι τι τὸ κρῖνον, ἔστι δὲ καὶ τὸ κρινόμενον, εἴη ἄν τι καὶ τὸ ἐκ τούτων ἀποτελούμενον, ὅπερ εἴποι ἄν τις κρίσιν. For a comparison of the criteria of truth of Potamo, Sextus, Alcinous, and Ptolemy, see Long, "Ptolemy on the Criterion," 154–62. Cf. Lehoux, "Observers, Objects," 461–62; Lucci, "Criterio e metodologia."

31. Sextus Empiricus *Adversus mathematicos* 7.35: λέγοντας τὸ μέν τι εἶναι κριτήριον ὡς ὑφ' οὗ, τὸ δὲ ὡς δι' οὗ, τὸ δὲ ὡς προσβολὴ καὶ σχέσις. Cf. *Theaetetus Commentary* 2.24–25, which refers to that through which an object is judged: τὸ [δ]ι' οὗ κ[ρίν]ομεν

Mapping Ptolemy's criterion onto his statement of how the soul's faculties and their locations may be determined, one notes that the objects being judged, or the soul's faculties and their locations, may be judged through perception of the movements the soul's faculties cause in and through the body, by the intellect in terms of what is reasonable and appropriate, and in a way that is concordant with a love of truth. While Ptolemy's dually rational and empirical criterion provides the foundation for his psychology, his psychology, in turn, elucidates the components of his criterion. After delineating the criterion, Ptolemy examines the relationships among its components: "Since sense perception and intellect are principles and elements, but the others are secondary to them as capacities, instruments, or activities, if we grasp both the similarities and differences between [sense perception and intellect], we shall have the whole procedure within our sight."[32] Thus, Ptolemy's expositions of the criterion of truth and the human soul mutually reinforce one another.

The Structure of the Soul in the *Harmonics*: The Aristotelian Account

Having examined *On the Kritêrion*, let us turn now to the *Harmonics* and assess its consistency with the former. A. A. Long has maintained that Ptolemy's divisions of the cognitive part of the soul in *Harmonics* 3.5 and *On the Kritêrion* are consistent, and, although he is correct in highlighting certain similarities, he has not mentioned the differences, which, as I aim to demonstrate, reveal a significant shift in the themes and methods Ptolemy employs in the two texts.[33] Most conspicuously, in *On the Kritêrion* Ptolemy presents a single account of the human soul, but in the *Harmonics* he outlines three: an Aristotelian, a Platonic, and a synthetic of the previous two. Presenting three alternative accounts, Ptolemy exhibits a particular attitude toward eclecticism. As in *On the Kritêrion*, he appropriates his terms from the Aristotelian, Platonic, and Stoic traditions, but in *Harmonics* 3.5 he does not choose between these traditions nor synthesize them into a single account as he does in *On the Kritêrion*. When transitioning from the Aristotelian to the Platonic account, Ptolemy merely remarks that the human soul is also divided in another way (κατ' ἄλλον τρόπον διαιρουμένης τῆς ψυχῆς ἡμῶν).[34] In other words, the three

---

32. Ptolemy, *On the Kritêrion*, La7, translation modified from Liverpool-Manchester Seminar, in Ptolemaeus, "On the Kriterion": ἐπεὶ δ' ἡ μὲν αἴσθησις καὶ ὁ νοῦς ἀρχαί τινές εἰσι καὶ στοιχεῖα, τὰ δ' ἄλλα τούτων παρακολουθήματα δυνάμεις ὄντα καὶ ὄργανα καὶ ἐνεργήματα, λαβοῦσιν ἡμῖν τὰς ἐκείνων ὁμοιότητάς τε καὶ διαφορὰς ὑπ' ὄψιν ἂν γένοιτο τὸ πᾶν.

33. Long, "Ptolemy on the Criterion," 170.

34. Ptolemy, *Harmonics* 3.5, D96.

accounts of the soul in *Harmonics* 3.5 are interchangeable; each describes the soul's structure.

In addition, Ptolemy uses different terminology in *On the Kritêrion* and the *Harmonics* to describe the soul's components. In the former, he discusses three faculties (δυνάμεις) of the soul; in the latter, he lists the soul's three parts (μέρη) and their forms, or species (εἴδη).[35] In order to explain this discrepancy, one might simply note that, as a result of his eclecticism, Ptolemy appropriated historically distinct sets of terminology, which he applied to the same set of phenomena. In this way, the terms "faculties" and "parts" would stand as simple alternatives, similar to how the terms for the soul's individual parts and species are alternatives in the Aristotelian, Platonic, and synthetic accounts of *Harmonics* 3.5. This interpretation, however, does not explain why Ptolemy chose to employ distinct terminology in *On the Kritêrion* and the *Harmonics*. I argue that this terminological distinction is a consequence of the texts' dissimilar themes. In *On the Kritêrion*, Ptolemy emphasizes the perceptibility of the movements the soul causes in and through the body, and he grounds his psychology in a dually rational and empirical criterion of truth. Ptolemy's empiricism is heavily dependent on Aristotle's theory of cognition in the *De anima*,[36] and the empirical emphasis of *On the Kritêrion* may have led Ptolemy to appropriate Aristotle's terminology in the *De anima* for the soul's faculties. In the *Harmonics*, on the other hand, Ptolemy appropriates the harmonic analogy of the soul's structure from the Platonic and Stoic traditions,[37] and, in order to analyze the soul's structure in harmonic terms, he uses language that is suitable for relationships among musical pitches and the soul's components: parts and species. Therefore, Ptolemy's choice of Aristotelian or Platonic terminology for the soul's components depends on his adherence to disparate textual themes.

In the *Harmonics*, Ptolemy explains that the human soul is tripartite because it has a harmonic form. As discussed in chapter 5, Ptolemy defines *harmonia* as a rational cause that produces harmonic ratios in movements and formal configurations, to some degree in all natural bodies but especially and to the greatest extent in those that have a more complete and rational nature—namely,

35. In harmonics, "species" describe the ways in which melodic intervals within a concord can be arranged. Each of the primary concords—the fourth, the fifth, and the octave—has a distinct number of species. Ptolemy applies this manner of distinguishing the concords, and dividing them into species, to analyzing the components of the human soul.

36. A. A. Long confirms Boll's assessment of Ptolemy's sources and points to a Peripatetic influence on Ptolemy's empiricism: Long, "Ptolemy on the Criterion," 163.

37. See Plato, *Republic* IV, 443d–e, and *Timaeus* 47d. On the Stoic use of the harmonic metaphor, see Long, *Stoic Studies*, 202–23.

musical pitches, heavenly bodies, and human souls. Like musical pitches and heavenly bodies, human souls experience only one type of change, the primary and most complete type of movement, circular motion from place to place, and their formal configurations reveal their harmonic structure. Ptolemy maps the most fundamental relationships in music—the octave and the concords of the fifth and the fourth—onto the three parts of the soul. Again, he presents these correspondences as follows:

> Well then, there are three primary parts of the soul—the intellectual, the perceptive, and the [part] that maintains a state—and there are three primary forms of homophones and concords—the homophone of the octave and the concords of the fifth and the fourth—and so the octave is attuned to the intellectual [part]—for in each there is the greatest simplicity, equality, and homogeneity—the fifth to the perceptive [part], and the fourth to the [part] that maintains a state.[38]

The soul's parts and species in the *Harmonics* differ in name and description from the soul's faculties in *On the Kritêrion*. According to this, the first of the three psychological accounts in *Harmonics* 3.5, the soul consists of three parts: the intellectual (νοερόν), the perceptive (αἰσθητικόν), and the part that maintains a state (ἑκτικόν). The terms for the first two parts are Aristotelian. The third is Stoic; it is the adjective derived from *hexis*, which refers to the function of *pneuma* in binding objects into a cohesive form.[39]

Each of the soul's parts has several species, the number of which is revealed by the part's correspondence to the octave or a concord in music. Just as the concord of the fourth has three species, the part that maintains a state (ἑκτικόν) has three species. Ptolemy explains, "One can say that the part of the soul that maintains a state has three species, equal in number to the species of the fourth, related respectively to growth, maturity, and decline—for these are its primary capacities . . ."[40] These species do not correspond in any way to the soul's components in *On the Kritêrion*, and, indeed, the terms do not even appear in the text. Although the term *hektikon*, the part that maintains a

38. Ptolemy, *Harmonics* 3.5, D95–96: Ἔστι τοίνυν τὰ μὲν πρῶτα τῆς ψυχῆς μέρη τρία, νοερόν, αἰσθητικόν, ἑκτικόν, τὰ δὲ πρῶτα τῶν ὁμοφώνων καὶ συμφώνων εἴδη τρία, τό τε διὰ πασῶν ὁμόφωνον καὶ σύμφωνα τό τε διὰ πέντε καὶ διὰ τεσσάρων, ὥστε ἐφαρμόζεσθαι τὸ μὲν διὰ πασῶν τῷ νοερῷ—πλεῖστον γὰρ ἐν ἑκατέρῳ τὸ ἁπλοῦν καὶ ἴσον καὶ ἀδιάφορον—τὸ δὲ διὰ πέντε τῷ αἰσθητικῷ, τὸ δὲ διὰ τεσσάρων τῷ ἑκτικῷ.

39. See Long, *Stoic Studies*, 230.

40. Ptolemy, *Harmonics* 3.5, D96: Καὶ μὴν τοῦ μὲν ἑκτικοῦ τῆς ψυχῆς τρία τις ἂν εἴποι τὰ εἴδη τοῖς τοῦ διὰ τεσσάρων ἰσάριθμα, τό τε κατὰ τὴν αὔξησιν καὶ τὴν ἀκμὴν καὶ τὴν φθίσιν—αὗται γὰρ αὐτοῦ πρῶται δυνάμεις

state, is Stoic, the species are Aristotelian, for Aristotle lists growth, maturity, and decline as definitional aspects of living beings in *De anima* I, 411a30,[41] and III, 434a24–25. In the latter passage, he declares, "Everything that lives and has a soul, then, must have the nutritive soul, from birth until death; for what has been born must have growth, maturity, and decline, and these are impossible without nourishment . . ."[42] Although Ptolemy does not utilize Aristotle's term for the most basic capacity of the soul, the nutritive soul (θρεπτικὴν ψυχήν), he appropriates Aristotle's description of it for the species of the part that maintains a state.

Corresponding to the octave, the intellectual part of the soul (νοερόν) has at most seven species, which Ptolemy lists as the following: *phantasia*,[43] intellect (νοῦς), conception (ἔννοια), thought (διάνοια), opinion (δόξα), *logos*, and knowledge (ἐπιστήμη). Although the account of the cognitive faculty of the soul in *On the Kritêrion* is more complex—in that thought is the principal movement of the faculty of thought, and opinion and knowledge are capacities of this faculty—in the *Harmonics* thought, opinion, and knowledge are simply species of the intellectual part of the soul. A more striking difference between *On the Kritêrion* and *Harmonics* 3.5 is their contradictory definitions of *phantasia*. Again, in *On the Kritêrion* Ptolemy defines *phantasia* as the medium that transmits sense impressions to the intellect. Albeit a medium, *phantasia* belongs solely to the faculty of sense perception and not to the faculty of thought. In *Harmonics* 3.5, however, Ptolemy defines *phantasia* as a species of the intellectual part of the soul, rather than the perceptive part as one might expect. Hence, *On the Kritêrion* and *Harmonics* 3.5 are in direct contradiction concerning the ascription of *phantasia* to a component of the soul, perceptive or cognitive.

I argue that Ptolemy lists *phantasia* as an intellectual, rather than a perceptive, species in the *Harmonics* because of the confines of the harmonic model and, in particular, the restriction to list a certain number of species for each of the soul's parts. To begin with, the seven species Ptolemy ascribes to the soul's intellectual part are components of the criterion of truth in *On the*

41. Boll identifies this passage as a source for Ptolemy's three species of the part of the soul that maintains a state: Boll, "Studien," 104. See also Düring, *Ptolemaios und Porphyrios*, 271; Barker, *Greek Musical Writings*, 375, n. 42.

42. Aristotle, *De anima* III, 434a22–25, translation modified from Hamlyn: Τὴν μὲν οὖν θρεπτικὴν ψυχὴν ἀνάγκη πᾶν ἔχειν ὅτι περ ἂν ζῇ καὶ ψυχὴν ἔχῃ, ἀπὸ γενέσεως καὶ μέχρι φθορᾶς· ἀνάγκη γὰρ τὸ γενόμενον αὔξησιν ἔχειν καὶ ἀκμὴν καὶ φθίσιν, ταῦτα δ' ἄνευ τροφῆς ἀδύνατον

43. See chapter 4 for a discussion of Ptolemy's conception of *phantasia*.

Kritêrion.[44] Providing a more detailed exposition than the initial list, examined above, Ptolemy enumerates the criterion's components: the objects of sense perceptions, the attributes of objects, the sense organs, sense perception, *phantasia*, intellect (νοῦς), conception (ἔννοια), *logos*, internal and thinking *logos* (ἐνδιάθετος καὶ διανοητικός), speech (διαλεκτικός), opinion (δόξα), knowledge (ἐπιστήμη), and truth (ἀλήθεια).

The seven species of the soul's intellectual part in *Harmonics* 3.5 correspond to seven of the criterion's components, including *phantasia* and six of the seven components associated with the faculty of thought. One might wonder why Ptolemy does not include speech as a species of the intellectual part of the soul, since it is the only cognitive component of the criterion that Ptolemy omits as an intellectual species. By listing speech rather than *phantasia* as an intellectual species, Ptolemy could have met the requirements of his harmonic model—including the necessity to list seven species of the intellectual part of the soul—while enumerating intellectual species that are indisputably cognitive. In addition, had Ptolemy listed speech rather than *phantasia* as a species of the soul's intellectual part, he could have reserved *phantasia* for the soul's perceptive part and thereby maintained its association with sense perception. Nevertheless, Ptolemy may have had sufficient reason to omit speech. As noted above, he argues in *On the Kritêrion* that speech makes no contribution to the process of judgment. As a repetition of thought, like an image to an original, it may not deserve classification as a species. Therefore, Ptolemy may have omitted speech deliberately and reasonably, but why did he allocate *phantasia* to the soul's intellectual part? If Ptolemy intended to maintain consistency between his criterion of truth and the soul's intellectual species, it is possible that once he eliminated speech he had no choice but to include *phantasia* as an intellectual species. After all, it seems that *phantasia* was the best candidate for the seventh intellectual species, for, of the components of the criterion associated with sense perception, it is the most closely related to cognitive function.

Moreover, Ptolemy's perceptive part of the soul could not accommodate *phantasia*. While the soul's intellectual part has seven species, the perceptive part has only four. Ptolemy explains, "the perceptive part has four [species], equal in number to those of the concord of the fifth, related respectively to sight, hearing, smell, and taste, if we regard the sense of touch as being common

---

44. Boll observes that Ptolemy lists these seven terms as components of his criterion of truth in *On the Kritêrion and Hêgemonikon*: Boll, "Studien," 105. Cf. Long, "Ptolemy on the Criterion," 170; Barker, "Ptolemy's Musical Models," 287.

to all, since it is by touching the perceptibles in some way or another that they produce our perceptions of them . . ."[45] The idea that touch need not be counted as a sense because the four senses have it in common is inconsistent with Ptolemy's description of the senses in *On the Kritêrion*, where he depicts touch accordingly: "of the senses, touch is more material and extends over the whole of the flesh and blood of the body, but the others [extend] only to the more easily penetrable and more moist [parts of the body] . . ."[46] While touch is unlike the other four senses in that it extends through the body's flesh and blood, Ptolemy still calls it a sense alongside the other four.

Andrew Barker has suggested a potential Stoic and/or Epicurean influence on the characterization of the senses in the *Harmonics*,[47] but, notwithstanding the corporealism of the Stoics, none argued that four of the senses are reducible to touch and, while the Epicureans may have advanced this position, Ptolemy generally incorporates few Epicurean concepts into his philosophical system. More likely is the explanation Franz Boll propounds. He finds a precedent for Ptolemy's association of sense perception with the number four in Pythagorean number symbolism, which drove Platonists, such as Theon of Smyrna in the early second century, to claim that touch is common to the four senses.[48] Theon states, "sense perception exists in four divisions; all the senses operate according to touch, since touch is common to all [the senses] in a fourfold manner."[49] Ptolemy appropriates this idea in order to justify his fitting of the perceptive faculty to a harmonic model. In order to maintain a strict correspondence between the soul's perceptive part and the concord of the fifth, Ptolemy portrays the perceptive part as having only four species. Classifying sight, hearing, taste, and smell as perceptive species, he could not add another, fifth, species for touch or, for that matter, *phantasia*. Consequently, Ptolemy claims that touch need not count as a species if one regards

---

45. Ptolemy, *Harmonics* 3.5, D96: τοῦ δὲ αἰσθητικοῦ τέσσαρα τοῖς τῆς διὰ πέντε συμφωνίας ἰσάριθμα, τό τε κατὰ τὴν ὄψιν καὶ τὴν ἀκοὴν καὶ τὴν ὄσφρησιν καὶ τὴν γεῦσιν, εἰ τὸ τῆς ἀφῆς ὥσπερ ἐπίκοινον θείημεν ἁπασῶν, ἐπεὶ τῷ ἄπτεσθαι τῶν αἰσθητῶν ὁπωσοῦν ποιοῦνται τὰς ἀντιλήψεις αὐτῶν

46. Ptolemy, *On the Kritêrion*, La20: τῶν αἰσθήσεων τὴν μὲν ἀφὴν ὑλικωτέραν οὖσαν παρ' ὅλον τὸ σαρκῶδες καὶ ἔναιμον τοῦ σώματος διατετάσθαι, τὰς δ' ἄλλας παρὰ μόνα τὰ εὐχωρότερα καὶ ὑγρότερα

47. Barker, *Greek Musical Writings*, 375, n. 43.

48. Boll, "Studien," 105.

49. Theon of Smyrna, *On Mathematics Useful for the Understanding of Plato* 98.4–7: ἡ δὲ αἴσθησις ὡς τετράς, ἐπειδὴ τετραπλῆ κοινῆς πασῶν οὔσης τῆς ἀφῆς κατ' ἐπαφὴν πᾶσαι ἐνεργοῦσιν αἱ αἰσθήσεις.

it as common to the four senses, and he classifies *phantasia* as an intellectual rather than a perceptive species of the soul.

The contradictory accounts of *phantasia* and touch in *On the Kritêrion* and *Harmonics* 3.5 follow from Ptolemy's commitment only in the latter to a restrictive, harmonic model, which dictates an exact number of species for each of the soul's parts. When Ptolemy mapped the harmonic model onto the psychological phenomena, *phantasia* and all five of the senses did not fit. In order to list seven intellectual species, Ptolemy chose to characterize *phantasia* as an intellectual rather than a perceptive species of the soul, and in order to list four species of the soul's perceptive part he claimed that touch is common to the four senses.

The Development of Ptolemy's Scientific Method

I argue that this shift in Ptolemy's psychological theory illuminates the manner in which Ptolemy developed his mature scientific method. Central to my argument is the chronology of the texts. The dates of composition of *On the Kritêrion* and the *Harmonics*, unfortunately, are unknown. The latter generally is considered an early text of Ptolemy because of the relation of *Harmonics* 3.14–16 to the *Canobic Inscription*.[50] Though lost, these chapters of the *Harmonics* apparently examined the relations between musical pitches and celestial bodies tabulated in the *Canobic Inscription*,[51] which, in turn, is believed to predate the *Almagest* because it contains numerical values that Ptolemy corrects in the *Almagest*.[52] *On the Kritêrion* is considered one of the earliest—perhaps the earliest—of Ptolemy's extant texts, and the most persuasive evidence for its early dating appeals to its relation to the *Harmonics*. It is more likely that Ptolemy wrote *On the Kritêrion* before the *Harmonics*, because (1) the criterion of truth presented in the *Harmonics* is indisputably more developed than the criterion in *On the Kritêrion*; and (2) Ptolemy merely summarizes his psychological accounts in the *Harmonics*, but he deliberates on the soul's nature and structure in *On the Kritêrion*.[53]

50. Byzantine scholars characterized the *Harmonics* as Ptolemy's final work, incomplete at the time of his death. See Boll, "Studien," 65. Cf. Wilson, *Scholars of Byzantium*, 267; Solomon, Ptolemy "Harmonics," xxx.

51. Swerdlow, "Ptolemy's Harmonics," 175.

52. See Hamilton, Swerdlow, and Toomer, "Canobic Inscription"; A. Jones, "Ptolemy's Canobic Inscription."

53. See Boll, "Studien," 105. A. A. Long concurs that the psychological exposition in *On the Kritêrion* seems to predate the *Harmonics*: Long, "Ptolemy on the Criterion," 170.

In spite of this underdetermined chronology, if one aims to offer a cogent explanation of the texts' incongruities, one must accept that Ptolemy composed *On the Kritêrion* before *Harmonics* 3.5. The opposite supposition would preclude any rigorous explanation for why Ptolemy does not ascribe *phantasia* to the faculty of thought or describe only four senses in *On the Kritêrion*. Unlike Ptolemy's harmonic model of the soul, the psychological account in *On the Kritêrion* is sufficiently unrestrictive that it could accommodate the classification of *phantasia* as cognitive and the elimination of touch as a sense, and these slight alterations to the existing text would have the added benefit of maintaining the consistency of *On the Kritêrion* with the *Harmonics*.

Comparison of the scientific methods Ptolemy employs in these two texts with the method he exercises in the remainder of his corpus provides additional evidence for this order of composition. Again, *On the Kritêrion* is the only text of Ptolemy devoid of mathematics, but in the *Harmonics* he applies a branch of mathematics, harmonics, to a physical science, psychology, and he applies mathematics to physics consistently in the rest of his corpus. As we saw in chapter 3, Ptolemy articulates and justifies the contribution of mathematics to physics in *Almagest* 1.1, and he applies geometry to element theory in several of his texts, including *Almagest* 1.1 and 1.7, *Planetary Hypotheses* 2.3, and his lost works *On the Elements* and *On Weights*. As we will see in the following chapter, Ptolemy applies astronomy to astrology—the science that studies and predicts physical changes in the sublunary realm caused by the powers emanating from celestial bodies—in the *Tetrabiblos,* and astronomy to cosmology in the *Planetary Hypotheses*.

With the exception of *On the Kritêrion*, Ptolemy consistently applies mathematics to physics in his natural philosophical investigations. Why, then, does he not apply mathematics to psychology in *On the Kritêrion*? I suggest that the only plausible explanation rests with the text's early composition. When Ptolemy composed *On the Kritêrion*, he had not yet formulated his mature scientific method, and the shift in psychological theory from *On the Kritêrion* to the *Harmonics* marks the point at which he devised it. When composing *On the Kritêrion*, Ptolemy demonstrated his commitment to his criterion of truth—the collaboration of reason and perception—but it was only after completing *On the Kritêrion*, but before completing the *Harmonics*, that he advanced the position that physics requires the contribution of mathematics. Thereafter, Ptolemy put this claim into practice by employing mathematics in each and every one of his natural philosophical investigations, including the psychological chapters of the *Harmonics*. Despite the contradiction of Ptolemy's description of certain psychological phenomena in the *Harmonics* with their earlier account in *On the Kritêrion,* by applying harmonics to psychology Ptolemy established his psychological theory on what he considered to be firmer epistemic ground. In this way, the psychological account in the *Harmonics* stands as an improvement on the corresponding account in *On the Kritêrion and Hêgemonikon*.

The Structure of the Soul in the *Harmonics*: The Platonic and Synthetic Accounts

Ptolemy presents not only the Aristotelian account of the soul in *Harmonics* 3.5 but also two more, alternative, accounts. As discussed in chapter 4, the second of the three portrays the soul as encompassing the following three parts: the rational (λογιστικόν), spirited (θυμικόν), and appetitive (ἐπιθυμητικόν). Ptolemy undoubtedly appropriated these terms from the Platonic tradition, as Socrates's account of the soul in *Republic* IV includes a rational (λογιστικόν), spirited (θυμοειδές), and appetitive part (ἐπιθυμητικόν).[54] Rather than employing Plato's term for the spirited part, *thumoeides*, Ptolemy follows the contemporary Platonic tradition in using the term *thumikon*, the same term Alcinous uses in *Didaskalikos* 17.4. While in the Aristotelian account Ptolemy derives the species of the parts of the soul from his criterion of truth, the senses, and Aristotle's account of the nutritive faculty of the soul, in the Platonic account the species are virtues, which I examined in chapter 4.

After delineating the species of virtue, Ptolemy comments on how the three parts of the soul relate. Addressing both his Aristotelian and Platonic accounts, he asserts, "so also in souls it is natural for the intellectual and rational parts to govern the others, which are subordinate, and they [i.e., the former] need greater accuracy in the imposition of correct ratio, since they are themselves responsible for the whole or the greater part of any error among the others."[55] Hence, the intellectual and rational parts of the soul—in the Aristotelian and Platonic accounts, respectively—govern the other two, subordinate, parts. Andrew Barker has noted that the idea of one part or faculty of the soul governing other parts was common in the Platonic tradition—as in *Republic* IV— as well as in the Aristotelian and Stoic traditions.[56] *Republic* IV no doubt also influenced, either directly or indirectly, Ptolemy's portrayal of the best condition of the soul, a concord among the soul's parts. Ptolemy explains, "The best condition of the soul as a whole, being justice, is as it were a concord between

---

54. Plato, *Republic* IV, 440e–441a.

55. Ptolemy, *Harmonics* 3.5, D97, trans. Barker, in *Greek Musical Writings*: οὕτω κἀν ταῖς ψυχαῖς ἄρχειν μὲν πέφυκε τὰ νοητικὰ καὶ λογιστικὰ μέρη τῶν λοιπῶν καὶ ὑποτεταγμένων, ἀκριβείας δὲ πλείονος δεῖται πρὸς τὸ κατὰ λόγον, ὡς καὶ τῆς ἐν ἐκείνοις ἁμαρτίας τὸ πᾶν ἢ τὸ πλεῖστον ἔχοντα παρ' ἑαυτοῖς.

56. See Plato, *Republic* IV, 441e; Barker, *Greek Musical Writings*, 377, n. 47.

the parts themselves in their relations to one another in correspondence with the ratio governing the principal parts . . ."[57] In *Republic* IV, Socrates defines justice as a relation among the parts of the soul: "Then, isn't to produce justice to establish the parts of the soul in a natural relation of control, one by another, while [to produce] injustice is to establish a relation of ruling and being ruled contrary to nature?"[58]

It is in the context of relating the best condition of the soul—a concord among its parts, justice—that Ptolemy presents his third, synthetic account. The soul is again tripartite, and it consists of (1) the part concerned with goodwill (εὔνοια) and right reckoning, or rationality (εὐλογιστία); (2) the part concerned with good perception (εὐαισθησία) and good health (εὐεξία), or, alternatively, courage (ἀνδρεία) and moderation (σωφροσύνη); (3) the part concerned with "the things that can produce and the things that participate in *harmoniai*" (τὰ ποιητικὰ καὶ τὰ μετέχοντα τῶν ἁρμονιῶν).[59] The terms Ptolemy employs for the three parts of the soul allude to the Aristotelian and Platonic accounts. With respect to the first part of the soul, the term "goodwill" (εὔνοια) implies some relation with the intellectual part of the soul (νοερόν) in the Aristotelian account. Likewise, the term "right reckoning" (εὐλογιστία) relates to the rational part of the soul (λογιστικόν) in the Platonic account. Concerning the second part of the soul, "good perception" (εὐαισθησία) refers to the perceptive part of the soul (αἰσθητικόν) in the Aristotelian account, and "good health" (εὐεξία) may refer to the health of the body in sustaining sense perception. Ptolemy provides an alternative description for this second part as concerning courage (ἀνδρεία) and moderation (σωφροσύνη). In his Platonic account, as we saw in chapter 4, Ptolemy portrays courage as a virtue of the spirited part of the soul and moderation as a virtue of the appetitive part. Correspondingly, in *Didaskalikos* 29.1 Alcinous portrays courage as the perfection of the spirited part of the soul and moderation as the perfection of the appetitive part. Ptolemy does not construct a hierarchy among the species of virtue, but, by highlighting courage and moderation in his third, synthetic, account, he implies that courage and moderation, as in the

57. Ptolemy, *Harmonics* 3.5, D97: καὶ ὅλως ἡ κρατίστη τῆς ψυχῆς διάθεσις, οὖσα δὲ δικαιοσύνη, συμφωνία τίς ἐστιν ὥσπερ τῶν μερῶν αὐτῶν πρὸς ἄλληλα κατὰ τὸν ἐπὶ τῶν κυριωτέρων προηγούμενον λόγον

58. Plato, *Republic* IV, 444d, translation modified from Grube, rev. Reeve: Οὐκοῦν αὖ, ἔφην, τὸ δικαιοσύνην ἐμποιεῖν τὰ ἐν τῇ ψυχῇ κατὰ φύσιν καθιστάναι κρατεῖν τε καὶ κρατεῖσθαι ὑπ' ἀλλήλων, τὸ δὲ ἀδικίαν παρὰ φύσιν ἄρχειν τε καὶ ἄρχεσθαι ἄλλο ὑπ' ἄλλου;

59. Ptolemy, *Harmonics* 3.5, D97. For another interpretation of how this account of the soul relates to the other two Ptolemy presents in *Harmonics* 3.5, see Barker, *Greek Musical Writings*, 377, n. 49.

contemporary Platonic tradition, are the principal virtues of the spirited and appetitive parts, respectively.

The third part of the soul in the synthetic account—which concerns the things that can produce and the things that participate in the *harmoniai*—may relate to the part of the soul that maintains a state (ἑκτικόν) in the Aristotelian account, as the latter is the part that preserves the form of an object or living being, whether or not its form consists of harmonic ratios. Ptolemy does not allude with this third part to the appetitive part in the Platonic account, for he already has brought the appetitive part into association with the spirited part in the synthetic account's second part of the soul. The integration of the spirited and appetitive parts is not surprising, as Ptolemy also joins these two parts when he deconstructs the faculty of impulse (ὁρμητικόν) in *On the Kritêrion*. Moreover, the combination of the spirited and appetitive parts into a single part in a bipartite model—consisting of a rational and an irrational part—was a common practice from the early Academy onward.[60] Ptolemy, however, does not join the spirited and appetitive parts into a single part in a bipartite model, but rather in a tripartite model, not only here—in *Harmonics* 3.5—but also in *On the Kritêrion* and the *Tetrabiblos*.[61]

Having enumerated the parts of the soul in his synthetic account, Ptolemy elaborates on the analogy between the harmonic relations in music and the relations among the soul's parts:

> the whole condition of a philosopher is like the whole *harmonia* of the complete *systêma*, comparisons between them, part by part, being made by reference to the concords and the virtues, while the most complete comparison is made by reference to what is, as it were, a concord of melodic concords and a virtue of the soul's virtues, constituted out of all the concords and all the virtues.[62]

The whole *harmonia* of the complete *systêma* of music consists in a concord of concords, and in a human being it is a virtue of the soul's virtues. It is only in the Platonic account of the soul that Ptolemy examines the virtues, and so whereas the synthetic account draws on both the Aristotelian and Platonic accounts for its terms, Ptolemy's characterization of the best condition of the soul appropriates the terms of the Platonic account alone. The best condition

---

60. See Rees, "Bipartition of the Soul."

61. Ptolemy, *On the Kritêrion*, La21; *Tetrabiblos* 3.14.1–2, H248–49.

62. Ptolemy, *Harmonics* 3.5, D97–98: ὅλης δὲ τῆς φιλοσόφου διαθέσεως ὅλῃ τῇ τοῦ τελείου συστήματος ἁρμονίᾳ, τῶν μὲν ἐπὶ μέρους παραβολῶν τασσομένων κατά τε τὰς συμφωνίας αὐτὰς καὶ τὰς ἀρετάς, τῆς δὲ τελειοτάτης κατὰ τὴν συνισταμένην ἐκ πασῶν τῶν συμφωνιῶν καὶ πασῶν τῶν ἀρετῶν, συμφωνίαν τινὰ καὶ ἀρετὴν ὥσπερ ἀρετῶν καὶ συμφωνιῶν, μελῳδικῶν τε καὶ ψυχικῶν.

of the soul is a concord among its parts, a relation among its species, which is a virtue of virtues, or justice.

After delineating the soul's parts and species in *Harmonics* 3.5, Ptolemy examines the genera in *Harmonics* 3.6. These genera are not of the soul's parts but of its species and, in particular, its virtues. The chapter is entitled "A comparison between the genera of attunement and the [genera] in relation to the primary virtues."[63] The principles of the genera are the theoretical and the practical, and Ptolemy divides each into three genera: the theoretical into the physical, mathematical, and theological, and the practical into the ethical, domestic, and political. He explains that for each principle the genera do not differ in capacity, for the same virtues apply in each genus: "these do not differ from one another in capacity, for the virtues of the three genera are shared and dependent on one another, but they do differ in magnitude and value and in the compass of their structure..."[64] In other words, when studying any of the theoretical sciences, for example, the soul takes on the same virtues. Ptolemy is unclear on whether all of the virtues obtain in all six genera, the theoretical and the practical, or whether the virtues in the theoretical domain are distinct from the ones in the practical. It may be the latter, since in *Harmonics* 3.5 Ptolemy defines wisdom as the virtue having to do with the theoretical and prudence as the virtue having to do with the practical. At least these two virtues are differentiated by their relation to distinct principles. Nevertheless, Ptolemy is concerned no longer with the individual species of the soul but, rather, their application in external situations, whether their exercise in the six genera, here in *Harmonics* 3.6, or in the crises of life in *Harmonics* 3.7.

Conclusion

With the development of his psychological theory from *On the Kritêrion and Hêgemonikon* to the *Harmonics*, Ptolemy sought to improve the study of the soul by mathematizing it. According to the *Harmonics*, the soul contains the same harmonic relations as exist among the movements of heavenly bodies and musical pitches. In his selection of terms for the faculties, parts, and species of the soul, Ptolemy reveals a particular attitude toward philosophical traditions. He does not appropriate terms indiscriminately from among the Aristotelian, Platonic, and, to a lesser extent, Stoic schools; rather, he selects terms in accordance with the themes of his texts. In *On the Kritêrion*, with its

---

63. Ptolemy, *Harmonics* 3.6, D98: Παραβολὴ τῶν τε τοῦ ἡρμοσμένου γενῶν καὶ τῶν κατὰ τοὺς πρώτας ἀρετάς.

64. Ibid.: τῇ μὲν δυνάμει τούτων μὴ διαφερόντων—κοιναὶ γὰρ αἱ τῶν τριῶν γενῶν ἀρεταὶ καὶ ἀλλήλων ἐχόμεναι—μεγέθει δὲ καὶ ἀξίᾳ καὶ τῇ περιβολῇ τῆς κατασκευῆς

emphasis on the perceptibility of the effects the soul causes in and through the body, Ptolemy employs Aristotelian terminology consistent with his dually empirical and rational criterion. In the *Harmonics*, he utilizes Platonic terminology to describe the parts and species not only in music but also in the human soul.

Moreover, in *Harmonics* 3.5 Ptolemy presents three alternative accounts of the soul, one that is primarily Aristotelian, another Platonic, and the third synthetic of the previous two. The fact that Ptolemy does not choose from among alternative accounts reveals his distinctly eclectic attitude. He does not adhere exclusively to either the Aristotelian or the Platonic tradition, and he here indicates that he has no desire to choose between them. These traditions provided him with alternative terminology to appropriate and blend as he saw fit. That in the *Harmonics* Ptolemy does not choose between alternative accounts of the soul likely also follows from his conviction that physics, even with the contribution of mathematics, remains conjectural. Ptolemy could not reliably choose from among the accounts and so he did not. That Ptolemy maintains this epistemic attitude toward the physical sciences not only in the *Almagest* and *Harmonics* but also in the *Tetrabiblos* and *Planetary Hypotheses*, and that he employs the scientific method of the *Harmonics'* psychological chapters in his studies of astrology and cosmology, we shall see in the following chapter.

8

Astrology and Cosmology

AFTER PTOLEMY composed *On the Kritêrion and Hêgemonikon*, he developed his mature scientific method. In addition to his criterion of truth, he mandated the application of mathematics to physics, a method he practices in every one of his subsequent natural philosophical studies, including the astrology and cosmology of the *Tetrabiblos* and *Planetary Hypotheses*, respectively. Astrology and cosmology are physical sciences. The former analyzes and predicts the physical effects stars' movements and configurations have on sublunary bodies and souls. The latter is the study of the physical properties of the heavenly bodies themselves. Both Ptolemy's astrology and his cosmology rely on astronomy. Indeed, Ptolemy suggests that their very study depends on an antecedent and complete examination of the stars' movements and configurations. The conclusions astrology and cosmology put forward, however, remain conjectural, and Ptolemy remarks in both the *Tetrabiblos* and *Planetary Hypotheses* on the conjectural nature of astrology, cosmology, and physics in general. Nevertheless, by placing his astrology and cosmology on an astronomical foundation, Ptolemy endeavors to produce the best guesses possible of the physical characteristics of aethereal bodies and the physical effects they have on the sublunary realm.

Astrological Conjecture

When Ptolemy defines astrology in the *Tetrabiblos*, he juxtaposes it with astronomy. He may have found it necessary to do so, for he does not use different terms for them as we do today. In the ancient Greek world, the terms *astronomia* and *astrologia* were used interchangeably, such that either one could denote astronomy and/or astrology.[1] According to the *Tetrabiblos*,

1. Cf. Ptolemy, *On the Kritêrion*, La11, where Ptolemy declares that, rather than analyzing the applicability of the terms "soul" and "body," he prefers to examine the actual differences between them. He takes a similar approach here, where, instead of discussing what to call

168

astrology and astronomy are both fields of inquiry with a predictive goal, and they procure this goal by means of *astronomia*. Ptolemy uses the term *astronomia* six times in the *Tetrabiblos*—but, rather surprisingly, never in the *Almagest*—and he consistently uses the syntax that appears in the first line of the text. The phrase *di' astronomias*, "by means of *astronomia*," characterizes a type of prognostication.[2] The term *astronomia* occurs in one other text of Ptolemy. In *Harmonics* 3.3, he juxtaposes astronomy with harmonics, *astronomia* with *harmonikê*.[3] Here, Ptolemy indicates that astronomy is one of the most rational of the sciences, the one that employs geometry as an indisputable instrument to study the quantity and quality of the movements from place to place of the bodies that are only visible, or celestial bodies (on the quality of movements, see chapter 2). In the *Tetrabiblos*, the term *astronomia* carries a similar meaning, and it serves as a means of procuring a prognostic goal.

Even though the same term denotes both astrology and astronomy in Ptolemy's corpus, the two sciences remain distinct. According to Ptolemy, astrology and astronomy use *astronomia* to different ends. Astrology employs *astronomia* to predict the qualitative changes produced in the sublunary realm from the stars' configurations, which result from their movements. Astronomy, on the other hand, uses *astronomia* to predict the stars' movements and configurations themselves. I suggest that it is astronomy's essentially predictive nature that mandates the study of astronomical models' quantitative aspects, which, as we saw in chapter 6, can be apprehended only approximately.

Despite astronomy's epistemic weaknesses, in the *Tetrabiblos* Ptolemy takes astronomy to be prior to astrology in both order (τάξις) and power (δύναμις). He bases this claim on a comparison of their methods and epistemic capabilities. Ptolemy's description of astronomy is consistent with his characterization of the mathematical sciences in *Almagest* 1.1, where he asserts that movements are the subject matter of mathematics. In the *Tetrabiblos*, astronomy predicts the configurations resulting from these movements. Astrology, on the other hand, is a physical science. Celestial bodies cause physical changes in the sublunary realm, and it is these physical changes that astrology aims to predict. Astronomy is prior to astrology, then, because, as branches of mathematics and physics, respectively, they have different claims to truth. In *Almagest* 1.1, Ptolemy claims that mathematics yields sure and incontrovertible knowledge but physics is conjectural, and in *Tetrabiblos* 1.1 he characterizes the kind of

---

astronomy and astrology, he simply characterizes them differently. On the relation between the terms "astronomy" and "astrology" in antiquity, see Hübner, *Begriffe "Astrologie" und "Astronomie."*

2. Ptolemy, *Tetrabiblos* 1.1.1, H3; 1.2.T, H5; 1.3.1, H14; 1.3.18, H21; 1.3.19, H21; 2.1.2, H88.

3. Ptolemy, *Harmonics* 3.3, D94.

apprehension that astronomy lays claim to as sure, but the claims of astrology as merely possible. Ptolemy makes this epistemic distinction when discussing the manner in which he will present his astrology:

> We shall now give an account concerning the second and not self-sufficient [means of procuring the prognostic goal; viz. astrology] in a way that is in tune with philosophy, so that one who above all has as an aim a love of truth may not compare its [kind of] apprehension with the sureness of the first and ever unchanging [viz. astronomy]—which does not lay claim to the weakness and indiscernibility of material quality that exists in many things—nor shrink from the investigation that is in the realm of possibility, when the greatest and more general occurrences clearly show their cause to be from that which encompasses [the earth; i.e., the ambient].[4]

Ptolemy distinguishes astronomy and astrology on three accounts. First, astronomy is self-sufficient and astrology is not.[5] Astrology studies qualitative changes in the sublunary realm caused by the superlunary movements and configurations studied by astronomy. Therefore, astrology is dependent on astronomy, but astronomy does not rely on astrology or any other science. Even geometry, which astronomy does employ, is not a science; it is a method or instrument of mathematical sciences like astronomy. Hence, astronomy is an independent science, and astrology is dependent. Second, Ptolemy sets out to present his astrology in a philosophical way, as opposed to the demonstrative way in which, as he explains earlier in the *Tetrabiblos*, he already has set forth his astronomy in the *Almagest*.[6] By proposing that the proper method of

---

4. Ptolemy, *Tetrabiblos* 1.1.2, H4: περὶ δὲ τοῦ δευτέρου καὶ μὴ ὡσαύτως αὐτοτελοῦς ἡμεῖς ἐν τῷ παρόντι ποιησόμεθα λόγον κατὰ τὸν ἁρμόζοντα φιλοσοφίᾳ τρόπον καὶ ὡς ἄν τις φιλαλήθει μάλιστα χρώμενος σκοπῷ μήτε τὴν κατάληψιν αὐτοῦ παραβάλλοι τῇ τοῦ πρώτου καὶ ἀεὶ ὡσαύτως ἔχοντος βεβαιότητι, τὸ ἐν πολλοῖς ἀσθενὲς καὶ δυσείκαστον τῆς ὑλικῆς ποιότητος μὴ προσποιούμενος, μήτε πρὸς τὴν κατὰ τὸ ἐνδεχόμενον ἐπίσκεψιν ἀποκνοίη, τῶν γε πλείστων καὶ ὁλοσχερεστέρων συμπτωμάτων ἐναργῶς οὕτως τὴν ἀπὸ τοῦ περιέχοντος αἰτίαν ἐμφανιζόντων. A textual problem affects how one may interpret Ptolemy's attitude toward the weakness and indiscernibility of material quality. According to the alternative reading—ἐπίπροσθεν ποιούμενος instead of μὴ προσποιούμενος—Ptolemy gives prior place to this aspect of material quality when expounding it in a philosophical way. Franz Boll notes that the discussion of physics in this passage recalls its description in *Almagest* 1.1: Boll, "Studien," 139.

5. Cf. Ptolemy, *Geography* 1.2.2, where he calls astronomical observation (μετεωροσκοπικόν) self-sufficient in contrast to land surveying.

6. Galen makes a similar distinction in *De animi cuiuslibet peccatorum dignotione et curatione*, 102. He contrasts the demonstrative method (λόγον ἀποδεικτικόν) that mathematicians use with philosophers' arguments, which are at most possible and likely (ἐνδεχομένους τε καὶ εἰκότας). Cf. ibid., 93.

astrology is philosophical, Ptolemy does not indicate, as one might suppose, that astrology is not empirical. After all, Ptolemy is not referring to the method of discovery but, rather, the method of presentation. The demonstrative method is appropriate to astronomy, because the geometrical demonstration, which the rigorous study of astronomy can and does employ, is indisputable. As a physical science, astrology cannot proceed by way of a demonstration. At best, its style of exposition is philosophical. These methods of presentation relate to the epistemic success achieved by each science, and so Ptolemy indicates that, third, as astronomy is a mathematical science and astrology is a physical science, the former furnishes judgments that are sure while the latter yields judgments that are merely possible. Given the epistemic insecurity of the quantitative aspects of Ptolemy's astronomy, one might wonder whether this distinction truly applies. Yet, even though Ptolemy reveals in the *Almagest* that the parameters and periods of the heavenly spheres are known only approximately, he does not deem them conjectural. Astrology, on the other hand, does not produce judgments that approximate the truth. Its judgments are merely possible. They are best guesses, but they could be wrong.

Ptolemy dedicates *Tetrabiblos* 1.2 to proving that the claims of astrology are possible, as opposed to impossible, and in the midst of this argument he refers to astrology as guessing well (εὐστόχως). When examining a hypothetical case in which a man knows the exact periods of the wandering stars, as well as the types of changes they cause in the sublunary realm, Ptolemy asserts the following:

> and if he is capable of determining in relation to such things, both naturally and by guessing well, the peculiar attribute of the quality [resulting] from the combination of all these things, [what is to prevent] him from being able to tell on each of the given occasions the peculiar qualities of the ambient from the condition of the phenomena at the time (such as that it will be warmer or wetter)[7]

Ptolemy adds that any field of inquiry concerned with the quality of matter is conjectural: "For in general, besides the fact that every theory concerning the quality of matter is conjectural [εἰκαστικήν] and not affirmative, and especially one that is mixed together out of many dissimilar things . . ."[8] The adjective

---

7. Ptolemy, *Tetrabiblos* 1.2.10, H8–9: ἱκανὸν δὲ πρὸς τοιαῦτα ὄντα φυσικῶς ἅμα καὶ εὐστόχως ἐκ τῆς συγκρίσεως πάντων τὸ ἴδιον τῆς ποιότητος διαλαβεῖν, ὡς δύνασθαι μὲν ἐφ' ἑκάστου τῶν διδομένων καιρῶν ἐκ τῆς τότε τῶν φαινομένων σχέσεως τὰς τοῦ περιέχοντος ἰδιοτροπίας εἰπεῖν (οἷον ὅτι θερμότερον ἢ ὑγρότερον ἔσται)

8. Ibid., 1.2.15, H10: καθόλου γάρ, πρὸς τῷ τὴν περὶ τὸ ποιὸν τῆς ὕλης θεωρίαν πᾶσαν εἰκαστικὴν εἶναι καὶ οὐ διαβεβαιωτικὴν καὶ μάλιστα τὴν ἐκ πολλῶν ἀνομοίων συγκιρναμένην

eikastikên recalls Ptolemy's use of the noun *eikasia* in *Almagest* 1.1 when labeling physics and theology conjecture.[9] Hence, in the *Tetrabiblos* as in *Almagest* 1.1, Ptolemy judges the claims of astrology, like the judgments of every other
physical science, conjectural, or merely possible.

Ptolemy delineates the several shortcomings that prevent astrology from
producing incontrovertible claims. In addition to the nature of its subject
matter—that is, the changing qualities of the sublunary realm—the primary
cause of these changes, the stars' movements and configurations, proves to be
an obstacle to astrology's epistemic endeavor. Recalling *Almagest* 3.1, Ptolemy
declares that the exact periods of celestial phenomena are indeterminable.
He affirms in *Tetrabiblos* 1.2 that the exact return of all heavenly bodies to an
identical configuration does not occur within the lifetime of human beings:

> and yet the ancient configurations of the wandering stars—on the basis of
> which we apply to the similar [configurations] of the present day the prog-
> nostications observed by our predecessors in theirs—can be more or less
> similar [to the present configurations] and [less similar] after long periods
> of time, and in no way precisely similar, since the precise joint return of
> everything in the heavens with the earth, unless one holds a vain opin-
> ion of his apprehension and knowledge of inapprehensible things, either
> is completed not entirely or not in the period of time that corresponds
> to what is perceptible by man; so, on account of this, the predictions also
> sometimes fail because of the dissimilarities existing among the founda-
> tional examples.[10]

Because the joint return of everything in the heavens has not been witnessed
by astrologers, the foundational examples that establish the correspondence
of celestial configurations with sublunary effects contain disparities and therefore are not completely reliable. Moreover, the exact periods of heavenly phenomena cannot be known. Because astrology depends on these astronomical
data, none of the claims of astrology constitute knowledge.

In addition to the limits placed on astrology by its dependence on astronomy, the preponderance of causes that effect changes in the sublunary realm

9. Ptolemy, *Almagest* 1.1, H6.

10. Ptolemy, *Tetrabiblos* 1.2.15–16, H10–11: ἔτι καὶ τοῖς παλαιοῖς τῶν πλανωμένων συσχηματισμοῖς,
ἀφ' ὧν ἐφαρμόζομεν τοῖς ὡσαύτως ἔχουσι τῶν νῦν τὰς ὑπὸ τῶν προγενεστέρων ἐπ' ἐκείνων
παρατετηρημένας προτελέσεις, παρόμοιοι μὲν [γὰρ] δύνανται γίνεσθαι μᾶλλον ἢ ἧττον καὶ οὗτοι
διὰ μακρῶν περιόδων, ἀπαράλλακτοι δὲ οὐδαμῶς, τῆς πάντων ἐν τῷ οὐρανῷ μετὰ τῆς γῆς κατὰ
τὸ ἀκριβὲς συναποκαταστάσεως, εἰ μή τις κενοδοξοίη περὶ τὴν τῶν ἀκαταλήπτων κατάληψιν καὶ
γνῶσιν, ἢ μηδ' ὅλως ἢ μὴ κατά γε τὸν αἰσθητὸν ἀνθρώπῳ χρόνον ἀπαρτιζομένης, ὡς διὰ τοῦτο καὶ
τὰς προρρήσεις ἀνομοίων ὄντων τῶν ὑποκειμένων παραδειγμάτων ἐνίοτε διαμαρτάνεσθαι.

prevents astrology from producing certain claims. Astrology takes into account only the stars' configurations when making predictions, but atmospheric phenomena as well as genetic, national, and cultural influences have an impact on a seed's development into a living being. Ptolemy attests to the epistemic limitation these causes—in addition to the celestial, or ambient, causes—place on astrology:

> Unless each of these is examined together with the causes from the ambient, even if the latter has the greatest power (the ambient is a contributory cause of these things themselves being such as they are, while not one of them [is a contributory cause] of [the ambient]), they can afford much difficulty for those who believe concerning such things that everything (even things not completely in its [domain]), can be discerned from the movement of the heavenly things alone.[11]

Although many causes contribute to the development of living beings, a hierarchy exists among the causes. The movements of celestial bodies have the greatest influence on sublunary events and, therefore, even though astrology does not account for every cause, because it takes into account the most influential causes, its claims remain possible.

Bounded by the inherent and practical limitations of astrology, the student of astrology still aims to make predictions that are as likely as possible. One way he achieves this goal is by maintaining his theories' conformity with nature. In other words, just as astrology is a physical science, the astrologer aims to follow a method that is *phusikos*. For instance, when discussing annual phenomena, Ptolemy remarks, "It seems more proper and more natural [φυσικώτερον] to me, however, to employ the four starting points for investigations in relation to the annual [phenomena] . . ."[12] Ptolemy rejects the practice of other astrologers and cultures of prioritizing one solstice or equinox over the others in an effort to establish a starting point of the year. He instead maintains that each solstice and equinox has a suitable claim. At the spring equinox, the daylight hours begin to exceed the night and, furthermore, this equinox occurs during the moist season, which signifies birth; the summer solstice has the longest days and, for the Egyptians, it signifies the flooding of

---

11. Ibid. 1.2.19, H12: ὧν ἕκαστον ἐὰν μὴ συνδιαλαμβάνηται ταῖς ἀπὸ τοῦ περιέχοντος αἰτίαις, εἰ καὶ ὅτι μάλιστα τὴν πλείστην ἔχει τοῦτο δύναμιν (τῷ τὸ μὲν περιέχον κἀκείνοις αὐτοῖς εἰς τὸ τοιοῖσδε εἶναι συναίτιον γίνεσθαι, τούτῳ δὲ ἐκεῖνα μηδαμῶς), πολλὴν ἀπορίαν δύνανται παρέχειν τοῖς ἐπὶ τῶν τοιούτων οἰομένοις ἀπὸ μόνης τῆς τῶν μετεώρων κινήσεως πάντα (καὶ τὰ μὴ τέλεον ἐπ᾽ αὐτῇ) δύνασθαι διαγινώσκειν.

12. Ibid. 2.11.5, H149: οἰκειότερον δέ μοι δοκεῖ καὶ φυσικώτερον πρὸς τὰς ἐνιαυσίους ἐπισκέψεις ταῖς τέτταρσιν ἀρχαῖς χρῆσθαι

the Nile as well as the rising of the star Sirius; by the fall equinox, farmers have harvested their crops and begun sowing seeds for future crops; with the winter solstice, the days once again begin to lengthen.

Ptolemy chooses to abide by a principle that Aristotle articulates in the *De caelo*—that a circle has no beginning or end[13]—and he applies this rule to the ecliptic. He asserts that each solsticial and equinoctial point is a natural starting point and no one is prior:

> In a circle, one in general simply can contrive and suppose not any one starting point, but in the [circle] through the middle of the zodiac [viz. the ecliptic], one suitably can take as the only starting points the points marked by the equinoctial [circle] and the tropical [circles], that is, the two equinoxes and the two solstices. Only then, however, one would be puzzled which of the four he should declare as going first. Certainly, in accordance with the simple and circular nature, none of these goes first, as in the case of one starting point[14]

Although in a circle no one point is prior to another, the ecliptic intersects with the celestial equator and therefore it has four natural starting points: the two points where it intersects the celestial equator and the two points where it is furthest from the celestial equator, the equinoctial and solsticial points, respectively. Still, because they are points on a circle, no one of these four points is prior to the others, and so the most natural method for the astrologer is to treat every one as a starting point.

Time and again, Ptolemy articulates a preference for natural theories and methods. When discussing the difficulty in establishing the degree of the horoscopic point, he asserts that the astrologer must use natural reasoning: "it would be necessary to relay in what way one can discover, in accordance with natural and consistent reason, the degree of the zodiac due to rise . . ."[15] When explaining his method for determining the length of life, he declares, "The method that is both especially agreeable to us and, above all, cleaves

---

13. Aristotle, *De caelo* I, 279b2–3.

14. Ptolemy, *Tetrabiblos* 2.11.2, H147–48: τίνα δ' ἄν τις ἀρχὴν ὑποστήσαιτο, ἐν κύκλῳ μὲν αὐτὸ μόνον ἁπλῶς οὐδ' ἂν ἐπινοήσειεν, ἐν δὲ τῷ διὰ μέσων τῶν ζῳδίων μόνας ἂν εἰκότως ἀρχὰς λάβοι τὰ ὑπὸ τοῦ ἰσημερινοῦ καὶ τῶν τροπικῶν ἀφοριζόμενα σημεῖα, τουτέστι τά τε δύο ἰσημερινὰ καὶ τὰ δύο τροπικά. ἐνταῦθα μέντοι τις ἀπορήσειεν ἂν ἤδη, τίνι τῶν τεσσάρων ὡς προηγουμένῳ χρήσαιτ' ἄν. κατὰ μὲν οὖν τὴν ἁπλῆν καὶ κυκλικὴν φύσιν οὐδὲν αὐτῶν ἐστιν ὡς ἐπὶ μιᾶς ἀρχῆς προηγούμενον

15. Ibid. 3.3.2, H173: ἀναγκαῖον ἂν εἴη προπαραδοθῆναι, τίνα ἄν τις τρόπον εὑρίσκοι τὴν ὀφείλουσαν ἀνατέλλειν μοῖραν τοῦ ζῳδιακοῦ κατὰ τὸν φυσικὸν καὶ ἀκόλουθον λόγον

to nature is the following."[16] Ptolemy also rejects methods that he considers unnatural. For instance, he spurns the so-called Chaldean system of Terms, one of the dignities alongside Houses, Triangles, and Exaltations.[17] He repudiates it because it is not logical (ἄλογον), not natural (οὐ φυσικόν), and conceited (κενόδοξον).[18] In each of these cases, Ptolemy affirms his preference for natural theories and methods, and he rejects the unnatural.

Ptolemy refines this position in *Tetrabiblos* 3.1. Recalling his exposition in *Tetrabiblos* 1.2 of the many causes that effect changes in the sublunary realm, he recognizes the impossibility of accounting for every single star:

Since it is our present purpose to supply also this part methodically in accordance with the reasoned principle laid down at the beginning of this composition concerning the possibility of such a prediction, we shall reject—on account of the difficulty both in using it and getting through it—the ancient method of predictions, the one corresponding to the form combining all or most of the stars (since it is manifold and nearly infinite, if one wishes to describe this [method] accurately in detail, and it is examined closely more in the particular attempts of those inquiring naturally than in the traditional powers). The very procedures, through which each of the kinds are comprehended according to the method that is quick to apprehend, and the productive powers of the stars, specifically and generally, in relation to each [of the kinds], we shall expound, as far as possible, both conscientiously and briefly, in accordance with natural conjecture[19]

16. Ibid. 3.11.2, H202: ἔστι δ' ὁ μάλιστά τε συμφωνῶν ἡμῖν καὶ ἄλλως ἐχόμενος φύσεως τρόπος τοιοῦτος.

17. On Terms in the *Tetrabiblos*, see Heilen, "Ptolemy's Doctrine." Heilen, following Neugebauer, takes the system Ptolemy ascribes to the Chaldeans to be Hellenistic in origin rather than Babylonian. See Neugebauer, *History of Ancient Mathematical Astronomy*, vol. 2, 690. See also ibid., 606.

18. Ptolemy, *Tetrabiblos* 1.22.1–2, H82.

19. Ibid. 3.2.5–6, H170–71: προθέσεως δὲ κατὰ τὸ παρὸν ἡμῖν οὔσης καὶ τοῦτο τὸ μέρος ἐφοδικῶς ἀναπληρῶσαι κατὰ τὸν ἐν ἀρχῇ τῆσδε τῆς συντάξεως ὑφηγημένον ἐπιλογισμὸν περὶ τοῦ δυνατοῦ τῆς τοιαύτης προγνώσεως, τὸν μὲν ἀρχαῖον τῶν προρρήσεων τρόπον τὸν κατὰ τὸ συγκρατικὸν εἶδος τῶν ἀστέρων πάντων ἢ τῶν πλείστων (πολύχουν τε ὄντα καὶ σχεδὸν ἄπειρον, εἴ τις αὐτὸν ἀκριβοῦν ἐθέλοι κατὰ τὴν διέξοδον, καὶ μᾶλλον ἐν ταῖς κατὰ μέρος ἐπιβολαῖς τῶν φυσικῶς ἐπισκεπτομένων ἢ ἐν ταῖς παραδόσεσι ἀναθεωρεῖσθαι δυναμένων) παραιτησόμεθα διά τε τὸ δύσχρηστον καὶ τὸ δυσδιέξοδον. τὰς δὲ πραγματείας αὐτάς, δι' ὧν ἕκαστα τῶν εἰδῶν κατὰ τὸν ἐπιβλητικὸν τρόπον συνορᾶται, καὶ τὰς κατὰ τὸ ἰδιότροπον καὶ ὁλοσχερὲς τῶν ἀστέρων πρὸς ἕκαστα ποιητικὰς δυνάμεις, ὡς ἕνι μάλιστα, παρακολουθητικῶς τε ἅμα καὶ ἐπιτετμημένως κατὰ τὸν φυσικὸν στοχασμὸν ἐκθησόμεθα

Although Ptolemy endeavors to utilize natural theories and methods, the practical limitations of astrology prevent him from doing so. The stars and their relations are so many that it is impossible to account for every one. The astrologer must rely on a more limited account of the phenomena, and, rather than employ the natural method, he must adopt the traditional allocation of the stars' powers.

Correspondingly, as Franz Boll observes,[20] Ptolemy treats Aries as the starting point of the zodiacal circle even though, as we saw above, he argues in favor of keeping each of the solstices and equinoxes as a starting point. In *Tetrabiblos* 2.12, he yields to practicability and lists the weather signs of Aries before the signs of the other zodiacal signs. Likewise, in *Almagest* 2.7 he designates Aries as the first zodiacal sign: "we call the first sign, [proceeding] from the spring equinox and toward the rear with respect to the motion of the universe, 'Aries,' the second 'Taurus,' and so on for the rest in accordance with the order handed down to us by tradition of the twelve zodiacal signs."[21] Thus, Ptolemy aspires to the most natural methods and theories but the practice of astrology forces him to admit the limited nature of astrology's claims, employ methods that are not absolutely natural, and aim only at producing predictions that are possible.

Celestial Powers and Rays

After defending astrology's possibility and utility, Ptolemy begins his analysis of astrology in *Tetrabiblos* 1.4.[22] He discusses the stars' powers (δυνάμεις), which he also refers to as productive powers (ποιητικαί) and activities (ἐνέργειαι). The stars' powers are not their essential properties or faculties inasmuch as they are aethereal bodies. Rather, they are the causes, transmitted by the stars' rays, of changes in sublunary bodies and souls. Ptolemy distinguishes between the underlying nature of stars and their powers in *Tetrabiblos* 1.2: "even if [he

20. Boll, "Studien," 166.

21. Ptolemy, *Almagest* 2.7, H117–18: τὸ μὲν ἀπὸ τῆς ἐαρινῆς ἰσημερίας ὡς εἰς τὰ ἑπόμενα τῆς τῶν ὅλων φορᾶς πρῶτον δωδεκατημόριον Κριὸν καλοῦντες, τὸ δὲ δεύτερον Ταῦρον, καὶ ἐπὶ τῶν ἑξῆς ὡσαύτως κατὰ τὴν παραδεδομένην ἡμῖν τάξιν τῶν ιβ̅ ζῳδίων.

22. Jaap Mansfeld has observed that in dedicating the first three chapters of the *Tetrabiblos* to defining astrology, defending its possibility, and examining its utility, Ptolemy follows the tripartite scheme—substance, possibility, and utility—that Albinus outlines for investigating any subject matter: ἀρέσκει δὲ τῷ φιλοσόφῳ περὶ παντὸς οὑτινοσοῦν τὴν σκέψιν ποιούμενον τὴν οὐσίαν τοῦ πράγματος ἐξετάζειν, ἔπειτα τί τοῦτο δύναται καὶ τί μή, πρὸς ὅ τε χρήσιμον πέφυκε καὶ πρὸς ὃ μή (*Introduction to the Dialogues of Plato* 1.5–8). Therefore, Ptolemy follows a Platonic introductory scheme not only in the first chapter of the *Almagest* but also in the first three chapters of the *Tetrabiblos*. See Mansfeld, *Prolegomena mathematica*, 72, n. 250.

may discern] not the actual underlying, but rather their potentially productive [powers] (such as the [productive power] of the sun that heats and the [productive power] of the moon that moistens, and so on for the rest) . . ."[23] The sun itself is not hot, and the moon is not moist, but their powers cause these effects in the sublunary realm, and, just as the sun heats and the moon moistens, every star affects sublunary bodies and souls. Ptolemy appropriates the traditional allocation of the wandering stars' powers, which manifest as two qualities from a set of four—hot, cold, wet, and dry—which are Aristotle's primary contraries in *On Generation and Corruption* and which, before Aristotle, played a role in the Hippocratic theory of the four humors, as in *The Nature of Man*.[24] Ptolemy explains that while the sun heats and dries the sublunary realm, the moon heats and humidifies, because of its proximity to the earth and the exhalations it emits. Jupiter and Venus, like the moon, heat and humidify, but Jupiter predominately heats while Venus chiefly humidifies. Mars, on the other hand, heats and dries, and Saturn dries and cools. In this way, Mars's and Saturn's powers are in opposition to the powers of Jupiter and Venus. Unlike the other wandering stars, Mercury is changeable in its power. Because of its proximity to both the sun and the moon—to the former in longitude and the latter in the order of the aethereal spheres—it both dries and humidifies, heats and cools.

A wandering star's power increases and diminishes as it makes its way through the zodiac and forms relations with other wandering stars and zodiacal signs. The sun's power affects the sublunary realm to the greatest degree, but each star's power plays a part. Ptolemy explains as follows:

> Furthermore, the aspects in their relations to one another, when the influences somehow come together and mix together, cause most complex changes; when the sun's power prevails over the general array of quality, the rest aid or oppose it in some way, and the moon more obviously and continuously—as in its new, quarter, and full [phases]—and the stars at greater intervals and more obscurely, as in their appearances, occultations, and approaches.[25]

23. Ptolemy, *Tetrabiblos* 1.2.10, H8: κἂν μὴ τὰς κατ᾽ αὐτὸ τὸ ὑποκείμενον, ἀλλὰ τάς γε δυνάμει ποιητικάς (οἷον ὡς τὴν τοῦ ἡλίου ὅτι θερμαίνει καὶ τὴν τῆς σελήνης ὅτι ὑγραίνει, καὶ ἐπὶ τῶν λοιπῶν ὁμοίως)

24. See Aristotle, *On Generation and Corruption* II, 329b15–331a6; *De natura hominis* 7: *Corpus Medicorum Graecorum* 1.1.3, 182.4–186.12. *The Nature of Man* is a composite work, but this chapter likely was composed by Polybus, Hippocrates's student and son-in-law.

25. Ptolemy, *Tetrabiblos* 1.2.5, H6–7: ἤδη δὲ καὶ οἱ πρὸς ἀλλήλους αὐτῶν συσχηματισμοί, συνερχομένων πως καὶ συγκιρναμένων τῶν διαδόσεων, πλείστας καὶ ποικίλας μεταβολὰς ἀπεργάζονται, κατακρατούσης μὲν τῆς τοῦ ἡλίου δυνάμεως πρὸς τὸ καθόλου τῆς ποιότητος

The powers of the stars increase and diminish as the wandering stars travel through the zodiac. They augment and curtail one another. Ptolemy summarizes how the stars' powers affect one another in the following:

> Each [of the wandering stars], then, accomplishes such [effects] when it attains its proper nature severally, but, when mixed one with another in accordance with the aspects, the alterations of the zodiacal signs, and the phases in relation to the sun, and in proportion admitting of the tempering in the activities, it produces the rendered character, which is complex, mixed from the natures that have taken part.[26]

The stars form aspects with one another—opposition, trine, quartile, and sextile—as well as disjunct relations. The stars have their own productive powers, but their configurations, as the wandering stars move through the zodiac, determine the complex effects they together have on sublunary events.

The stars transmit their powers through the heavens and into the sublunary realm by means of rays (ἀκτῖνες). Ptolemy nowhere states what these rays consist of, if anything, or how the stars transmit them. Yet, these rays have the ability to bring the stars' powers into contact with one another as well as with sublunary bodies and souls. In *Tetrabiblos* 1.24, Ptolemy juxtaposes the bodily conjunction of stars—when they occupy the same meridian—with the convergence of stars' rays: "Such is accepted whether it happens bodily or according to some of the traditional configurations [. . .]. In relation to [conjunctions and separations] by aspects, however, such [a practice] is superfluous, since all the rays always move and similarly come together from every direction toward the same (that is to say, to the center of the earth)."[27] Ptolemy indicates that a star's rays travel away from the star, through the heavens, into the sublunary realm, and toward the center of the earth.

Ptolemy depicts the rays as traveling through the zodiac in *Tetrabiblos* 3.11. Here he juxtaposes the bodily conjunction of stars with their projection of

---

τεταγμένης, συνεργούντων δὲ ἢ ἀποσυνεργούντων κατά τι τῶν λοιπῶν, καὶ τῆς μὲν σελήνης ἐκφανέστερον καὶ συνεχέστερον ὡς ἐν ταῖς συνόδοις καὶ διχοτόμοις καὶ πανσελήνοις, τῶν δὲ ἀστέρων περιοδικώτερον καὶ ἀσημότερον, ὡς ἐν ταῖς φάσεσι καὶ κρύψεσι καὶ προσνεύσεσιν.

26. Ibid. 2.9.19, H142–43: ἰδίως μὲν οὖν τῆς οἰκείας φύσεως ἐπιτυχὼν ἕκαστος τὰ τοιαῦτα ἀποτελεῖ, συγκιρνάμενος δὲ ἄλλος ἄλλῳ κατά τε τοὺς συσχηματισμοὺς καὶ τὰς τῶν ζῳδίων ἐναλλοιώσεις καὶ τὰς πρὸς ἥλιον φάσεις, ἀναλόγως τε καὶ τὴν ἐν τοῖς ἐνεργήμασι σύγκρασιν λαμβάνων μεμιγμένην ἐκ τῶν κεκοινωνηκυιῶν φύσεων τὴν περὶ τὸ ἀποτελούμενον ἰδιοτροπίαν ποικίλην οὖσαν ἀπεργάζεται.

27. Ibid. 1.24.1–2, H85: παραλαμβάνεται δὲ τὸ τοιοῦτον ἐάν τε σωματικῶς ἐάν τε καὶ κατά τινα τῶν παραδεδομένων σχηματισμῶν συμβαίνῃ [. . .]. πρὸς δὲ τὰς διὰ τῶν συσχηματισμῶν περισσόν ἐστι τὸ τοιοῦτον, πασῶν ἀεὶ τῶν ἀκτίνων ἐπὶ τὰ αὐτά (τουτέστιν ἐπὶ τὸ κέντρον τῆς γῆς) φερομένων καὶ ὁμοίως πανταχόθεν συμβαλλουσῶν.

rays when they are in the process of prorogation, an astrological technique of continuous horoscopy: "In accordance with the prorogation toward the rear of the zodiacal signs, the places of the maleficent [wandering stars] destroy (Saturn and Mars: whether they are meeting bodily or projecting rays from any place whatsoever in quartile or in opposition, and sometimes in sextile, as well, upon the 'hearing' or 'seeing' [zodiacal signs] according to equivalence of power) . . ."[28] Ptolemy likewise describes stars projecting rays through the zodiac in *Tetrabiblos* 3.10:

> In these circumstances, if the luminaries happen to be engaged in a separation from one of the beneficent [wandering stars], or even if they are otherwise in some aspect to them (they project their rays, however, to the parts that precede them), the [child] that was born will survive so long as the months or days or even hours equal to the number of degrees between the prorogator and the nearest rays of the maleficent [wandering stars], in proportion to the magnitude of the affliction and the power of the [wandering stars] producing the cause. But if the rays of the maleficent [wandering stars] fall on the luminaries toward the front, and the [rays fall] on the beneficent [wandering stars] toward the rear, the exposed child will be adopted and will live[29]

Thus, Ptolemy portrays the stars as projecting their rays through the aether to various zodiacal signs and the stars in them, as well as down through the sublunary realm toward the center of the earth.

Ptolemy mostly uses the term *aktines* for the stars' rays, but he also uses two other terms to denote the rays of the sun and moon in particular. When a planet is weak and invisible in relation to the sublunary realm because of its bodily proximity to the sun, it falls under the sun's *augai*. In *Tetrabiblos* 2.6, Ptolemy says, "inasmuch as the [wandering stars] that are rising or are stationary produce intensifications of the events, but the ones setting and under the rays [αὐγάς] [of the sun] or the ones leading, making evening risings, bring about

28. Ibid. 3.11.12, H210: κατὰ δὲ τὴν εἰς τὰ ἑπόμενα τῶν ζῳδίων ἄφεσιν ἀναιροῦσιν οἵ τε τῶν κακοποιῶν τόποι (Κρόνου καὶ Ἄρεως· ἤτοι σωματικῶς ὑπαντώντων ἢ ἀκτῖνα ἐπιφερόντων ὅθενδήποτε τετράγωνον ἢ διάμετρον, ἐνίοτε δὲ καὶ ἐπὶ τῶν ἀκουόντων ἢ βλεπόντων κατ' ἰσοδυναμίαν ἑξαγώνων)

29. Ibid. 3.10.5–6, H200–1: τούτων δὲ οὕτως ἐχόντων, ἐὰν μὲν ἀπόρροιαν ἀπό τινος τῶν ἀγαθοποιῶν ἔχοντα τὰ φῶτα τυγχάνῃ ἢ καὶ ἄλλως αὐτοῖς ἢ συνεσχηματισμένα (τοῖς προηγουμένοις ἑαυτῶν μέρεσι μέντοι γε τὰς ἀκτῖνας αὐτῶν ἐπιφερόντων), ἐπιζήσεται τὸ τεχθὲν ἄχρι τοῦ τῶν μεταξὺ τῆς τε ἀφέσεως καὶ τοῦ τῶν ἐγγυτέρων τῶν κακοποιῶν ἀκτίνων ἀριθμοῦ τῶν μοιρῶν τοὺς ἴσους μῆνας ἢ ἡμέρας ἢ καὶ ὥρας πρὸς τὸ μέγεθος τῆς κακώσεως καὶ τὴν δύναμιν τῶν τὸ αἴτιον ποιούντων. ἐὰν δὲ αἱ μὲν τῶν κακοποιῶν ἀκτῖνες εἰς τὰ προηγούμενα φέρωνται τῶν φώτων, αἱ δὲ τῶν ἀγαθοποιῶν εἰς τὰ ἑπόμενα, τὸ γεννώμενον ἐκτεθὲν ἀναληφθήσεται καὶ ζήσεται

an abatement of the things produced."[30] The term *augai* for the sun's rays always appears in the plural, and so individual rays remain undistinguished. The *augai* flow like a looming shield and make the effects of the planets they cover weak.

In the case of the moon, Ptolemy refers to its effluence, *aporroia*, in *Tetrabiblos* 1.2: "The moon, too, since it is nearest the earth, passes on to us on the earth the greatest effluence [ἀπόρροιαν], as most [things on earth], both inanimate and animate, are sympathetic and change with it . . ."[31] Side by side Ptolemy presents a causal theory, in which the rays directly affect the materials they come into contact with, and a theory of sympathy, where objects affect one another from afar. These two kinds of explanation rarely co-occur in ancient Greek scientific texts.[32] It could be that Ptolemy employs them both in order to explain the causal phenomena underlying sympathetic relationships. In other words, it may appear that the moon influences terrestrial events from afar, but in fact its effluence comes into contact with, and in this way causes changes in, the sublunary realm.

As for the rays' composition, Ptolemy nowhere states whether they are material or immaterial and, if material, whether they consist of aether or some other element or combination of elements. I suggest that because the stars' rays affect sublunary bodies and souls, and bodies and souls are, according to Ptolemy, material, it is likely that he interpreted the stars' rays also to be material. After all, in *On the Kritêrion* Ptolemy makes plain his affinity for the Hellenistic view that every object that can act or be acted upon is a body, and he even describes the soul as material. He portrays the soul as consisting of finer particles than constitute the body, and he remarks that the body, as opposed to the soul, is characterized by the thick consistency of its matter.[33]

---

30. Ibid. 2.7.4, H128: ἐπειδήπερ ἀνατέλλοντες μὲν ἢ στηρίζοντες ἐπιτάσεις ποιοῦνται τῶν συμπτωμάτων, δύνοντες δὲ καὶ ὑπὸ τὰς αὐγὰς ὄντες ἢ ἀκρονύκτους ποιούμενοι προηγήσεις ἄνεσιν τῶν ἀποτελουμένων ποιοῦσιν. Cf. ibid. 3.11.15, H212; 4.5.1, H307; 4.5.19, H317. This concept is often, although not in the *Tetrabiblos*, denoted by the term ὕπαυγος, "exposed to rays." See Bouché-Leclercq, *L'astrologie grecque*, 112. On αὐγαί and its derivatives, see Orlando and Torre, "Lessico astronomico-astrologico greco," 297–99.

31. Ptolemy, *Tetrabiblos* 1.2.3, H6: ἥ τε σελήνη πλείστην ὡς περιγειοτάτη διαδίδωσιν ἡμῖν ἐπὶ τὴν γῆν τὴν ἀπόρροιαν, συμπαθούντων αὐτῇ καὶ συντρεπομένων τῶν πλείστων καὶ ἀψύχων καὶ ἐμψύχων
Cf. ibid. 3.11.4, H205. Ptolemy more often uses the term ἀπόρροια as a technical term in astrology meaning "separation," where a more swiftly moving star, usually the moon, moves away from a slower one.

32. Daryn Lehoux remarks on the co-occurrence of Ptolemy's theories of stellar radiation and sympathy in *What Did the Romans Know?*, 161–62.

33. Ptolemy, *On the Kritêrion*, La11–12.

In *Tetrabiblos* 3.12, Ptolemy likewise maintains that the body is more material (ὑλικώτερον) than the soul.[34] As the rays act upon material objects, bodies and souls, they too must be material. Moreover, it is likely that Ptolemy imagined the material of the rays to be aether. Although sublunary bodies experience the stars' powers with respect to the four sublunary qualities—hot, cold, wet, and dry—it is unlikely that Ptolemy would have conceived of the stars as emitting any other substance than that of which they consist. It may appear problematic that stars' aethereal rays travel into the sublunary realm, but Ptolemy reveals in *On the Kritêrion* that aether does exist below the moon. The human soul's faculty of thought, for instance, is composed of aether, and, if the human soul contains aether, then it is possible that other substances in the sublunary realm, like the stars' rays, are aethereal.[35]

If the rays are material, what type of movement do they experience? Although it is likely that they are aethereal, it seems unlikely that, traveling toward the center of the earth, they would move only circularly. Therefore, I suggest that, outside of the star's system from which they emanate, they experience only rectilinear motion. My interpretation relates to Simplicius's description of Ptolemy's element theory. In his commentary on Aristotle's *De caelo*, Simplicius states that Ptolemy—like Xenarchus, the first-century BCE Peripatetic, before him and Plotinus after him—claimed in his *On the Elements* and *Optics* that elements experience rectilinear motion only when outside their natural places. When in their natural places, they either rest or move circularly.[36] Simplicius portrays this view as concerning only earth, water, air, and fire, as he contextualizes it within Xenarchus's larger argument, where the elements are not fully realized—they do not attain their proper form—until they reach their natural places. Xenarchus put forward this argument in his *Against the Fifth Substance*, aimed at dismantling Aristotle's argument for the existence of the aether. Plotinus, too, did not accept aether theory,

34. See Ptolemy, *Tetrabiblos* 3.12.1, H224.

35. Symeon Seth, an eleventh-century Byzantine teacher and philosopher, attributes a theory of sublunary aether to Ptolemy. He states in *Conspectus rerum naturalium* 4.74, "Ptolemy says in the *Optics* that the optical *pneuma* is something aethereal and composed of the fifth substance" (φησὶ δὲ ὁ Πτολεμαῖος ἐν τοῖς ὀπτικοῖς ὅτι αἰθερῶδές τί ἐστι καὶ τῆς πέμπτης οὐσίας τὸ ὀπτικὸν πνεῦμα). The *Optics* attributed to Ptolemy contains no such theory. The text is incomplete, and the first book, now lost, may have contained such a theory, but Harald Siebert has argued that, even if the theory had been articulated in the first book, it would be inconsistent with the surviving text: Siebert, *Die ptolemäische "Optik,"* 201–39. Nevertheless, if Ptolemy did compose an optical text—different or not from the one attributed to him in the manuscript tradition—it is possible that he portrayed the rays as aethereal, as Symeon Seth maintains.

36. See Simplicius, *In de caelo* 1.2.20.10–25. Cf. Proclus, *In Platonis Timaeum commentaria* III, 4.114.31–115.4.

but, as Marwan Rashed has shown, neither did he appropriate Xenarchus's theory that elements moving rectilinearly are still coming to be.[37] In fact, the only direct evidence we have for Plotinus entertaining a view like the one attributed to him by Simplicius consists in his response to the question of why fire moves circularly, and does not stay still, when it arrives at its proper place. Plotinus's answer explains fire's motion in relation to the universal soul, and so Simplicius's ascription to Plotinus of Xenarchus's element theory contains only a kernel of truth.[38]

Ptolemy, on the other hand, does not articulate the view Simplicius ascribes to him in any of his surviving texts. Again, Simplicius cites Ptolemy's *On the Elements* and *Optics*, but *On the Elements* is no longer extant, and the surviving, incomplete text of the *Optics* attributed to him does not contain this element theory. Indeed, nowhere in Ptolemy's extant corpus does he propose this view. What's more, the very reason Xenarchus and Plotinus consider fire's circular motion does not apply to Ptolemy. Unlike Xenarchus and Plotinus, Ptolemy was a staunch proponent of the fifth substance; according to Ptolemy, the heavens consist of aether, not fire. How, then, could the claim that elements rest or move circularly in their proper places cohere with Ptolemy's element theory? As in the case of Plotinus, where only an aspect of the theory Simplicius attributes to him is consistent with his philosophy, perhaps Simplicius's discussion also contains only an element of truth in Ptolemy's case. When Simplicius says that according to Ptolemy, Plotinus, and Xenarchus "motion in a straight line belongs to the elements that are still coming to be and that are in a non-natural place,"[39] I suggest that only the latter restriction, when "in a non-natural place," applies in Ptolemy's element theory. In other words, when all five of the elements are outside their natural places, they move rectilinearly and, when in their natural places, they either rest or move circularly. Although this interpretation is speculative, it would serve to explain how the stars' rays move toward the center of the earth in Ptolemy's astrology. When the aethereal rays are in the sublunary realm, they move rectilinearly.

As for the rays' movement within the heavens, if they moved only circularly, then they would never leave the system of spheres belonging to the star from which they emanate. It is possible to imagine that when rays leave a star's system, and travel to the aethereal bodies of other stars, they move rectilinearly. In this interpretation, the natural place of a star's rays is not the entire heavens but only the system of spheres belonging to that star, just as in the

37. Rashed, "Contre le mouvement rectiligne."

38. Plotinus, *Enneads* 2.2.1.19–51. Cf. ibid. 2.1.8.3–28.

39. Simplicius, *In de caelo* 1.2.20.13–14, trans. Falcon, in *Aristotelianism*: τὴν μὲν ἐπ' εὐθείας κίνησιν τῶν στοιχείων γινομένων ἔτι καὶ ἐν τῷ παρὰ φύσιν ὄντων τόπῳ

sublunary realm the natural place of each of the four elements is not the entire sublunary region but only the element's sphere within the four nested spheres of earth, water, air, and fire. Although the aethereal bodies of a star's system remain in their natural place, the star's rays leave their proper system and move rectilinearly though the heavens. This interpretation is highly unorthodox, even according to Ptolemy's extant accounts of aethereal physics, but it would explain how stars' rays move throughout the entire cosmos, as Ptolemy indicates they do in the *Tetrabiblos*. Even more, it would explain why it is that adjacent zodiacal signs, and the stars in them, are disjunct (ἀσύνδετα). Ptolemy remarks in *Tetrabiblos* 1.17 that signs that are one and five apart are "as it were, turned away from one another" (ἀπέστραπταί τε ὥσπερ ἀλλήλων).[40] If the rays only moved circularly through the heavens, then all signs would be able to "see" one another; all relations in the zodiac would be aspects. Only the rectilinear motion of the rays, when outside of their proper systems, can account for these astrological relationships.

When the stars' rays enter the sublunary realm—however they move and of whatever they consist—they affect all sublunary bodies, starting with the elemental layers of fire and air that reside at the periphery of the sublunary region. Ptolemy portrays this sequence of effects in *Tetrabiblos* 1.2:

> Well then, it would appear most clear to everyone even in few words that some power from the aethereal and eternal substance spreads through and penetrates all of the region about the earth, which through the whole of it is subject to change, since of the primary sublunary elements fire and air are encompassed and changed by the movements in accordance with the aether, and they also encompass and together change all the rest, earth and water and the plants and animals therein.[41]

Ptolemy does not state here that the stars' rays, specifically, cause changes in the sublunary realm. He could be suggesting simply that the aether's circular motion affects the sublunary region in the same manner as it does in Aristotle's *On Generation and Corruption* and *Meteorologica*. In the former, Aristotle portrays the sun's annual motion along the ecliptic as causing generation and corruption in the sublunary world.[42] In the latter, friction between the aether

---

40. Ptolemy, *Tetrabiblos* 1.17.2, H57.

41. Ibid. 1.2.1, H5: Ὅτι μὲν τοίνυν διαδίδοται καὶ διικνεῖταί τις δύναμις ἀπὸ τῆς αἰθερώδους καὶ ἀϊδίου φύσεως ἐπὶ πᾶσαν τὴν περίγειον καὶ δι᾽ ὅλων μεταβλήτην, τῶν ὑπὸ τὴν σελήνην πρώτων στοιχείων πυρὸς καὶ ἀέρος περιεχομένων μὲν καὶ τρεπομένων ὑπὸ τῶν κατὰ τὸν αἰθέρα κινήσεων, περιεχόντων δὲ καὶ συντρεπόντων τὰ λοιπὰ πάντα, γῆν καὶ ὕδωρ καὶ τὰ ἐν αὐτοῖς φυτὰ καὶ ζῷα, πᾶσιν ἂν ἐναργέστατον καὶ δι᾽ ὀλίγων φανείη.

42. Aristotle, *On Generation and Corruption* II, 336a13–337a33.

and the layer of fire below it produces meteorological phenomena including comets, meteors, and the Milky Way.[43] Nonetheless, because Ptolemy advocates a theory of rays, which transmit the powers of the stars, it is likely that the stars' rays, specifically, affect the layers of fire and air and thereafter the lower regions of the terrestrial realm.

Ptolemy describes a causal relationship between the stars and meteorological events in *Tetrabiblos* 2.13:

> For in relation to such positions of these [stars], the abatements and intensifications of the weather change hour by hour, just as in relation to the [phases] of the moon the ebb and flow of the tides [change], and the changes of the air currents especially are produced in connection with such positions of the luminaries at the cardinal points, in the direction of those winds, toward the same as the latitude of the moon is found to be inclining.[44]

As the many stars move through the heavens, their powers, by means of rays, affect the sublunary realm, from the sphere of fire down to the earth. Just as the moon's phases cause the ebb and flow of tides, the many stars cause hourly alterations in the weather.

The stars' powers also affect living things, including, of course, human beings. Ptolemy summarizes the effects the stars' rays have on human bodies and souls in *Tetrabiblos* 1.3:

> Therefore, it has been made clear by us, more or less summarily, in what way prognostication by means of *astronomia* is possible and that it can extend only as far as the occurrences in the ambient itself and the effects on human beings from such a cause (these would be in respect of the original tendencies of faculties and actions of the body and soul, and the occasional diseases of these for long and short durations, and, still, as many of the external matters that have a directive and natural connection to the original combination [of body and soul], as property and marriage are in relation to the body, and honor and dignity are in relation to the soul, and the occasional matters of chance of these things).[45]

43. Aristotle, *Meteorologica* I, 341b1–346b15.

44. Ptolemy, *Tetrabiblos* 2.13.8, H157–58: πρὸς γὰρ τὰς τοιαύτας αὐτῶν σχέσεις αἱ καθ' ὥραν ἀνέσεις καὶ ἐπιτάσεις τῶν καταστημάτων μεταβάλλουσι, καθάπερ πρὸς τὰς τῆς σελήνης αἵ τε ἀμπώτεις καὶ αἱ παλίρροιαι, καὶ αἱ τῶν πνευμάτων δὲ τροπαὶ μάλιστα περὶ τὰς τοιαύτας τῶν φώτων κεντρώσεις ἀποτελοῦνται, πρὸς οὓς ἂν τῶν ἀνέμων ἐπὶ τὰ αὐτὰ τὸ πλάτος τῆς σελήνης τὰς προσνεύσεις ποιούμενον καταλαμβάνηται.

45. Ibid. 1.3.1, H14: Τίνα μὲν οὖν τρόπον δυνατὸν γίνεται τὸ δι' ἀστρονομίας προγνωστικὸν καὶ ὅτι μέχρι μόνων ἂν φθάνοι τῶν τε κατ' αὐτὸ τὸ περιέχον συμπτωμάτων καὶ τῶν ἀπὸ τῆς τοιαύτης

In *Tetrabiblos* 3.2, Ptolemy explains that the configuration of the stars at a human being's conception determines the qualities of his body and soul for the rest of his life:

> Since the chronological starting point of human childbearing naturally starts at the time of the very conception, but potentially and accidentally the moment of birth, in the cases when the exact time of conception truly is known coincidentally or even by observation, it is more proper to follow it [i.e., the time of conception] with a view to the properties of the body and of the soul, examining the productive [cause] of the configuration of the stars at that time; for in the beginning the semen is given some sort of quality once and for all by the influences of the ambient, and even though it may become different in accordance with the time the body successively grows, in growth it successively adds only matter proper to it by nature and it resembles even more the specific form of its initial quality.[46]

The configuration of the stars at any one time determines the combined effect all of the stars' powers have on the sublunary realm, including human beings. For determining what a human being's body and soul will be like for the rest of his life, the stars' configuration at his conception is crucial. Over time, a human being develops but does not change substantially. As a human being grows, he simply adds material to himself of like kind to his initial composition, and he comes to resemble even more the form molded by the stars' powers at his conception.

In *Tetrabiblos* 3.14, Ptolemy explains how the stars and their movements affect the human soul's quality (ποιότης) and peculiar properties (ἰδιώματα). In so doing, he introduces terms for the parts of the soul, some of which are

---

αἰτίας τοῖς ἀνθρώποις παρακολουθούντων (ταῦτα δ' ἂν εἴη περί τε τὰς ἐξ ἀρχῆς ἐπιτηδειότητας δυνάμεων καὶ πράξεων σώματος καὶ ψυχῆς καὶ τὰ κατὰ καιροὺς αὐτῶν πάθη πολυχρονιότητάς τε καὶ ὀλιγοχρονιότητας, ἔτι δὲ καὶ ὅσα τῶν ἔξωθεν κυρίαν τε καὶ φυσικὴν ἔχει πρὸς τὰ πρῶτα συμπλοκὴν ὡς πρὸς τὸ σῶμα μὲν ἡ κτῆσις καὶ ἡ συμβίωσις, πρὸς δὲ τὴν ψυχὴν ἥ τε τιμὴ καὶ τὸ ἀξίωμα, καὶ τὰς τούτων κατὰ καιροὺς τύχας), σχεδὸν ὡς ἐν κεφαλαίοις γέγονεν ἡμῖν δῆλον.

46. Ibid. 3.2.1–2, H168–69: Ἀρχῆς δὲ χρονικῆς ὑπαρχούσης τῶν ἀνθρωπίνων τέξεων φύσει μὲν τῆς κατ' αὐτὴν τὴν σποράν, δυνάμει δὲ καὶ κατὰ τὸ συμβεβηκὸς τῆς κατὰ τὴν ἀποκύησιν ἐκτροπῆς, ἐπὶ μὲν τῶν ἐγνωκότων τὸν τῆς σπορᾶς καιρὸν ἤτοι συμπτωματικῶς ἢ καὶ παρατηρητικῶς, ἐκείνῳ μᾶλλον προσήκει πρός τε τὰ τοῦ σώματος καὶ τὰ τῆς ψυχῆς ἰδιώματα κατακολουθεῖν τὸ ποιητικὸν τοῦ κατ' αὐτὸν τῶν ἀστέρων σχηματισμοῦ διασκεπτομένους· ἅπαξ γὰρ ἐν ἀρχῇ τὸ σπέρμα ποιόν πως γενόμενον ἐκ τῆς τοῦ περιέχοντος διαδόσεως, κἂν διάφορον τοῦτο γίνηται κατὰ τοὺς ἐφεξῆς τῆς σωματοποιήσεως χρόνους, αὐτὸ τὴν οἰκείαν μόνην ὕλην φυσικῶς προσεπισυγκρίνον ἑαυτῷ κατὰ τὴν αὔξησιν ἔτι μᾶλλον ἐξομοιοῦται τῇ τῆς πρώτης ποιότητος ἰδιοτροπίᾳ. On the whole field of astrological influences on human conception and birth, see Frommhold, *Bedeutung und Berechnung*. On Ptolemy, see ibid. 10–27.

identical to the terms he employs in *On the Kritêrion* and the *Harmonics*. Ptolemy begins his account of the astrological effects on the human soul with the following:

> of the qualities of the soul, those in respect of the rational and intellectual [λογικὸν καὶ νοερόν] part are apprehended by means of the position of Mercury observed on each occasion, and those in respect of the perceptive and irrational [αἰσθητικὸν καὶ ἄλογον] [part are apprehended] from the one of the luminaries that is more corporeal (that is, the moon) and from the stars similarly situated with it in its separations and conjunctions.[47]

Distinguishing the rational part of the soul from the irrational, Ptolemy applies two terms to the former and two to the latter. The former is the part of the soul that is rational (λογικόν) and intellectual (νοερόν). These terms are significant, as Ptolemy describes a rational (λογικόν) faculty in *On the Kritêrion* and an intellectual (νοερόν) part of the soul in the *Harmonics*.[48] Moreover, Ptolemy's choice of terms in the *Tetrabiblos* replicates his tendency in *On the Kritêrion* and the *Harmonics* to utilize both Platonic and Aristotelian terminology for the soul's parts. In the case of the irrational part, the term "perceptive" (αἰσθητικόν) appropriates the Aristotelian faculty of the soul, which Ptolemy describes in the *Harmonics* and *On the Kritêrion*, and the term "irrational" (ἄλογον) recalls Plato's two irrational parts.[49] Moreover, as in both *On the Kritêrion* and the *Harmonics*, Ptolemy combines the spirited and appetitive parts of the soul into a single part.[50] In *On the Kritêrion*, he calls it the faculty of impulse (ὁρμητικόν), and in *Tetrabiblos* 3.14, after discussing the soul's sensory and irrational part, Ptolemy remarks on the soul's impulses: "But since the kinds of impulses of the soul are manifold . . ."[51] Hence, Ptolemy amalgamates the irrational components of the soul into a single part in the *Tetrabiblos*, *On the Kritêrion*, and *Harmonics*, and he associates impulses with this resultant part in both the *Tetrabiblos* and *On the Kritêrion*.[52]

---

47. Ptolemy, *Tetrabiblos* 3.14.1, H248–49: τῶν δὲ ψυχικῶν ποιοτήτων αἱ μὲν περὶ τὸ λογικὸν καὶ νοερὸν μέρος καταλαμβάνονται διὰ τῆς κατὰ τὸν τοῦ Ἑρμοῦ ἀστέρα θεωρουμένης ἑκάστοτε περιστάσεως, αἱ δὲ περὶ τὸ αἰσθητικὸν καὶ ἄλογον ἀπὸ τοῦ σωματωδεστέρου τῶν φώτων (τουτέστι τῆς σελήνης) καὶ τῶν πρὸς τὰς ἀπορροίας ἢ καὶ τὰς συναφὰς αὐτῆς συνεσχηματισμένων ἀστέρων.

48. Ptolemy, *On the Kritêrion*, La6; *Harmonics* 3.5, D95.

49. Ptolemy, *Harmonics* 3.5, D95–96; *On the Kritêrion*, La13.

50. Ptolemy, *On the Kritêrion*, La21; *Harmonics* 3.5, D97.

51. Ptolemy, *Tetrabiblos* 3.14.2, H249: πολυτροπωτάτου δ' ὄντος τοῦ κατὰ τὰς ψυχικὰς ὁρμὰς εἴδους

52. Ptolemy appropriates this grouping of the soul's two irrational parts—the spirited and appetitive—in contrast to the rational part from the tradition following Plato. On the bipartite model of the soul in Plato, Aristotle, and the early Academy, see Rees, "Bipartition of the Soul";

Of further note is Ptolemy's characterization in *Tetrabiblos* 3.15 and 4.10 of the rational part of the soul as the thinking part (διανοητικόν).[53] In the former, he contrasts the thinking part with a passive part of the soul (παθητικόν); in the latter he associates it with the logical part (λογικόν). In *On the Kritêrion*, Ptolemy associates the faculty of thought (διανοητικόν) with the rational faculty (λογικόν) and, although he does not call a part of the soul passive, he does indicate that the faculty of sense perception is around the body's passive elements, earth and water.[54] Thus, in the *Tetrabiblos* Ptolemy uses several of the same terms for the parts of the soul as he employs in *On the Kritêrion* and the *Harmonics*, and he repeats certain methodological choices, such as employing both Platonic and Aristotelian terminology for the soul's parts and combining two irrational parts of the soul into a single part.

Celestial Souls and Bodies

Human souls are not the only type of soul Ptolemy examines. Just as significant in his cosmology are celestial souls, which he discusses in Book 2 of the *Planetary Hypotheses*. Unfortunately, only part of the first book of the *Planetary Hypotheses* exists in Greek. The remainder of the first book and the entirety of the second book exist only in a ninth-century Arabic translation, as well as a fourteenth-century Hebrew translation from the Arabic. In addition to the obvious difficulties involved in interpreting a text that exists only in translation, the poor quality of the Arabic translation constrains its study. Andrea Murschel has observed that the second book does not follow a logical order, as if it were translated out of sequence.[55] As I do not read Arabic, my own study of Book 2 of the *Planetary Hypotheses* relies on Murschel's synopsis and the German translation of the Arabic by Ludwig Nix, completed by F. Buhl and P. Heegard after Nix's death, and published in Heiberg's edition.[56] As Nix's translation itself is not exact, my examination will set aside any attempt at philological rigor and aim instead to tease out the concepts Ptolemy puts forward in the text and bring them into conversation with the rest of his philosophy as well as the prevailing philosophical traditions from which he drew.

In the *Planetary Hypotheses*, Ptolemy introduces a physical representation of his astronomical models. In so doing, he extends the project he starts in

---

C. Gill, *Structured Self*, 211–12. For Ptolemy, the spirited and appetitive parts combine into a single part within a tripartite rather than a bipartite model.

53. Ptolemy, *Tetrabiblos* 3.15.2, H276; 4.10.7, H348.

54. Ptolemy, *On the Kritêrion*, La6 and La20.

55. Murschel, "Structure and Function," 37.

56. Heiberg, *Opera astronomica minora*, 111–45.

his astronomical texts. In the *Almagest*, he calculates the relative distances of the stars and the absolute distances of the sun and moon. In Book 1 of the *Planetary Hypotheses*, he states the absolute distances from the earth of every star's system, as well as the stars' sizes. Although he modifies the *Almagest's* astronomical models, other than in his latitude theory, which differs drastically between the texts, the mathematical models in the *Planetary Hypotheses* are similar to the models he puts forward in the *Almagest*. In Book 2 of the *Planetary Hypotheses*, Ptolemy describes the physical characteristics of the heavenly bodies and, in general, he expounds a theory of aethereal physics.

This sequence, from the astronomy of Book 1 to the aethereal physics of Book 2, is not accidental. In basing his physical account of the heavens on an astronomical foundation, Ptolemy employs the same method he uses in the *Harmonics* and *Tetrabiblos*. To examine the physical aspects of a set of bodies, he first studies them mathematically. In the case of the *Planetary Hypotheses*, in order to discern the physical qualities of aethereal bodies, Ptolemy first constructs astronomical models. Indeed, in *Planetary Hypotheses* 2.2 Ptolemy distinguishes these two types of study. He maintains that the phenomena are analyzable in two respects: physically and mathematically. By the end of Book 1, he has completed the mathematical account; thereafter, he presents the physical study, which depends on the mathematical account's prior completion.

In *Planetary Hypotheses* 2.2, Ptolemy also characterizes physical accounts of heavenly bodies as hypothetical rather than certain. In her examination of the text, Liba Taub references Stephen Toulmin and Ernan McMullin as suggesting "that this caution may be a veiled reference to the 'likely account' described by Timaeus at 29d."[57] I suggest that it is also possible, and more likely, that it alludes to Ptolemy's epistemological assessment of physics, relayed in *Almagest* 1.1 and *Tetrabiblos* 1.2, that physics is conjectural. That Ptolemy should present the same epistemological position here, when embarking on a physical account of the heavens, would be consistent.

In *Planetary Hypotheses* 2.3, Ptolemy presents an account of superlunary aether that is largely consistent with Aristotle's *De caelo* and *Metaphysics* Λ8. According to Ptolemy, every heavenly movement belongs to a distinct aethereal body, and every aethereal body is unique in its periodicity and position in the heavens. Despite the existence of several aethereal bodies, they do not hinder one another. The only change they experience is their own regular circular motion from place to place. Ptolemy contrasts the aether and its uniform circular motion with the four sublunary elements and their motions. When displaced from their natural places but unobstructed, the sublunary elements

57. Taub, *Ptolemy's Universe*, 167, n. 15.

move briskly toward their natural places. When in their natural places, they rest. This element theory stands in contrast to that of *On the Elements*, wherein Ptolemy is purported by Simplicius to have stated that the elements either rest or move circularly in their natural places.

Recalling *Almagest* 1.3, Ptolemy maintains in the *Planetary Hypotheses* that the aether consists of parts that are similar to one another. Yet, his search for a cause of the various spheres' order, orientations, and velocities leads him to posit some physical variation in the aether. In *Planetary Hypotheses* 2.5, he explains that this variation is not in density but in power. Ptolemy asserts that the stars are ensouled and the heavenly bodies move voluntarily by means of the stars' powers.[58] No doubt the term Ptolemy used for the stars' powers is the same as in the *Tetrabiblos*: *dynameis*. In support of this interpretation is Simplicius's quotation of Book 2 of the *Planetary Hypotheses*: "So, to be sure, it is more reasonable that each of the stars causes motion in its own place, since this is both their power and activity [δύναμις καὶ ἐνέργεια], and about its center, again, regularly and circularly; for it is right that what produces it [i.e., the motion] in the surrounding structures belongs to the [star] first."[59] While in the *Tetrabiblos* Ptolemy portrays celestial powers as affecting sublunary bodies, in the *Planetary Hypotheses* the stars' powers cause the movements of aethereal bodies.

When advancing his animistic theory of aethereal motion, Ptolemy sets out to disprove a mechanistic account. He claims in *Planetary Hypotheses* 2.2 that it is not his present purpose to correct the hypotheses of the ancients, but he devotes a considerable portion of Book 2 to disproving Aristotle's mechanistic model of the heavens, such as conveyed in *Metaphysics* Λ8. In *Planetary Hypotheses* 2.5, Ptolemy cites Aristotle by name and attributes to him the theory that the motion of the heavenly spheres has its cause in the spheres' poles, which are fixed on the spheres surrounding them. Indeed, Aristotle mentions the

58. Both Plato and Aristotle characterize the stars as ensouled. See Plato, *Timaeus* 38e–42e; Aristotle, *De caelo* II, 292a18–b2. Cf. Pseudo-Plato, *Epinomis* 982c–984b. Commenting on the dissertation from which this book grew, Stephan Heilen remarked that Ptolemy's ascription of souls to celestial bodies may explain why Ptolemy continues the Platonic manner of denoting planets by the phrase "the star of the [god]," which is first attested in Plato, *Timaeus* 38d, while other astrologers had by Ptolemy's time adopted the simpler, and more materialistic, way of denoting the planets simply with the gods' names. Cf. Cumont, "Noms des planètes."

59. Simplicius, *In de caelo* 2.8.456.23–27, translation modified from Mueller, in Simplicius, *On Aristotle's "On the Heavens 2.1–9"*: "ὥστε εὐλογώτερον εἶναι τὸ κινεῖν μὲν τῶν ἄστρων ἕκαστον, ὅτι τοῦτό ἐστι καὶ δύναμις καὶ ἐνέργεια αὐτῶν, κατὰ τὸν ἴδιον μέντοι τόπον καὶ περὶ τὸ αὐτοῦ μέσον ὁμαλῶς πάλιν καὶ ἐγκυκλίως· ὑπάρχειν γὰρ αὐτῷ πρώτῳ δίκαιον, ὃ καὶ ἐν ταῖς περιεχούσαις αὐτὸ συστάσεσι περιποιεῖ."

poles (πόλοι) of the heavenly spheres in *Metaphysics* Λ8.[60] Ptolemy argues that it is unnecessary to account for the spheres' rotation in terms of the poles, and, further, that it is impossible to give an account of the poles—where they act as causes of the spheres' rotation—that is both consistent with the principles of aethereal physics and explains how the poles would be physically different from the spheres of which they are a part, how they would be attached to other spheres, and how they would cause the spheres' motion.

Having rejected the poles as at all significant in aethereal physics, Ptolemy proposes that some aethereal bodies do not even have poles. These bodies are the sawn-off pieces he introduces in *Planetary Hypotheses* 2.4. Ptolemy posits two kinds of aethereal bodies, each capable of producing the same phenomena. The first kind is the complete sphere, which is either hollow, like the spheres that surround other aethereal spheres or the earth, or solid, like the epicyclic spheres. The second kind is a thin, equatorial section of a sphere, or a sawn-off piece. Like a complete sphere, a sawn-off piece is either solid or hollow. When solid, its shape is like a tambourine; when hollow, its shape is similar to a belt, a ring, or a whorl. While Ptolemy cites Aristotle when rejecting the causal function of the poles, he cites Plato when approving a theory of heavenly whorls. Plato describes a cosmological whorl in the *Republic*'s myth of Er:

> The nature of the whorl was as follows: its shape was just such as is of this world, but, from what he said, we must consider it to be as follows, as if in one big whorl, hollow and having been scooped out thoroughly, lay another, smaller such [whorl] fitted into it, just like the jars that fit one in another, and similarly another third and a fourth and four others. For there were eight whorls altogether, lying inside one another, their rims appearing from above as circles, while from the back they made one continuous whorl around the spindle; this [spindle] had been driven right through the middle of the eighth.[61]

Ptolemy adopts a Platonic account of heavenly bodies, where some of the bodies are sawn-off pieces. While the sawn-off pieces rotate, the remainders of

---

60. See Aristotle, *Metaphysics* XII, 1073b28.

61. Plato, *Republic* X, 616d–e: τὴν δὲ τοῦ σφονδύλου φύσιν εἶναι τοιάνδε· τὸ μὲν σχῆμα οἵαπερ ἡ τοῦ ἐνθάδε, νοῆσαι δὲ δεῖ ἐξ ὧν ἔλεγεν τοιόνδε αὐτὸν εἶναι, ὥσπερ ἂν εἰ ἐν ἑνὶ μεγάλῳ σφονδύλῳ κοίλῳ καὶ ἐξεγλυμμένῳ διαμπερὲς ἄλλος τοιοῦτος ἐλάττων ἐγκέοιτο ἁρμόττων, καθάπερ οἱ κάδοι οἱ εἰς ἀλλήλους ἁρμόττοντες, καὶ οὕτω δὴ τρίτον ἄλλον καὶ τέταρτον καὶ ἄλλους τέτταρας. ὀκτὼ γὰρ εἶναι τοὺς σύμπαντας σφονδύλους, ἐν ἀλλήλοις ἐγκειμένους, κύκλους ἄνωθεν τὰ χείλη φαίνοντας, νῶτον συνεχὲς ἑνὸς σφονδύλου ἀπεργαζομένους περὶ τὴν ἡλακάτην· ἐκείνην δὲ διὰ μέσου τοῦ ὀγδόου διαμπερὲς ἐληλάσθαι.

the spheres of which they are a part presumably remain at rest. Therefore, not all of the aether experiences uniform circular motion. Only the spheres and parts of spheres that the stars' powers move experience circular motion. The remaining parts of the sawn-off pieces' spheres stay at rest in their natural places, and so Ptolemy's element theory of *On the Elements* seems to be at work here. When in their natural places, some of the aethereal bodies move circularly and others remain at rest.

In addition to critiquing Aristotle's poles, Ptolemy rejects the counter-rolling spheres proposed by Aristotle and his followers.[62] In *Metaphysics* Λ8, Aristotle argues that because, in his mechanistic model, outer spheres cause inner spheres to rotate, counter-rolling spheres must exist in order that a wandering star's individual movements do not affect the movements of the stars below it. The counter-rolling spheres cancel out a star's movements such that only the diurnal rotation remains. Aristotle explains the need for counter-rolling spheres accordingly:

> But it is necessary, if all [the spheres] are going to be combined to render the phenomena, that in accordance with each of the wandering stars there be other, fewer by one, counteracting spheres that restore to the same position the first sphere which in every case is situated below the star; for only in this way do all these things allow the motion of the wandering [stars] to be produced.[63]

In Ptolemy's anti-mechanistic theory, counter-rolling spheres are unnecessary. Because the aethereal spheres do not impede one another's motion, counter-rolling spheres would have no purpose. Moreover, positing counter-rolling spheres produces a significant increase in the number of aethereal bodies and, by extension, the size of the cosmos. In *Metaphysics* Λ8, Aristotle declares that, with the counter-rolling spheres, the heavens consist of up to fifty-five spheres.[64]

Ptolemy, on the other hand, expresses a preference for economy. The superfluity, as well as uselessness, of the counter-rolling spheres is reason to reject them. In the case of the spheres that do account for the stars' motions, Ptolemy must decide whether all of the aethereal bodies are complete spheres

---

62. Alan Bowen suggests that Ptolemy may be critiquing here the homocentric theory of Sosigenes: Bowen, *Simplicius on the Planets*, 282.

63. Aristotle, *Metaphysics* XII, 1073b38–1074a5: ἀναγκαῖον δέ, εἰ μέλλουσι συντεθεῖσαι πᾶσαι τὰ φαινόμενα ἀποδώσειν, καθ' ἕκαστον τῶν πλανωμένων ἑτέρας σφαίρας μιᾷ ἐλάττονας εἶναι τὰς ἀνελιττούσας καὶ εἰς τὸ αὐτὸ ἀποκαθιστάσας τῇ θέσει τὴν πρώτην σφαῖραν ἀεὶ τοῦ ὑποκάτω τεταγμένου ἄστρου· οὕτω γὰρ μόνως ἐνδέχεται τὴν τῶν πλανήτων φορὰν ἅπαντα ποιεῖσθαι.

64. Ibid., 1074a11–12.

or whether some of them are sawn-off pieces. He faces a similar problem in *Almagest* 3.4 with the equivalency of the eccentric and epicyclic models. Both hypotheses account for the sun's anomaly but only one is needed. Ptolemy chooses the eccentric hypothesis because it is "more simple" (ἁπλούστερα) than the epicyclic; it accomplishes the sun's apparent irregularity by means of one movement rather than two.[65] Correspondingly, Ptolemy argues in *Planetary Hypotheses* 2.6 that it is more likely that—other than the sphere of fixed stars, which must be a complete sphere, as the stars are scattered visibly around it—all the aethereal bodies that bring about the stars' movements are sawn-off pieces. To hypothesize that complete spheres, enormous as they must be, rotate through the heavens in order to produce the movements of the wandering stars near the ecliptic is to suppose that large portions of the cosmos move unnecessarily. To hypothesize that only thin, equatorial sections of the spheres move reduces the amount of aether responsible for producing the wandering stars' movements. The movements of these equatorial sections are necessary for the movements of the wandering stars and in this way they serve a purpose. Consequently, Ptolemy calls this latter hypothesis of the sawn-off pieces "the better and the natural" choice.[66] Similarly to how in the *Tetrabiblos* he is concerned with maintaining the conformity of his astrological theories with nature, here in the *Planetary Hypotheses* Ptolemy decides between two cosmological theories—a superlunary realm of only complete rotating spheres or one that includes sawn-off pieces—and he chooses the theory consistent with nature, where aether does not move idly but only purposefully— i.e., to move the stars through the heavens. Thus, Ptolemy posits the existence of twenty-nine aethereal bodies—complete spheres, sawn-off pieces, and bodies of resting aether, which remain after the complete spheres are truncated into sawn-off pieces—and he advances a model with nearly half the number of aethereal bodies as Aristotle's cosmology.[67]

This choice of theory—of sawn-off pieces over only complete spheres—has the additional advantage of facilitating *sphairopoïïa*, or sphere making. Ptolemy indicates at the beginning of the *Planetary Hypotheses* that he has two audiences for the text: "ourselves" (ἡμῶν αὐτῶν)—presumably not just Ptolemy and his dedicatee, Syrus, but also readers like them, mathematicians and philosophers alike—and instrument makers.[68] Ptolemy notes that it is customary for craftsmen to depict in models of the heavens only the phenomena, the resultant appearances of the stars' movements, rather than the

65. Ptolemy, *Almagest* 3.4, H232.

66. Murschel, "Structure and Function," 51.

67. Ibid., 50.

68. Ptolemy, *Planetary Hypotheses* 1.1, H70.

underlying reality that produces these appearances, and he is critical of this convention. He entreats instrument makers to construct models that display the regular circular motions, eccentric and epicyclic, that together produce the stars' apparently anomalous movements. If the heavens were to contain sawn-off pieces rather than only complete spheres, then these instruments would be even more accurate. Systems of sawn-off pieces lend themselves to their representation in concrete models. It is easy to peer through a system of rings but—given the materials of ancient Greek instrument making—it is impossible to see inside a series of nested complete spheres, and so while an accurate representation of a system of nested complete spheres is impossible, a system of sawn-off pieces, stripped of the complete spheres and bodies of resting aether, is possible. Therefore, in addition to maintaining consistency with Ptolemy's theory of nature, the hypothesis of sawn-off pieces facilitates accurate modeling.

Ptolemy also endeavors to eliminate useless aspects of his cosmological model when establishing the order of the wandering stars. In particular, he argues that Mercury and Venus lie between the moon and the sun, even though, as he admits, no observations had yet been made of their occultation of the sun. In *Almagest* 9.1, Ptolemy advances an aesthetic argument in support of this planetary order. If Mercury and Venus lie between the moon and sun, then the sun separates those planets that always remain in its vicinity, Mercury and Venus, from the planets that have any and all elongations. In Book 1 of the *Planetary Hypotheses*, Ptolemy presents the same planetary order and adds that, according to this model, the two wandering stars with the most complicated astronomical systems, Mercury and the moon, are adjacent to one another and closest to the sublunary sphere of ever-changing bodies, while the simplest system, which enacts the movements of the fixed stars, resides at the periphery of the cosmos.

Ptolemy further supports this order of wandering stars with his model of nested spheres, where the minimum absolute distance of a star's system is equal to the maximum distance of the star's system below it. As a result, each star's system is adjacent to the next, and no gaps exist among the (parts of) spheres enacting the stars' movements. If Mercury and Venus were above the sun, so Ptolemy argues, a large region of useless aether would reside between the moon's and sun's systems. It is not because this region would contain aether at rest that it would be useless; it is because it would contain only aether at rest. No aether between the sun's and moon's systems would contribute to any star's movements. By locating Mercury and Venus between the moon and the sun, however, the gap disappears and the hypothesis of nested planetary spheres prevails. In *Planetary Hypotheses* 2.6, Ptolemy revisits his choice to place Mercury and Venus between the moon and sun, and he affirms that this model eliminates gaps in the aether that would be useless, as if forgotten and deserted by nature. Again, Ptolemy chooses the cosmological theory in conformity with nature.

Consistent with his esteem for the economical and his rejection of the useless is Ptolemy's preference for simplicity, which he articulates in *Almagest* 13.2. Faced with the complexity of his latitude theory, Ptolemy argues that the correct judgment of what is simple differs according to whether one is analyzing superlunary or sublunary phenomena. Because heavenly bodies experience no hindrance, they glide past each other without hindering one another's motion. Sublunary instruments designed to represent heavenly movements, on the other hand, cannot demonstrate this lack of friction, as their parts hinder one another's motion. Hence, Ptolemy maintains that his models are, despite appearances, simple:

> Rather, it is more proper to judge the simplicity itself of heavenly things not from what appears to be this way [i.e., simple] to us, when in our case the same thing is not equally simple for all; for, to those examining in this way, none of the things existing in the heavens would seem simple, not even the unchangeableness of the first motion, since this very existence in the same way for all time is for us not [merely] difficult but rather altogether impossible; but [it is more proper to judge the simplicity itself of heavenly things] from the unchangeableness of the natures of things in the heavens themselves and their movements; for in this way everything will appear simple and more so than what seem to us to be so, since no one can conceive of no labor or difficulty in the case of their circuits.[69]

This difference in the nature of super- and sublunary substances entails that one should not infer the nature and effects of heavenly bodies from the corresponding aspects of sublunary bodies. Ptolemy denounces the use of mundane analogies as the bases of astronomical hypotheses earlier in *Almagest* 13.2:

> Let no one considering the difficulty of our devices deem such of the hypotheses to be overworked; for it is not proper to compare human [devices]

---

69. Ptolemy, *Almagest* 13.2, HII533–34: μᾶλλον δὲ καὶ αὐτὸ τὸ ἁπλοῦν τῶν οὐρανίων οὐκ ἀπὸ τῶν παρ' ἡμῖν οὕτως ἔχειν δοκούντων προσήκει κρίνειν, ὁπότε μηδ' ἐφ' ἡμῶν τὸ αὐτὸ πᾶσιν ὁμοίως ἐστὶν ἁπλοῦν· οὕτω γὰρ σκοποῦσιν οὐδὲν ἂν δόξειε τῶν κατὰ τὸν οὐρανὸν γινομένων ἁπλοῦν οὐδ' αὐτὸ τὸ τῆς πρώτης φορᾶς ἀμετάστατον, ἐπειδὴ καὶ τοῦτο αὐτὸ τὸ πάντα τὸν χρόνον ὡσαύτως ἔχειν ἐφ' ἡμῶν ἐστιν οὐ δύσκολον, ἀλλὰ παντάπασιν ἀδύνατον· ἀπὸ δὲ τῆς τῶν ἐν αὐτῷ τῷ οὐρανῷ φύσεων καὶ τῆς τῶν κινήσεων ἀμεταβλησίας· οὕτω γὰρ ἂν πᾶσαι καταφανείησαν ἁπλαῖ καὶ μᾶλλον ἢ τὰ παρ' ἡμῖν οὕτως ἔχειν δοκοῦντα μηδενὸς πόνου μηδὲ δυσχερείας τινὸς περὶ τὰς περιόδους αὐτῶν ὑπονοηθῆναι δυναμένων.

with the divine, nor to form beliefs about such great things based on most dissimilar examples; for what is more dissimilar from the eternal and unchanging than the things that are never [unchanging], or the things that can be hindered by everything than the things [that can be hindered] not even by themselves?[70]

Correspondingly, in the *Planetary Hypotheses* Ptolemy censures Peripatetics for inferring the physical properties of heavenly bodies from sublunary bodies.

Yet, Ptolemy himself employs several mundane analogies in the *Planetary Hypotheses*. I contend, however, that he does not use them in the same way that he criticizes. Rather, he introduces them only after he has deduced the heavenly bodies' various physical properties from his general principles of aethereal physics. For instance, in *Planetary Hypotheses* 2.5 Ptolemy argues that aethereal bodies do not vary in density. He claims that if the poles of aethereal bodies differed from their spheres in density, then the aether that composes them would sink and move rapidly toward the center of the cosmos. In other words, the disparity in density would cause the denser, and therefore more heavy, aether to have a natural motion downward toward the center of the cosmos. Only once Ptolemy has established that the aether has uniform density does he offer three mundane analogies. First, he discusses birds as an example of bodies that move at high elevations; however, he admits that they are not a perfect analogy for aethereal bodies because, unlike aethereal poles, birds differ from their environment in density. Ptolemy then presents two sets of sublunary bodies that, unlike birds, do not vary in density from their surroundings. In dry weather, so Ptolemy claims, clouds differ from the air surrounding them only in color, and two liquids that have the same density may vary only in color. Hence, only after arguing that the aether has uniform density does Ptolemy present three mundane analogies to illustrate this property of superlunary bodies.

Similarly, in *Planetary Hypotheses* 2.3 Ptolemy compares aethereal bodies to the parts of a universal animal. He proclaims that the stars move the aethereal bodies in their systems by means of their powers, and thereafter he likens the heavens to an entire, or universal, animal (*al-ḥayawân al-kullî*). The analogy between the heavens and an animal was common in ancient Greek philosophy.

---

70. Ibid., HII532: καὶ μηδεὶς τὰς τοιαύτας τῶν ὑποθέσεων ἐργώδεις νομισάτω σκοπῶν τὸ τῶν παρ' ἡμῖν ἐπιτεχνημάτων κατασκελές· οὐ γὰρ προσήκει παραβάλλειν τὰ ἀνθρώπινα τοῖς θείοις οὐδὲ τὰς περὶ τῶν τηλικούτων πίστεις ἀπὸ τῶν ἀνομοιοτάτων παραδειγμάτων λαμβάνειν· τί γὰρ ἀνομοιότερον τῶν ἀεὶ καὶ ὡσαύτως ἐχόντων πρὸς τὰ μηδέποτε καὶ τῶν ὑπὸ παντὸς ἂν κωλυθησομένων πρὸς τὰ μηδ' ὑφ' αὑτῶν;

In the *Timaeus*, for example, Plato portrays the entire cosmos as a living thing that encompasses all other living things:

> rather, let us lay it down that [the cosmos] is most like, of all things, that [Living Thing] of which all other living things are parts, both individually and generically. For, indeed, it [viz. the Living Thing] comprehends within itself all intelligible living things, just as this cosmos is composed of us and all the other visible creatures. For since the god wanted most of all to make [the cosmos] like the most beautiful of the intelligible things, complete in every way, [he made it] a single visible living thing, which contains within itself all living things who by nature share its kind.[71]

Timaeus depicts the cosmos as a living thing, or an animal (ζῷον). It has a soul, the world soul, composed of Being, the Same, and the Different.[72] Aristotle portrays the heavens animistically in the *Nicomachean Ethics* and *De caelo*. In the former, he implies that the heavenly bodies—"the [bodies] of which the cosmos is composed"—are animals, or living things (ζῴων).[73] In the latter, commenting on the Pythagorean convention of assigning a left and a right side to the heavens, he explains, "Since it has been determined by us already that such functions belong in things that have a principle of movement, and that the heavens are ensouled [ἔμψυχος] and have a principle of movement, it is clear that [the heavens] also have up and down, right and left."[74] For Ptolemy, the heavens are not a single living thing, but the concept of a universal animal makes for a useful analogy. The aethereal bodies relate to the heavens

71. Plato, *Timaeus* 30c–31a, translation modified from Zeyl, in Plato, *Complete Works*: οὗ δ' ἔστιν τἆλλα ζῷα καθ' ἓν καὶ κατὰ γένη μόρια, τούτῳ πάντων ὁμοιότατον αὐτὸν εἶναι τιθῶμεν. τὰ γὰρ δὴ νοητὰ ζῷα πάντα ἐκεῖνο ἐν ἑαυτῷ περιλαβὸν ἔχει, καθάπερ ὅδε ὁ κόσμος ἡμᾶς ὅσα τε ἄλλα θρέμματα συνέστηκεν ὁρατά. τῷ γὰρ τῶν νοουμένων καλλίστῳ καὶ κατὰ πάντα τελέῳ μάλιστα αὐτὸν ὁ θεὸς ὁμοιῶσαι βουληθεὶς ζῷον ἓν ὁρατόν, πάνθ' ὅσα αὐτοῦ κατὰ φύσιν συγγενῆ ζῷα ἐντὸς ἔχον ἑαυτοῦ, συνέστησε.

72. Ibid. 35a.

73. Aristotle, *Nicomachean Ethics* VI, 1141a33–b2: εἰ δ' ὅτι βέλτιστον ἄνθρωπος τῶν ἄλλων ζῴων, οὐδὲν διαφέρει· καὶ γὰρ ἀνθρώπου ἄλλα πολὺ θειότερα τὴν φύσιν, οἷον φανερώτατά γε ἐξ ὧν ὁ κόσμος συνέστηκεν.

74. Aristotle, *De caelo* II, 285a27–31: Ἡμῖν δ' ἐπεὶ διώρισται πρότερον ὅτι ἐν τοῖς ἔχουσιν ἀρχὴν κινήσεως αἱ τοιαῦται δυνάμεις ἐνυπάρχουσιν, ὁ δ' οὐρανὸς ἔμψυχος καὶ ἔχει κινήσεως ἀρχήν, δῆλον ὅτι ἔχει καὶ τὸ ἄνω καὶ τὸ κάτω καὶ τὸ δεξιὸν καὶ τὸ ἀριστερόν. In *De caelo* I, 278b9–24, Aristotle delineates three meanings of οὐρανός, the last of which is the cosmos as a whole, τὸ πᾶν. In *De caelo* II, 285a32, amidst his discussion of directionality in the heavens (οὐρανός), he mentions the shape of the whole: τὸ σχῆμα τοῦ παντός. Therefore, Aristotle here seems to be using οὐρανός to signify the cosmos as a whole.

as a whole in a similar way to how an animal's parts relate to the animal in its entirety. Thus, Ptolemy appropriates a common analogy, which he introduces after asserting an animistic theory of the aethereal bodies' motion.

In *Planetary Hypotheses* 2.7, Ptolemy compares aethereal bodies' motion to the movements of a flock of birds. The juxtaposition of heavenly bodies and birds also appears in Plato's *Timaeus*. According to Timaeus, four types of living creatures inhabit the cosmos: "Indeed, there are four [of these kinds]: first the heavenly race of gods, another that is winged and traverses the air, third the kind that lives in water, and fourth [the kind] that has feet and lives on land."[75] One type of living thing corresponds to each element. In relation to fire and air, the two outermost of Plato's elements, are celestial bodies, or gods, and the animals that fly through the air, or birds. Timaeus again makes reference to birds when examining the metempsychosis of human souls into animals: "and the race of birds was transformed—growing feathers instead of hair—from innocent but insubstantial men, who were skilled in the study of heavenly phenomena but because of their simplicity believed that the surest demonstrations concerning them were by sight."[76] In other words, astronomers who limit their comprehension of the heavens to the phenomena are reborn as birds. Ptolemy appropriates this association of birds with the higher reaches of the cosmos for his own purpose, the illustration of heavenly bodies' movements.

Ptolemy explicates how the movement of a bird within a flock comes about. The movement has its source in the bird's vitality, and from this vitality arises an impulse. This impulse moves the bird's nerves, and the nerves move the bird's feet and wings. These movements end at the bird's extremities, as the bird's vitality does not cause the air surrounding it nor the other birds in the flock to move. According to Ptolemy, a star's system of aethereal bodies functions in a similar way. Every star is animate, it has a soul, and the celestial soul's power causes the star to project rays, which transmit its power and cause aethereal bodies in its system to move. First the rays cause the aether closest to the star, the epicycle, to move. Then they cause the movements of the eccentric body and the body whose center point is the center of the entire cosmos.

Ptolemy's discussion of rays in the *Planetary Hypotheses* may illuminate his theory of rays in the *Tetrabiblos*. In the *Planetary Hypotheses*, the stars emit

---

75. Plato, *Timaeus* 39e–40a: εἰσὶν δὴ τέτταρες, μία μὲν οὐράνιον θεῶν γένος, ἄλλη δὲ πτηνὸν καὶ ἀεροπόρον, τρίτη δὲ ἔνυδρον εἶδος, πεζὸν δὲ καὶ χερσαῖον τέταρτον.

76. Ibid. 91d–e: τὸ δὲ τῶν ὀρνέων φῦλον μετερρυθμίζετο, ἀντὶ τριχῶν πτερὰ φύον, ἐκ τῶν ἀκάκων ἀνδρῶν, κούφων δέ, καὶ μετεωρολογικῶν μέν, ἡγουμένων δὲ δι' ὄψεως τὰς περὶ τούτων ἀποδείξεις βεβαιοτάτας εἶναι δι' εὐήθειαν.

rays, which travel through the aethereal bodies of a star's system. Their ability to cross through many aethereal bodies entails that they do not move uniformly and circularly in the aethereal body closest to the star. Indeed, Ptolemy indicates in *Planetary Hypotheses* 2.3 that the rays disperse all around without being hindered or influenced. The aethereal bodies, too, experience no hindrance, and so the rays affect them not mechanistically, or by force, but only animistically. In the *Tetrabiblos*, the rays travel further abroad. They disperse through the heavens to other stars' systems and down into the sublunary realm. Although it is possible that the rays carve a circular path as they pass through the aethereal bodies of their own star's system, when they cross into other stars' systems and down to the earth, it seems that they travel rectilinearly. Moreover, they do not affect the aethereal bodies of other stars' systems. All aether remains unhindered, and, furthermore, Ptolemy indicates that a star's rays cause motion only in the (parts of) spheres of their own system. Just as a bird's vitality does not cause other birds to move, a star's power only causes the aethereal bodies of its own system to move. Outside of their systems, the rays only cause changes in the sublunary realm, where mechanistic causation does effect change.

To illuminate this animistic theory of the heavens, Ptolemy compares the powers of celestial and human souls. In so doing, he mentions the human soul's powers of intellect and impulse, which recall his account in *On the Kritêrion* of the soul's three faculties: the faculties of thought (διανοητικόν), sense perception (αἰσθητικόν), and impulse (ὁρμητικόν). The juxtaposition of the intellectual and impulsive powers in the *Planetary Hypotheses* suggests that Ptolemy is contrasting the first faculty of the soul—labeled variously as the intellectual part, the rational part, and the faculty of thought in the *Harmonics* and *On the Kritêrion*—with the third, lowest faculty of the soul, the faculty of impulse of *On the Kritêrion*. In the *Planetary Hypotheses*, these two human faculties correspond to the two powers of celestial souls that cause aethereal bodies' motion. Moreover, in *Planetary Hypotheses* 2.3 Ptolemy compares the stars' rays to human souls' powers of sight and mind. By likening celestial powers and rays to human souls' powers of intellect, sight, and impulse, Ptolemy indicates that celestial souls have three faculties, which correspond to the three faculties of human souls.

Furthermore, Ptolemy claims that the aether's uniform circular motion pertains to reason and desire, where no change of intent occurs. In other words, Ptolemy describes the stars as not only ensouled but also desiring. Considering that Ptolemy alludes to the Prime Mover in *Almagest* 1.1, it is plausible that the ensouled stars and their aethereal bodies move uniformly and circularly because of the celestial souls' desire for the Prime Mover. In Aristotle's *Metaphysics* Λ, the Prime Mover acts as a final cause, and so it is possible that it serves as a final cause in Ptolemy's cosmology as well.[77] After all, if the stars desire, they must desire something. Aristotle declares in the *De anima* that desire is a function of the soul and that every desire must be for the sake of something.[78] If the celestial souls' desire must be for the sake of some end, I suggest that this end is the Prime Mover.

Ptolemy's cosmology, then, would function in a somewhat similar way to that propounded by Alexander of Aphrodisias in *Quaestio* 1.25 (40.8–30).[79] According to Alexander, the heavenly spheres, rather than the stars, are ensouled, and—recalling the plurality of unmoved movers in Aristotle's *Metaphysics* Λ8—each heavenly sphere's soul desires an unmoved mover. More specifically, the sphere-souls desire to be as similar as possible to their corresponding unmoved movers, and this desire results in the diurnal rotation of the several heavenly spheres.[80] In Ptolemy's cosmology, I contend, the stars' souls desire to be like the Prime Mover—the one unmoved mover Ptolemy describes—and to this end they cause the motion of aethereal bodies in their respective systems. Ptolemy's appropriation of the Prime Mover as a final cause of heavenly motion also would explain why he calls it "the first cause of

77. Murschel and Taub both argue that Ptolemy implicitly rejects the Prime Mover as a cause of aethereal motion, but in the *Planetary Hypotheses* Ptolemy only rejects mechanistic causes in the heavens and does not explicitly address final causes: Murschel, "Structure and Function," 39; Taub, *Ptolemy's Universe*, 116.

78. See Aristotle, *De anima* III, 432b3–8; 433a15–16.

79. I follow Robert W. Sharples's interpretation of this passage. See Sharples, "Divine Providence: Two Problems," 209–10. Cf. Alexander, *On the Cosmos* 96, where Alexander of Aphrodisias also discusses the spheres' souls.

80. István Bodnár presents a reading of *Quaestio* 1.25 where the desire of the first, outermost sphere of fixed stars for the unmoved mover brings about the first sphere's diurnal rotation, and this westward motion is transmitted from the first sphere to the planetary spheres. The desire of each planetary sphere for an unmoved mover causes the planetary spheres' individual eastward motions. On this reading, Alexander's cosmology is similar to that of Adrastus of Aphrodisias, the Aristotelian philosopher from earlier in the second century. According to Adrastus, the celestial bodies, rather than the spheres, are ensouled, but still the outermost sphere of fixed stars—which, in this case, rotates as a consequence of the choice of the fixed stars' souls—transmits the diurnal rotation down to the planetary spheres. The souls of the sun, moon, and five planets cause, by choice, the movements other than the diurnal rotation. Ptolemy, on the other hand, portrays the celestial systems as independent of one another. In his anti-mechanistic theory, the sphere enacting the diurnal rotation of the fixed stars does not transmit its motion outside the fixed stars' aethereal system. Instead, each wandering star has an aethereal body in its own system that enacts the diurnal rotation. For Bodnár's argument, see Bodnár, "Alexander of Aphrodisias." For a discussion of Adrastus's and Alexander's cosmologies, see Sorabji, "Adrastus."

the first motion of the universe" in the *Almagest*.[81] The Prime Mover is the first cause inasmuch as it is the final cause. The celestial souls love and wish to be like the Prime Mover, just as, in *Almagest* 1.1, the souls of mathematicians love and strive to resemble the heavenly divine. For human beings the Prime Mover remains an object of conjecture, but perhaps celestial souls know and desire it.

Conclusion

In *Almagest* 1.1, Ptolemy claims that physics is conjectural but that mathematics contributes to physics significantly. To guess well at the physical nature of any set of bodies, one must first study them mathematically. After Ptolemy composed *On the Kritêrion*, he mandated the application of mathematics to physics, which every one of his subsequent natural philosophical studies enacts. Geometry reveals the elements' fundamental qualities in several of Ptolemy's texts, harmonics serves as the foundation of the *Harmonics'* account of the human soul, astronomy makes astrology possible in the *Tetrabiblos*, and astronomy comes before cosmology in the *Planetary Hypotheses*. These applications of mathematics, however, do not entail that mathematics is propaedeutic to physics. For Ptolemy, mathematics is a necessary forerunner to any physical science, but mathematics is not for the sake of, it is not subordinate to, the physical sciences. Instead, mathematics is the highest science, both epistemologically and ethically. Only mathematics yields knowledge; only mathematics furnishes the good life. Even so, Ptolemy was not so single-minded as to study mathematics and only mathematics. He also pursued the sciences where mathematics' contribution proved promising. Even if the physical sciences must remain conjectural—which Ptolemy confirms in the *Tetrabiblos* and *Planetary Hypotheses*—mathematics makes it possible to guess well at the physical nature of the soul, the physical effects of the stars' movements and configurations on the sublunary realm, and the aethereal physics of the heavens.

---

81. Ptolemy, *Almagest* 1.1, H5: τὸ μὲν τῆς τῶν ὅλων πρώτης κινήσεως πρῶτον αἴτιον

9

Conclusion

ALMAGEST 1.1 functions as an epitome of Ptolemy's philosophical system. He structured it in such a way as to mimic the opening chapters of philosophical handbooks, a contemporary genre used to introduce the philosophical ideas of authoritative figures like Plato. Ptolemy employs this format to establish the authority of his own philosophy, where mathematics reigns supreme. Like so-called middle Platonists, Ptolemy appropriated concepts from the Aristotelian and Platonic traditions as well as, to a lesser extent, the Stoic and Epicurean, but the end product of this amalgamation is not a derivative philosophy or one of optimum agreement among schools of thought. Ptolemy's philosophy is subversive. He puts forward claims that would have been extraordinarily controversial at the time, and that discredit attempts by philosophers to answer the very questions of philosophy. According to Ptolemy, philosophers will never attain knowledge, they will never agree on the nature of theological and physical objects, because theology and physics are conjectural. Ptolemy argues that mathematics alone yields knowledge and that, furthermore, it is the only path to the good life.

Despite the singular status of mathematics in his philosophy, Ptolemy does not dispense with the other sciences. He improves them by means of mathematics. He argues that mathematics contributes to theology and physics and that, moreover, mathematics is foundational to practical philosophy and even the ordinary affairs of life. Mathematics reveals the ultimate goal of human life: to be like the divine, mathematical objects of the heavens, the movements and configurations of the stars. The best life is one where everything an individual does is guided by and in the service of this mathematical-ethical objective. Positioning mathematics at the foundation of all philosophy, theoretical and practical, as well as each and every one of the ordinary activities of life, Ptolemy propounds the mathematical way of life.

Ptolemy's corpus suggests that he lived in accordance with this way of life. Every one of his extant texts is a study or an application of mathematics, except *On the Kritêrion and Hêgemonikon*, which I argue he composed first of all of his extant texts and before he formulated his mature scientific method, including the persistent and necessary application of mathematics to physics and theology. In his corpus, Ptolemy applies geometry to element theory; harmonics to psychology; and astronomy to astrology, cosmology, and theology. Besides *On the Kritêrion*, every one of his texts is an inquiry into or an implementation of mathematics. Yet, for Ptolemy mathematics is not propaedeutic; it is not preparatory to another, higher science. Mathematics is the highest science, and Ptolemy dedicated his life to it. If we take him at his word, he did so with the aim of transforming his soul into a condition that resembles divine, astronomical objects, which are constant, well ordered, commensurable, and calm. Ptolemy's ethics justifies his study of mathematics and provides the foundation for his many philosophical and scientific contributions.

Ptolemy defines mathematics alongside the other two theoretical sciences in *Almagest* 1.1. Although he cites Aristotle, Ptolemy's definitions of physics, mathematics, and theology are not Aristotle's. They are Aristotelian. Ptolemy distinguishes the sciences according to their objects of study, and he classifies the three fundamental types of objects in the cosmos according to epistemic criteria. In other words, he incorporates epistemology at a foundational level to his metaphysics. More specifically, whether and how an object is perceptible to human beings defines what type of object it is. The Prime Mover, the object of theology, is imperceptible; physical objects are special sensibles, meaning they are perceptible by only one of the five senses; mathematical objects are common sensibles, perceptible by more than one sense.

Ptolemy fuses an Aristotelian theory of perception with the Platonic concern for distinguishing knowledge from opinion, which he associates with conjecture. According to Ptolemy, whether a human being is able to construct knowledge or mere conjecture when studying one of the sciences depends entirely on the properties of the objects the science studies. These properties are epistemic and ontological: perceptibility, clarity, and stability. A human being can have knowledge only of something perceptible. Because the Prime Mover is imperceptible, it is ungraspable and its study, theology, is conjectural. Although physical objects are perceptible, they are unstable and unclear. Their instability and lack of clarity prevent the human intellect from making clear and skillful judgments from their sense impressions, and therefore physics, too, is conjectural. Ptolemy calls physics conjectural not only in *Almagest* 1.1 but also in the *Tetrabiblos* and *Planetary Hypotheses*. Nevertheless, Ptolemy mitigates the epistemic deficiency of physics and theology by systematically placing them on a mathematical foundation. He first studies entities mathematically before investigating their physical or theological nature.

The objects of mathematics, on the other hand, are stable and clear. A human being can skillfully examine clear sense impressions of them and thereby create knowledge. Among the mathematical sciences, harmonics and astronomy are complementary. Harmonics studies the relations among musical pitches, which are perceptible only by the sense of hearing; astronomy studies the movements and configurations of celestial bodies, which are perceptible only by sight. Moreover, harmonics and astronomy each employ an indisputable method, or instrument, of mathematics: arithmetic or geometry. Ptolemy indicates that harmonics generates sure and incontrovertible knowledge—as he claims all mathematics does in *Almagest* 1.1—but astronomy does so only inasmuch as it both employs Ptolemy's criterion of truth—the interplay of reason and perception—and relies on geometry, one of the indisputable instruments of mathematics. In the *Almagest*, Ptolemy takes the existence of eccentric and epicyclic spheres to be certain, but the quantitative aspects of his astronomical models are not exact. They are approximate rather than precise. The eccentric and epicyclic hypotheses follow from his criterion of truth and derive from geometry, but the spheres' parameters and periods of revolution do not follow from geometry and they depend solely on observation for their determination. Hence, Ptolemy concludes that mathematicians cannot know and should not claim that the quantitative features of astronomical models are true. They forever remain approximate.

Why, then, does Ptolemy even attempt to quantify his astronomical models? I suggest that he does so because astronomy, like astrology, is by definition a predictive science. Ptolemy defines astronomy in juxtaposition with astrology in the first chapter of the *Tetrabiblos*. Consistent with *Almagest* 1.1, he claims that the judgments of astronomy are sure, whereas astrology, like every other study of material quality, is conjectural. Yet, the claims of astrology are still possible, as opposed to impossible, and therefore they merit study. Astrology relies on the prior, complete study of astronomy, because it examines the effects the stars' movements and configurations have on sublunary bodies and souls. The stars' powers affect sublunary events by means of rays, which travel through the heavens and into the sublunary realm. Ptolemy also discusses the stars' powers and rays in the *Planetary Hypotheses*, and just as his astrology depends on astronomy, so, too, does his cosmology. In the *Planetary Hypotheses*, Ptolemy presents his astronomical models before he propounds his aethereal physics. According to Ptolemy, the heavens are animate. Celestial souls instigate and maintain the uniform circular motion of the spheres and parts of spheres that constitute their aethereal systems. Moreover, celestial souls cause these movements, I argue, as a result of their desire for the Prime Mover, which Ptolemy calls in *Almagest* 1.1 "the first cause of the first motion of the universe."[1]

---

1. Ptolemy, *Almagest* 1.1, H5: τὸ μὲν τῆς τῶν ὅλων πρώτης κινήσεως πρῶτον αἴτιον

At the foundation of every one of Ptolemy's studies is his ethics. It is because mathematics furnishes the good life that Ptolemy pursues its study and application. I suggest that in Ptolemy's philosophical system both astronomy and harmonics produce the virtuous transformation of the human soul, just as in Plato's *Timaeus* the study of the heavens and harmony facilitate the ordering of the human soul's orbits.[2] For Ptolemy, it is because *harmonia* causes the existence of harmonic ratios among the constituent parts, movements, and configurations of the most complete and rational objects in the cosmos—human souls, musical systems, and heavenly bodies—that harmonic and astronomical objects serve as exemplars for human souls. As corruptible, human souls fall out of attunement. They require an exemplar, an instantiation of harmonic ratios in a body physically unlike themselves on which to model their formal configuration. By studying and practicing harmonics and/or astronomy, human beings attain good order in their souls. Through astronomy, they come to love the heavenly divine and transform their souls into the constant, well-ordered, commensurable, and calm condition of the stars' movements and configurations. Restoring the harmonic structure of their souls, they resemble the divine. They become godlike.

By late antiquity, Ptolemy was celebrated as an authoritative figure. He acquired the epithets "wonderful," "excellent," and even "divine." Both Simplicius and Philoponus call him "the wonderful Ptolemy" (ὁ θαυμαστὸς Πτολεμαῖος), John Lydus—the sixth-century CE bureaucrat and author of a compilation of celestial and meteorological omen literature—refers to "the most divine Ptolemy" (ὁ θειότατος Πτολεμαῖος), and Synesius remarks on "the excellent Ptolemy and his divine company of successors" (Πτολεμαίου τοῦ πάνυ καὶ τοῦ θεσπεσίου θιάσου τῶν διαδεξαμένων).[3] They use these epithets when discussing Ptolemy's contributions in astronomy, astrology, and physics, but did Ptolemy's reputation extend to his more traditionally philosophical contributions? Were his metaphysics, epistemology, and ethics influential?

I already have pointed toward some areas of possible influence. Similarly to how Ptolemy identifies mathematical objects with common sensibles, Philoponus and Sophonias, the late thirteenth- to early fourteenth-century Constantinopolitan monk, address the nature of mathematical objects in their studies of Aristotle's *De anima*. They define mathematical objects as forms, abstracted by thought from matter, that are common sensibles.[4] In addition, Syrianus, the fifth-century Neoplatonic philosopher, addresses the possibility that mathe-

2. Plato, *Timaeus* 47c–e.

3. Simplicius, *In de caelo* 1.1.9.21; Philoponus, *In libros de generatione animalium commentaria* 183.4; Joannes Lydus, *De ostentis* 2.17–18; Synesius, *De dono* 5.7–8.

4. Philoponus, *In de anima* 15.57.28–30; Sophonias, *In de anima* 9.28–31.

matical objects are common sensibles abstracted by thought from perceptible objects.[5] It is possible that Ptolemy's identification of mathematical objects with common sensibles influenced Syrianus, Philoponus, and thereafter Sophonias.

The influence of Ptolemy's epistemology was evidently long lasting. In the early fourteenth century, the statesman and scholar Theodorus Metochites examines mathematics in comparison to the other species of theoretical philosophy in the opening chapters of his *Stoicheiôsis astronomikê*, and he calls mathematical objects the only type of objects that are "really knowable" (ὄντως ἐπιστητά).[6] Moreover, he describes mathematics as, with the exception of theology, superior to the other species of theoretical philosophy.[7] Similarly, in the sixteenth century the French mathematician Oronce Fine characterizes mathematics, as well as its relationship to physics and metaphysics, in a manner that recalls *Almagest* 1.1: "Mathematics is intermediate between the natural or physical investigation and the supernatural or metaphysical investigation (which deserve to be called conjecture rather than knowledge), taking part, with the natural, in matter, and joining the supernatural in the fact that it considers the same things as if they were separated from matter."[8] Fine appropriates both Ptolemy's argument for the intermediate status of mathematics and his epistemological assessment of the three theoretical sciences. Albert the Great in the thirteenth century cites Ptolemy in his commentary on Euclid's *Elements*. He calls Ptolemy "great in all divisions of learning" and paraphrases the epistemological argument of *Almagest* 1.1, which judges physics and theology to be conjectural and mathematics alone as productive of knowledge.[9] Albert's student, Thomas Aquinas repeatedly cites Ptolemy in his commentary on Boethius's *De trinitate*. When affirming the trichotomy of the theoretical sciences, he appeals to Aristotle and Ptolemy side by side, and he quotes Ptolemy not only with respect to the sciences' epistemological status, as knowledge or conjecture, but also with respect to the indisputability of mathematical demonstrations: "But as Ptolemy says in the beginning of the

5. Syrianus, *In metaphysica* 95.13–17.

6. Theodorus Metochites, *Stoicheiôsis astronomikê* 1.2.8.91. I would like to thank Anne-Laurence Caudano for bringing Metochites's epistemology to my attention.

7. Ibid. 1.3.1.4–6.

8. Oronce Fine, *Protomathesis* 1532B, sig. AA2ʳ, translation modified from Axworthy, "Epistemological Foundations": Sunt enim Mathematicae mediae inter naturalem seu Physicam auscultationem, & supernaturalem siue Metaphysicam (quae coniecturae potius, quàm scientiae dici meruerunt) cum naturali participantes in materia, & cum supernaturali conuenientes in eo, quoniam res easdem, perinde acsi forent à materia seiunctae considerant. Axworthy argues for the influence of Ptolemy on Fine: ibid., 36.

9. See Lo Bello, "Albertus Magnus and Mathematics," 5–6. I must thank Angela Axworthy for bringing to my attention these references to Ptolemy by Fine, Albert, and Aquinas.

Almagest, 'Mathematics alone, if one applies himself diligently to it, will give the inquirer after knowledge firm and unshaken certitude by demonstrations carried out with unquestionable methods.'"[10]

Ptolemy's and Hero of Alexandria's claim that geometrical demonstration is indisputable echoes through the Greek corpus in the works of at least ten historical figures from the third to the fourteenth century. Notably, Proclus, in the fifth century, claims that Euclid improved on the work of his predecessors by reforming their propositions into irrefutable demonstrations (ἀνελέγκτους ἀποδείξεις), and he appropriates the geometrical style of proof in his studies of theology and physics, the *Elements of Theology* and *Elements of Physics*, respectively.[11] In a way, Proclus combines Ptolemy's characterization of geometrical demonstration as indisputable with his mandate to apply mathematics to theology and physics. For Ptolemy, this application means inferring the nature of the Prime Mover by way of an analogy with mathematical, specifically astronomical, objects and studying bodies mathematically before investigating their physical properties. For Proclus, mathematics' application entails composing discourses on nonmathematical objects in the style of geometry. It is the demonstration of theological propositions on the procession and characteristics of the various classes of gods as well as the demonstration from first principles of the existence of an unmoved cause of motion and change in the world. For Proclus, the irrefutability of geometrical demonstration extends to other fields of inquiry when their discourses are constructed in a geometrical style.

The appropriation of the geometrical style for nonmathematical studies flourished in the seventeenth century.[12] Prompted by Mersenne, Descartes, for example, presents the central arguments of his *Meditations* in a geometrical form at the end of the *Second Replies*. He divides them into definitions, postulates, axioms or common notions, propositions and their demonstrations. Similarly, Spinoza employs the geometrical style in several works, most significantly in his *Ethics Demonstrated in Geometrical Order*. Thomas Hobbes

10. Aquinas, *Expositio super librum Boethii De trinitate*, q. 6, a. 1, translation modified from Maurer, in Aquinas, *Division and Methods*: Sed contra, disciplinaliter procedere est demonstrative procedere et per certitudinem. Sed, sicut Ptolemaeus in principio Almagesti dicit, "solum mathematicum genus, si quis huic diligentiam exhibeat inquisitionis, firmam stabilemque fidem intendentibus notitiam dabit, velut demonstratione per indubitabiles vias facta." See also ibid., q. 5, a. 1, 3.

11. Proclus, *In primum Euclidis elementorum* 68.10. Cf. 3.9, 11.21–22, 70.17. After discussing Ptolemy's epistemology in *Almagest* 1.1, J. L. Berggren states that the view of the mathematical argument as a source of certainty influenced Proclus, but he does not go so far as to claim that Ptolemy's philosophy specifically influenced Proclus: Berggren, "Mathematics and Religion," 17.

12. For a history of the geometrical order, see Schüling, *Geschichte der axiomatischen Methode*.

embraces the geometrical style in his *Leviathan*. Contributing to the *Quaestio de certitudine mathematicarum*, the sixteenth- and seventeenth-century debate on the scientific status and certainty of mathematics, he calls geometry "the onely Science that it hath pleased God hitherto to bestow on mankind," and he observes, concerning philosophers, "For there is not one of them that begins his ratiocination from the Definitions, or Explications of the names they are to use; which is a method that hath been used onely in Geometry; whose Conclusions have thereby been made indisputable."[13] Could the appropriation of the geometrical demonstration and the extension of its indisputability to nonmathematical studies have its roots in Ptolemy's epistemology?

The influence of Ptolemy's ethics likewise requires further study. Notably, in his commentary on the *Almagest* Theon of Alexandria proclaims that the overall message of *Almagest* 1.1 is ethical. He paraphrases Ptolemy as claiming that the human being who lives well maintains the fine and well-ordered state.[14] It is possible that the authors of commentaries on Ptolemy's *Almagest*, including Theon, adopted Ptolemy's ethical theory. At the very least, we know that later philosophers ascribed some ethical value to mathematics, as a type of mental or spiritual exercise. Matthew Jones has shown that Descartes, Pascal, and Leibniz considered mathematics to be avenues to the good life.[15] According to these philosophers, mathematics cultivates the mind. It exercises the intelligence and reveals the powers and limits of human reason. Yet, none of these philosophers took mathematics to be necessary or sufficient for the good life. For Ptolemy it is. Ptolemy's ethics is distinctly mathematical. It is only through mathematics that a human being achieves the good life.

Is Ptolemy alone in his devotion to the mathematical way of life? At the very least, we know that he was not alone in ascribing some ethical benefit to mathematics. Ptolemy's ethics is, after all, Platonic, and it resonates with mathematical practice well into the seventeenth century. What requires further study is why mathematicians throughout history chose to be mathematicians, or why individuals who concentrated their studies on mathematics did just that. In the modern world we ascribe a high value to mathematics and the mathematical sciences—particularly with respect to the economic rise of the so-called STEM fields of science, technology, engineering, and mathematics—but a benefit supplemental to this economic payout and the value in the liberal, economically independent, study of mathematics are not self evident. To choose to devote one's studies and the better part of one's life

---

13. Hobbes, *Leviathan*, 28, 34. On the relation of Hobbes's work to the *Quaestio*, see especially Mancosu, "Aristotelian Logic," 255–58.

14. Theon, *In Ptolemaei syntaxin mathematicam* 320.1–3.

15. M. L. Jones, *Good Life*.

to mathematics, an individual requires some reason or purpose for doing so, and, according to Ptolemy, the benefits mathematics provides are epistemological and ethical. Mathematics is the only field of inquiry through which human beings can acquire knowledge, and it is the only path to the good life. In Ptolemy's philosophy, the best way an individual can live his life is to live the mathematical way of life.

BIBLIOGRAPHY

Aaboe, Asger. *Episodes from the Early History of Mathematics.* New York: L. W. Singer, 1964.

Acerbi, Fabio, Nicolas Vinel, and Bernard Vitrac. "Les *Prolégomènes à l'Almageste.* Une édition à partir des manuscrits les plus anciens: Introduction générale—parties I–III." *SCIAMVS* 11 (2010): 53–210.

Alcinous. *The Handbook of Platonism.* Translated by John Dillon. Oxford: Clarendon, 1993.

Alexander of Aphrodisias. *On Aristotle Metaphysics 1.* Translated by William E. Dooley. Ithaca, NY: Cornell University Press, 1989.

———. *On Aristotle Metaphysics 4.* Translated by Arthur Madigan. London: Duckworth, 1993.

———. *On Aristotle On Sense Perception.* Translated by Alan Towey. London: Duckworth, 2000.

———. *"Quaestiones" 1.1–2.15.* Translated by Robert W. Sharples. London: Duckworth, 1992.

Allan, Donald J. "The Fine and the Good in the *Eudemian Ethics.*" In *Untersuchungen zur Eudemischen Ethik,* edited by Paul Moraux and Dieter Harlfinger, 63–72. Berlin: De Gruyter, 1971.

Aquinas, Thomas. *The Division and Methods of the Sciences: Questions V and VI of His Commentary on the "De Trinitate" of Boethius.* Translated by Armand Maurer. 4th rev. ed. Toronto: Pontifical Institute of Mediaeval Studies, 1986.

Aristotle. *Aristotle's "Metaphysics": Books "M" and "N."* Translated by Julia Annas. Oxford: Clarendon, 1976.

———. *The Complete Works of Aristotle: The Revised Oxford Translation.* 2 vols. Edited by Jonathan Barnes. Princeton, NJ: Princeton University Press, 1984.

———. *De anima: Books II and III, with Certain Passages from Book I.* Translated by D. W. Hamlyn. Oxford: Clarendon, 1968.

———. *Meteorologica.* Translated by H.D.P. Lee. Cambridge, MA: Harvard University Press, 1952.

———. *The Nicomachean Ethics.* Translated by H. Rackham. Rev. ed. Cambridge, MA: Harvard University Press, 1934.

———. *On the Heavens.* Translated by W.K.C. Guthrie. Cambridge, MA: Harvard University Press, 1939.

Asmis, Elizabeth. *Epicurus' Scientific Method.* Ithaca, NY: Cornell University Press, 1984.

Aspasius. *On Aristotle's "Nicomachean Ethics 1–4, 7–8."* Translated by David Konstan. Ithaca, NY: Cornell University Press, 2006.

Asper, Markus. "Mathematik, Milieu, Text. Die frühgriechische(n) Mathematik(en) und ihr Umfeld." *Sudhoffs Archiv* 87, no. 1 (2003): 1–31.

Asper, Markus, in collaboration with Anna-Maria Kanthak, eds. *Writing Science: Medical and Mathematical Authorship in Ancient Greece.* Berlin: De Gruyter, 2013.

Athanassiadi, Polymnia, and Michael Frede, eds. *Pagan Monotheism in Late Antiquity*. New York: Oxford University Press, 1999.

Axworthy, Angela. "The Epistemological Foundations of the Propaedeutic Status of Mathematics according to the Epistolary and Prefatory Writings of Oronce Fine." In *The Worlds of Oronce Fine: Mathematics, Instruments, and Print in Renaissance France*, edited by Alexander Marr, 31–51. Donington, UK: Shaun Tyas, 2009.

Barker, Andrew, ed. *Greek Musical Writings*. Vol. 2, *Harmonic and Acoustic Theory*. Cambridge: Cambridge University Press, 1989.

———. "Mathematical Beauty Made Audible: Musical Aesthetics in Ptolemy's *Harmonics*." *Classical Philology* 105, no. 4 (October 2010): 403–20.

———. "Ptolemy's Musical Models for Mind-Maps and Star-Maps." In *Rationality and Reality: Conversations with Alan Musgrave*, edited by Colin Cheyne and John Worrall, 273–91. Dordrecht: Springer, 2006.

———. "Ptolemy's Pythagoreans, Archytas, and Plato's Conception of Mathematics." *Phronesis* 39, no. 2 (1994): 113–35.

———. *The Science of Harmonics in Classical Greece*. Cambridge: Cambridge University Press, 2007.

———. *Scientific Method in Ptolemy's "Harmonics."* Cambridge: Cambridge University Press, 2000.

Berggren, John Lennart. "Mathematics and Religion in Ancient Greece and Medieval Islam." In *The Alexandrian Tradition: Interactions between Science, Religion, and Literature*, edited by Luis Arturo Guichard, Juan Luis García Alonso, and María Paz de Hoz, 11–34. Bern: Peter Lang, 2014.

Berggren, John Lennart, and Alexander Jones. *Ptolemy's "Geography": An Annotated Translation of the Theoretical Chapters*. Princeton, NJ: Princeton University Press, 2000.

Bernard, Alain. "The Significance of Ptolemy's *Almagest* for Its Early Readers." *Revue de synthèse* 131, no. 4 (2010): 495–521.

Bodnár, István M. "Alexander of Aphrodisias on Celestial Motions." *Phronesis* 42, no. 2 (1997): 190–205.

Boll, Franz. "Studien über Claudius Ptolemäus: Ein Beitrag zur Geschichte der griechischen Philosophie und Astrologie." *Jahrbücher für classische Philologie*, supplement 21 (1894): 49–243.

Bouché-Leclercq, Auguste. *L'astrologie grecque*. Paris: E. Leroux, 1899.

Bowen, Alan C. "The Demarcation of Physical Theory and Astronomy by Geminus and Ptolemy." *Perspectives on Science* 15, no. 3 (2007): 327–58.

———. "Review of *Ptolemy's Universe: The Natural Philosophical and Ethical Foundations of Ptolemy's Astronomy* by Liba Chaia Taub." *Isis* 85, no. 1 (March 1994): 140–41.

———. *Simplicius on the Planets and Their Motions: In Defense of a Heresy*. Leiden: Brill, 2013.

Bowen, Alan C., and Robert B. Todd. *Cleomedes' Lectures on Astronomy: A Translation of "The Heavens" with an Introduction and Commentary*. Berkeley: University of California Press, 2004.

Bowen, Alan C., and Christian Wildberg, eds. *New Perspectives on Aristotle's "De caelo."* Leiden: Brill, 2009.

Britton, John Phillips. *Models and Precision: The Quality of Ptolemy's Observations and Parameters*. New York: Garland, 1992.

Bydén, Börje. *Theodore Metochites' "Stoicheiosis Astronomike" and the Study of Natural Philosophy and Mathematics in Early Palaiologan Byzantium.* Gothenburg: Acta Universitatis Gothoburgensis, 2003.

Cicero. *De natura deorum: Academica.* Translated by H. Rackham. London: W. Heinemann, 1933.

Creese, David. *The Monochord in Ancient Greek Harmonic Science.* Cambridge: Cambridge University Press, 2010.

Cumont, Franz. "Les noms des planètes et l'astrolâtrie chez les Grecs." *Antiquité Classique* 4 (1935): 5–43.

Cuomo, Serafina. *Ancient Mathematics.* London: Routledge, 2001.

Dambska, Izydora. "La théorie de la science dans les oeuvres de Claude Ptolémée." *Organon* 8 (1971): 109–22.

De Lacy, Phillip. "Galen's Platonism." *American Journal of Philology* 93 (1972): 27–39.

Delambre, J.B.J. *Histoire de l'astronomie ancienne.* 2 vols. Paris: Courcier, 1817.

———. *Histoire de l'astronomie du moyen âge.* Paris: Courcier, 1819.

de Pace, Anna. "Elementi Aristotelici nell'*Ottica* di Claudio Tolomeo." *Rivista Critica di Storia della Filosofia* 36 (1981): 123–38; 37 (1982): 243–76.

Dillon, John M. "Commentary." In Alcinous, *Handbook of Platonism,* 49–211.

———. "Introduction." In Alcinous, *Handbook of Platonism,* ix–xliii.

———. *The Middle Platonists: A Study of Platonism, 80 B.C. to A.D. 220.* London: Duckworth, 1977.

Dillon, John M., and A. A. Long, eds. *The Question of "Eclecticism": Studies in Later Greek Philosophy.* Berkeley: University of California Press, 1988.

Drake, Stillman. "Ptolemy, Galileo, and Scientific Method." *Studies in History and Philosophy of Science* 9, no. 2 (1978): 99–115.

Dreyer, J.L.E. *History of the Planetary Systems from Thales to Kepler.* Cambridge: Cambridge University Press, 1906.

Duhem, Pierre. "ΣΩΖΕΙΝ ΤΑ ΦΑΙΝΟΜΕΝΑ: Essai sur la notion de théorie physique de Platon à Galilée." *Annales de philosophie chrétienne* 156 (1908): 113–38, 277–302, 352–77, 482–514, 576–92.

———. *ΣΩΖΕΙΝ ΤΑ ΦΑΙΝΟΜΕΝΑ: Essai sur la notion de théorie physique de Platon à Galilée.* Paris: A. Hermann, 1908.

———. *To Save the Phenomena: An Essay on the Idea of Physical Theory from Plato to Galileo.* Translated by Edmund Doland and Chaninah Maschler. Chicago: University of Chicago Press, 1969.

Düring, Ingemar. *Die Harmonielehre des Klaudios Ptolemaios.* Gothenburg: Elanders Boktryckeri Aktiebolag, 1930.

———. *Ptolemaios und Porphyrios über die Musik.* Gothenburg: Elanders Boktryckeri Aktiebolag, 1934.

Evans, James. *The History and Practice of Ancient Astronomy.* New York: Oxford University Press, 1998.

———. "On the Function and the Probable Origin of Ptolemy's Equant." *American Journal of Physics* 52, no. 12 (December 1984): 1080–89.

Evans, James, and John Lennart Berggren. *Geminos's "Introduction to the Phenomena": A Translation and Study of a Hellenistic Survey of Astronomy.* Princeton, NJ: Princeton University Press, 2006.

Falcon, Andrea. *Aristotelianism in the First Century BCE: Xenarchus of Seleucia.* Cambridge: Cambridge University Press, 2012.

———, ed. *Brill's Companion to the Reception of Aristotle in Antiquity.* Leiden: Brill, 2016.

Feke, Jacqueline. "Mathematizing the Soul: The Development of Ptolemy's Psychological Theory from *On the Kritērion and Hēgemonikon* to the *Harmonics.*" *Studies in History and Philosophy of Science* 43, no. 4 (December 2012): 585–94.

———. "Meta-mathematical Rhetoric: Hero and Ptolemy against the Philosophers." *Historia Mathematica* 41, no. 3 (2014): 261–76.

———. "Ptolémée d'Alexandrie (Claude)." In *Dictionnaire des philosophes antiques.* Vol. 5, 2nd part, *De Plotina à Rutilius Rufus*, edited by Richard Goulet, 1718–33. Paris: CNRS Éditions, 2012.

———. "Ptolemy in Philosophical Context: A Study of the Relationships between Physics, Mathematics, and Theology." PhD diss., University of Toronto, 2009.

———. "Ptolemy's Defense of Theoretical Philosophy." *Apeiron* 45, no. 1 (January 2012): 61–90.

———. "Ptolemy's Philosophy of Geography." In Claudio Ptolomeo, *Geografía (Capítulos teóricos)*, edited by René Ceceña, 281–326. México: Universidad Nacional Autónoma de México, 2018.

———. "What Can We Know of What the Romans Knew? Comments on Daryn Lehoux's *What Did the Romans Know? An Inquiry into Science and Worldmaking.*" *Expositions* 6, no. 2 (2012): 23–32.

Feke, Jacqueline, and Alexander Jones. "Ptolemy." In *Cambridge History of Philosophy in Late Antiquity*, vol. 1, edited by Lloyd P. Gerson, 197–209. Cambridge: Cambridge University Press, 2010.

Finamore, John F., and John M. Dillon. *Iamblichus "De anima": Text, Translation, and Commentary.* Leiden: Brill, 2002.

Fletcher, Richard. *Apuleius' Platonism: The Impersonation of Philosophy.* Cambridge: Cambridge University Press, 2014.

Fowler, Ryan C. *Imperial Plato: Albinus, Maximus, Apuleius.* Las Vegas: Parmenides, 2016.

Frommhold, Katrin. *Die Bedeutung und Berechnung der Empfängnis in der Astrologie der Antike.* Orbis antiquus, vol. 38. Münster, Germany: Aschendorff, 2004.

Furley. David J. *The Greek Cosmologists.* Cambridge: Cambridge University Press, 1987.

Gee, Teri. "Strategies of Defending Astrology: A Continuing Tradition." PhD diss., University of Toronto, 2012.

Genequand, Charles. *Alexander of Aphrodisias on the Cosmos.* Leiden: Brill, 2001.

———. "Quelques aspects de l'idée de la nature, d'Aristote à al-Ghazâlî." *Revue de Théologie et de Philosophie* 116 (1984): 105–29.

Gerson, Lloyd P. *Aristotle and Other Platonists.* Ithaca, NY: Cornell University Press, 2005.

———. *God and Greek Philosophy: Studies in the Early History of Natural Theology.* London: Routledge, 1990.

Gill, Christopher. *The Structured Self in Hellenistic and Roman Thought.* Oxford: Oxford University Press, 2006.

Gill, Mary Louise. *Aristotle on Substance: The Paradox of Unity*. Princeton, NJ: Princeton University Press, 1989.

Gill, Mary Louise, and James G. Lennox, eds. *Self-Motion: From Aristotle to Newton*. Princeton, NJ: Princeton University Press, 1994.

Gill, Mary Louise, and Pierre Pellegrin, eds. *A Companion to Ancient Philosophy*. Malden, MA: Blackwell, 2006.

Goldstein, Bernard R. "The Arabic Version of Ptolemy's *Planetary Hypotheses*." *Transactions of the American Philosophical Society* 57, no. 4. (1967): 3–55.

Goldstein, Bernard R., and Alan C. Bowen. "A New View of Early Greek Astronomy." *Isis* 74, no. 3 (September 1983): 330–40.

Gourinat, Jean-Baptiste. "The Stoics on Matter and Prime Matter: 'Corporealism' and the Imprint of Plato's *Timaeus*." In *God and Cosmos in Stoicism*, edited by Ricardo Salles, 46–70. Oxford: Oxford University Press, 2009.

Graeser, Andreas, ed. *Mathematics and Metaphysics in Aristotle*. Bern: P. Haupt, 1987.

Grasshoff, Gerd. *The History of Ptolemy's Star Catalogue*. New York: Springer, 1990.

Griffin, Michael J. *Aristotle's "Categories" in the Early Roman Empire*. Oxford: Oxford University Press, 2015.

Guichard, Luis Arturo, Juan Luis García Alonso, and María Paz de Hoz, eds. *The Alexandrian Tradition: Interactions between Science, Religion, and Literature*. Bern: Peter Lang, 2014.

Guillaumin, Jean-Yves. "L'éloge de la géométrie dans la préface du livre 3 des *Metrika* d'Héron d'Alexandrie." *Revue des Études Anciennes* 99 (1997): 91–99.

Hadot, Pierre. *Philosophy as a Way of Life*. Edited by Arnold I. Davidson; translated by Michael Chase. Oxford: Blackwell, 1995.

Hagel, Stefan. *Ancient Greek Music: A New Technical History*. Cambridge: Cambridge University Press, 2009.

Hahm, David E. *The Origins of Stoic Cosmology*. Columbus, OH: Ohio State University Press, 1977.

Hamilton, N. T., N. M. Swerdlow, and G. J. Toomer. "The *Canobic Inscription*: Ptolemy's Earliest Work." In *From Ancient Omens to Statistical Mechanics: Essays on the Exact Sciences Presented to Asger Aaboe*, edited by J. Lennart Berggren and Bernard R. Goldstein, 55–73. Copenhagen: University Library, 1987.

Hamm, Elizabeth Anne. "Ptolemy's Planetary Theory: An English Translation of Book One, Part A of the *Planetary Hypotheses* with Introduction and Commentary." PhD diss., University of Toronto, 2011.

Hankinson, R. J. *Cause and Explanation in Ancient Greek Thought*. Oxford: Clarendon, 1998.

———. "Galen." In *The Cambridge History of Philosophy in Late Antiquity*, vol. 1, edited by Lloyd P. Gerson, 210–31. Cambridge: Cambridge University Press, 2010.

———. "Galen and the Best of All Possible Worlds." *Classical Quarterly* 39, no. 1 (1989): 206–27.

———. "Usage and Abusage: Galen on Language." In *Companions to Ancient Thought*, vol. 3, *Language*, edited by Stephen Everson, 166–87. Cambridge: Cambridge University Press, 1994.

Harari, Orna. "John Philoponus and the Conformity of Mathematical Proofs to Aristotelian Demonstrations." In *The History of Mathematical Proof in Ancient Traditions*, edited by Karine Chemla, 206–27. Cambridge: Cambridge University Press, 2012.

Harvey, F. D. "Two Kinds of Equality." *Classica et mediaevalia* 26 (1965): 101–46.

Hatzimichali, Myrto. *Potamo of Alexandria and the Emergence of Eclecticism in Late Hellenistic Philosophy.* Cambridge: Cambridge University Press, 2011.

Heath, T. L. *The Thirteen Books of Euclid's "Elements": Translated from the Text of Heiberg with Introduction and Commentary.* Cambridge: Cambridge University Press, 1908.

Heiberg, J. L., ed. *Claudii Ptolemaei: Opera quae exstant omnia.* Vol. 2, *Opera astronomica minora.* Leipzig: Teubner, 1907.

Heilen, Stephan. "Ptolemy's Doctrine of the Terms and Its Reception." In *Ptolemy in Perspective: Use and Criticism of His Work from Antiquity to the Nineteenth Century,* edited by Alexander Jones, 45–93. Dordrecht: Springer, 2010.

Hobbes, Thomas. *Leviathan.* Edited by Richard Tuck. Revised student ed. Cambridge: Cambridge University Press, 1996.

Hübner, Wolfgang. *Die Begriffe "Astrologie" und "Astronomie" in der Antike: Wortgeschichte und Wissenschaftssystematik mit einer Hypothese zum Terminus "Quadrivium."* Abhandlungen der Akademie der Wissenschaften und der Literatur Mainz. Geistes- und Sozialwissenschaftliche Klasse 1989/7. Wiesbaden, Germany: Franz Steiner, 1989.

———, ed. *Claudii Ptolemaei: Opera Quae Exstant Omnia.* Vol. 3.1, *Apotelesmatika.* 2nd ed. Stuttgart: Teubner, 1998.

Huffman, Carl. "Response to Barker." *Classical Philology* 105, no. 4 (October 2010): 420–25.

Hutchinson, D. S., and Monte Ransome Johnson. "Authenticating Aristotle's *Protrepticus.*" *Oxford Studies in Ancient Philosophy* 29 (2005): 193–294.

Inwood, Brad, ed. *The Cambridge Companion to the Stoics.* Cambridge: Cambridge University Press, 2003.

———. *Ethics after Aristotle.* Cambridge, MA: Harvard University Press, 2014.

———. *Ethics and Human Action in Early Stoicism.* Oxford: Clarendon, 1985.

Inwood, Brad, and L. P. Gerson, eds. *The Epicurus Reader: Selected Writings and Testimonia.* Indianapolis: Hackett, 1994.

———, trans. *Hellenistic Philosophy: Introductory Readings.* Indianapolis: Hackett, 1988.

Johansen, Thomas K. *Plato's Natural Philosophy: A Study of the "Timaeus-Critias."* Cambridge: Cambridge University Press, 2004.

Jones, Alexander. "In order That We Should Not Ourselves Appear to Be Adjusting Our Estimates . . . to Make Them Fit Some Predetermined Amount." In *Wrong for the Right Reasons,* edited by Jed Z. Buchwald and Allan Franklin, 17–39. Dordrecht: Springer, 2005.

———. "Ptolemy." In *New Dictionary of Scientific Biography,* vol. 6, edited by Noretta Koertge, 173–78. Detroit: Scribner, 2008.

———. "Ptolemy's *Canobic Inscription* and Heliodorus' Observation Reports." *SCIAMVS* 6 (2005): 53–97.

———. "Ptolemy's First Commentator." *Transactions of the American Philosophical Society,* 80, no. 7. (1990): i–vi, 1–61.

———. "Ptolemy's Mathematical Models and their Meaning." In *Mathematics and the Historian's Craft: The Kenneth O. May Lectures,* edited by Glen van Brummelen and Michael Kinyon, 23–42. New York: Springer, 2005.

Jones, Matthew L. *The Good Life in the Scientific Revolution: Descartes, Pascal, Leibniz, and the Cultivation of Virtue.* Chicago: University of Chicago Press, 2006.

Karamanolis, George E. *Plato and Aristotle in Agreement? Platonists on Aristotle from Antiochus to Porphyry.* Oxford: Clarendon, 2006.

Kattsoff, Louis O. "Ptolemy and Scientific Method: A Note on the History of an Idea," *Isis* 38, no. 1/2 (November 1947): 18–22.

Kosman Aryeh. "Divine Being and Divine Thinking in *Metaphysics* Lambda." In *Proceedings of the Boston Area Colloquium in Ancient Philosophy*, vol. 3, edited by John J. Cleary, 165–88. Lanham, MD: University Press of America, 1987.

Lammert, Friedrich. "Eine neue Quelle für die Philosophie der mittleren Stoa." *Wiener Studien* 41 (1919): 113–21; 42 (1920–21): 36–46.

Lear, Jonathan. *Aristotle: The Desire to Understand.* Cambridge: Cambridge University Press, 1988.

Lee, Mi-Kyoung. *Epistemology after Protagoras: Responses to Relativism in Plato, Aristotle, and Democritus.* Oxford: Clarendon, 2005.

Lehoux, Daryn. "Observation and Prediction in Ancient Astrology." *Studies in History and Philosophy of Science* 35, no. 2 (2004): 227–46.

———. "Observers, Objects, and the Embedded Eye; or, Seeing and Knowing in Ptolemy and Galen." *Isis* 98, no. 3 (2007): 447–67.

———. *What Did the Romans Know? An Inquiry into Science and Worldmaking.* Chicago: University of Chicago Press, 2012.

Lilla, Salvatore R. C. *Clement of Alexandria: A Study in Christian Platonism and Gnosticism.* Oxford: Oxford University Press, 1971.

Lloyd, G.E.R. *Aristotelian Explorations.* Cambridge, Cambridge University Press, 1996.

———. "*Metaphysics* Λ 8." In *Aristotle's "Metaphysics" Lambda: Symposium Aristotelicum*, edited by Michael Frede and David Charles, 245–73. Oxford: Clarendon, 2000.

———. *Methods and Problems in Greek Science.* Cambridge: Cambridge University Press, 1991.

———. "The Pluralism of Greek 'Mathematics.'" In *The History of Mathematical Proof in Ancient Traditions*, edited by Karine Chemla, 294–310. Cambridge: Cambridge University Press, 2012.

———. "Saving the Appearances." *Classical Quarterly* 28, no. 1 (1978): 202–22.

———. "The Theories and Practices of Demonstration." In *Aristotelian Explorations*, 7–37.

Lo Bello, Anthony. "Albertus Magnus and Mathematics: A Translation with Annotations of Those Portions of the Commentary on Euclid's *Elements* Published by Bernhard Geyer." *Historia Mathematica* 10, no. 1 (February 1983): 3–23.

Long, A. A. "Astrology: Arguments Pro and Contra." In *Science and Speculation: Studies in Hellenistic Theory and Practice*, edited by Jonathan Barnes, 165–92. Cambridge: Cambridge University Press, 1982.

———. *From Epicurus to Epictetus: Studies in Hellenistic and Roman Philosophy.* Oxford: Clarendon, 2006.

———. *Hellenistic Philosophy: Stoics, Epicureans, Sceptics.* London: Duckworth, 1974.

———. "Ptolemy on the Criterion: An Epistemology for the Practising Scientist." In *The Criterion of Truth: Essays Written in Honour of George Kerferd together with a Text and Translation (with Annotations) of Ptolemy's "On the Kriterion and Hegemonikon,"* edited by Pamela Huby and Gordon Neal, 151–78. Liverpool: Liverpool University Press, 1989.

———. "Stoic Psychology." In *The Cambridge History of Hellenistic Philosophy*, edited by Keimpe Algra, Jonathan Barnes, Jaap Mansfeld, and Malcom Schofield, 560–84. Cambridge: Cambridge University Press, 1999.

————. *Stoic Studies*. Cambridge: Cambridge University Press, 1996.

Long, A. A., and D. N. Sedley. *The Hellenistic Philosophers*. 2 vols. Cambridge: Cambridge University Press, 1987.

Lucci, Gualberto. "Criterio e metodologia in Sesto Empirico e Tolomeo." *Annali dell'Istituto di Filosofia di Firenze* 2 (1980): 23–52.

Mancosu, Paolo. "Aristotelian Logic and Euclidean Mathematics: Seventeenth-Century Developments of the *Quaestio de certitudine mathematicarum*." *Studies in History and Philosophy of Science* 23, no. 2 (1992): 241–65.

————. *Philosophy of Mathematics and Mathematical Practice in the Seventeenth Century*. New York: Oxford University Press, 1996.

Mansfeld, Jaap. *Prolegomena Mathematica: From Apollonius of Perga to Late Neoplatonism*. Leiden: Brill, 1998.

————. *Prolegomena: Questions to Be Settled Before the Study of an Author, or a Text*. Leiden: Brill, 1994.

Manuli, Paola. "Claudio Tolomeo: Il criterio e il principio." *Rivista Critica di Storia della Filosofia* 36, no. 1 (1981): 64–88.

Miller, Jon, and Brad Inwood, eds. *Hellenistic and Early Modern Philosophy*. Cambridge: Cambridge University Press, 2003.

Morison, Ben. "Language." In *The Cambridge Companion to Galen*, edited by R. J. Hankinson, 116–56. Cambridge: Cambridge University Press, 2008.

Mueller, Ian. "Aristotle's Doctrine of Abstraction in the Commentators." In Sorabji, *Aristotle Transformed*, 463–80.

Murschel, Andrea. "The Structure and Function of Ptolemy's Physical Hypotheses of Planetary Motion." *Journal for the History of Astronomy* 26, no. 1 (February 1995): 33–61.

Netz, Reviel. "The Bibliosphere of Ancient Science (Outside of Alexandria)." *N.T.M. Zeitschrift für Geschichte der Wissenschaften, Technik und Medizin* 19, no. 3 (2011): 239–69.

————. *The Shaping of Deduction in Greek Mathematics: A Study in Cognitive History*. Cambridge: Cambridge University Press, 1999.

————. "What Did Greek Mathematicians Find Beautiful?" *Classical Philology* 105, no. 4 (October 2010): 426–44.

Neugebauer, Otto. *The Exact Sciences in Antiquity*. Princeton, NJ: Princeton University Press, 1952.

————. *A History of Ancient Mathematical Astronomy*. 3 vols. Berlin: Springer, 1975.

————. "Über eine Methode zur Distanzbestimmung Alexandria-Rom bei Heron I." *Det Kongelige Danske Videnskabernes Selskab* 26, no. 2 (1938): 3–26.

Newton, Robert R. *The Crime of Claudius Ptolemy*. Baltimore: Johns Hopkins University Press, 1977.

Nussbaum, Martha Craven. *Aristotle's "De motu animalium": Text with Translation, Commentary and Interpretive Essays*. Princeton, NJ: Princeton University Press, 1978.

Nussbaum, Martha Craven, and Amélie Oksenberg Rorty, eds. *Essays on Aristotle's "De anima."* Oxford: Oxford University Press, 1992.

O'Meara, Dominic J. *Pythagoras Revived: Mathematics and Philosophy in Late Antiquity*. Oxford: Clarendon, 1989.

Orlando, Carmela, and Rita Torre, "Lessico astronomico-astrologico greco." In *Atti del I Seminario di studi sui lessici tecnici greci e latini (Messina, 8–10 marzo 1990)*, edited by Paola Radici Colace and Maria Caccamo Caltabiano, 291–309. Messina: Accademia Peloritana dei Pericolanti, 1991.

Pedersen, Olaf. *A Survey of the "Almagest," with Annotation and New Commentary by Alexander Jones*. New York: Springer, 2011.

Plato. *Complete Works*. Edited by John M. Cooper, associate edited by D. S. Hutchinson. Indianapolis: Hackett, 1997.

———. *Meno*. Translated by G.M.A. Grube. 2nd ed. Indianapolis: Hackett, 1981.

———. *Republic*. Translated by G.M.A. Grube. Revised by C.D.C. Reeve. Indianapolis: Hackett, 1992.

———. *Timaeus*. Translated by Donald J. Zeyl. Indianapolis: Hackett, 2000.

Ptolemaeus, Claudius. "On the Kriterion and Hegemonikon," edited by Liverpool-Manchester Seminar on ancient Greek philosophy. In *The Criterion of Truth: Essays Written in Honour of George Kerferd Together with a Text and Translation (with Annotations) of Ptolemy's "On the Kriterion and Hegemonikon,"* edited by Pamela Huby and Gordon Neal, 179–230. Liverpool: Liverpool University Press, 1989.

Ptolemy. *Ptolemy's "Almagest."* Translated by G. J. Toomer. Princeton, NJ: Princeton University Press, 1998.

———. *Tetrabiblos*. Translated by F. E. Robbins. Cambridge, MA: Harvard University Press, 1940.

Raffa, Massimo. *La scienza armonica di Claudio Tolemeo: Saggio critico, traduzione e commento*. Messina: A. Sfameni, 2002.

Rashed, Marwan. "Contre le mouvement rectiligne naturel: Trois adversaires (Xénarque, Ptolémée, Plotin) pour une thèse." In *Physics and Philosophy of Nature in Greek Neoplatonism: Proceedings of the European Science Foundation Exploratory Workshop (Il Ciocco, Castelvecchio Pascoli, June 22–24, 2006)*, edited by Riccardo Chiaradonna and Franco Trabattoni, 17–42. Leiden: Brill, 2009.

Rees, D. A. "Bipartition of the Soul in the Early Academy." *Journal of Hellenic Studies* 77, no. 1 (1957): 112–18.

Riedweg, Christoph. *Pythagoras: His Life, Teaching, and Influence*. Translated by Steven Rendall in collaboration with Christoph Riedweg and Andreas Schatzmann. Ithaca, NY: Cornell University Press, 2005.

Rist, John M. *Stoic Philosophy*. London: Cambridge University Press, 1969.

———, ed. *The Stoics*. Berkeley: University of California Press, 1978.

Ross, W. D. *Aristotle*. London: Methuen, 1923.

Samburit, S. *The Physical World of Late Antiquity*. London: Routledge and Kegan Paul, 1962.

Sandbach, F. H. *The Stoics*. New York: Norton, 1975.

Schüling, Hermann. *Die Geschichte der axiomatischen Methode im 16. und beginnenden 17. Jahrhundert (Wandlung der Wissenschaftsauffassung)*. Hildesheim, Germany: G. Olms, 1969.

Sextus Empiricus. *Sextus Empiricus*. 4 vols. Translated by R. G. Bury. Cambridge, MA: Harvard University Press, 1949–71.

Sharples, Robert W. "Alexander of Aphrodisias on Divine Providence: Two Problems." *Classical Quarterly* 32, no. 1 (1982): 198–211.

————. *Alexander of Aphrodisias "On Fate": Text, Translation, and Commentary.* London: Duckworth, 1983.

Sidoli, Nathan. "Heron of Alexandria's Date." *Centaurus* 53 (2010): 55–61.

————. "Mathematical Tables in Ptolemy's *Almagest.*" *Historia Mathematica* 41 (2014): 13–37.

————. "Ptolemy's Mathematical Approach: Applied Mathematics in the Second Century." PhD diss., University of Toronto, 2004.

Sidoli, Nathan, and Glen Van Brummelen, eds. *From Alexandria, through Baghdad: Surveys and Studies in the Ancient Greek and Medieval Islamic Mathematical Sciences in Honor of J. L. Berggren.* Berlin: Springer, 2014.

Siebert, Harald. *Die ptolemäische "Optik" in Spätantike und byzantinischer Zeit: Historiographische Dekonstruktion, textliche Neuerschliessung, Rekontextualisierung.* Stuttgart: Franz Steiner, 2014.

Simplicius. *On Aristotle's "On the Heavens 1.1–4."* Translated by R. J. Hankinson. Ithaca, NY: Cornell University Press, 2002.

————. *On Aristotle's "On the Heavens 2.1–9."* Translated by Ian Mueller. Ithaca, NY: Cornell University Press, 2004.

Smith, A. Mark. "The Physiological and Psychological Grounds of Ptolemy's Visual Theory: Some Methodological Considerations." *Journal of the History of the Behavioral Sciences* 34, no. 3 (1998): 231–46.

————. "Ptolemy's Search for a Law of Refraction: A Case-Study in the Classical Methodology of 'Saving the Appearances' and its Limitations." *Archive for History of Exact Sciences* 26, no. 3 (1982): 221–40.

————. "Ptolemy's Theory of Visual Perception: An English Translation of the *Optics* with Introduction and Commentary." *Transactions of the American Philosophical Society* 86, no. 2. (1996): vii–xi, 1–300.

————. "Saving the Appearances of the Appearances: The Foundations of Classical Geometrical Optics." *Archive for History of Exact Sciences* 24, no. 2 (1981): 73–99.

Solomon, Jon. *Ptolemy "Harmonics": Translation and Commentary.* Leiden: Brill, 2000.

Sorabji, Richard. "Adrastus: Modifications to Aristotle's Physics of the Heavens by Peripatetics and Others, 100 BC to 200 AD." In Sorabji and Sharples, *Greek and Roman Philosophy,* vol. 2, 575–94.

————, ed. *Aristotle Transformed: The Ancient Commentators and Their Influence.* Ithaca, NY: Cornell University Press, 1990.

————, ed. *The Philosophy of the Commentators, 200–600 AD: A Sourcebook.* London: Duckworth, 2004.

Sorabji, Richard, and Robert W. Sharples, eds. *Greek and Roman Philosophy 100 BC–200 AD.* 2 vols. London: Institute of Classical Studies, University of London, 2007.

Steenbakkers, Piet. "The Geometrical Order in the *Ethics.*" In *The Cambridge Companion to Spinoza's "Ethics,"* edited by Olli Koistinen, 42–55. Cambridge: Cambridge University Press, 2009.

Striker, Gisela. *Essays on Hellenistic Epistemology and Ethics.* Cambridge: Cambridge University Press, 1996.

Swerdlow, N. M. "Ptolemy's *Harmonics* and the 'Tones of the Universe' in the *Canobic Inscription.*" In *Studies in the History of the Exact Sciences in Honour of David Pingree,* edited by

Charles Burnett, Jan P. Hogendijk, Kim Plofker, and Michio Yano, 137–80. Leiden: Brill, 2004.

Szabó, Árpád. *The Beginnings of Greek Mathematics*. Translated by A. M. Ungar. Dordrecht: D. Reidel, 1978.

Tarantino, Piero. "L'applicazione della dottrina aristotelica della scienza all'armonica." *Rivista di Filosofia Neo-Scolastica* 2/3 (2012): 289–309.

Tarrant, Harold. *Thrasyllan Platonism*. Ithaca, NY: Cornell University Press, 1993.

Taub, Liba Chaia. *Ptolemy's Universe: The Natural Philosophical and Ethical Foundations of Ptolemy's Astronomy*. Chicago: Open Court, 1993.

Thesaurus Linguae Graecae: A Digital Library of Greek Literature. www.tlg.uci.edu. University of California, Irvine.

Tieleman, Teun. *Chrysippus' "On Affections": Reconstruction and Interpretation*. Leiden: Brill, 2003.

———. *Galen and Chrysippus on the Soul: Argument and Refutation in the "De placitis," Books II–III*. Leiden: Brill, 1996.

Tihon, Anne. "Alexandrian Astronomy in the 2nd Century AD: Ptolemy and His Times." In *The Alexandrian Tradition: Interactions between Science, Religion, and Literature*, edited by Luis Arturo Guichard, Juan Luis García Alonso, and María Paz de Hoz, 73–95. Bern: Peter Lang, 2014.

Todd, Robert B. *Alexander of Aphrodisias on Stoic Physics: A Study of the De Mixtione with Preliminary Essays, Text, Translation and Commentary*. Leiden: Brill, 1976.

Tolsa, Cristian. "The 'Ptolemy' Epigram: A Scholion on the Preface of the *Syntaxis*." *Greek, Roman, and Byzantine Studies* 54, no. 4 (2014): 687–97.

Toomer, G. J. "Ptolemy." In *Dictionary of Scientific Biography*, vol. 11, edited by Charles Coulston Gillispie, 186–206. New York: Scribner, 1975.

Tuplin, C. J., and T. E. Rihll, eds. *Science and Mathematics in Ancient Greek Culture*. Oxford: Oxford University Press, 2002.

Tybjerg, Karin. "Doing Philosophy with Machines: Hero of Alexandria's Rhetoric of Mechanics in Relation to the Contemporary Philosophy." PhD diss., University of Cambridge, 2000.

———. "Hero of Alexandria's Mechanical Treatises: Between Theory and Practice." In *Physik/Mechanik: Geschichte der Mathematik und der Naturwissenschaften in der Antike*, vol. 3, edited by Astrid Schürmann, 204–26. Stuttgart, Franz Steiner, 2005.

———. "Wonder-Making and Philosophical Wonder in Hero of Alexandria." *Studies in History and Philosophy of Science* 34, no. 3 (2003): 443–66.

Vitrac, Bernard. "Les préfaces des textes mathématiques grecs anciens." In *Liber amicorum Jean Dhombres*, edited by P. Radelet-de-Grave in collaboration with C. Brichard, 518–56. Collection Réminiscences, no. 8. Turnhout, Belgium: Brepols, 2008.

Watson, Gerald. *Phantasia in Classical Thought*. Galway: Galway University Press, 1988.

West, M. L. *Ancient Greek Music*. Oxford: Clarendon, 1992.

White, Michael J. "Stoic Natural Philosophy (Physics and Cosmology)." In Inwood, *Cambridge Companion*, 124–52.

Whittaker, John. "Numenius and Alcinous on the First Principle." *Phoenix* 32, no. 2 (1978): 144–54.

Wilson, Malcolm. *Aristotle's Theory of the Unity of Science*. Toronto: University of Toronto Press, 2000.

Wilson, Nigel Guy. *Scholars of Byzantium*. Baltimore: Johns Hopkins University Press, 1983.

Witt, R. E. *Albinus and the History of Middle Platonism*. Cambridge: Cambridge University Press, 1937.

Zeller, Eduard. *A History of Eclecticism in Greek Philosophy*. London: Longmans, 1883.

INDEX

Academica (Cicero), 149

accuracy, 96, 120–24

activity/passivity, 17n19, 49, 80n4, 82, 148–50, 153

Adrastus of Aphrodisias, 199n80

Adversus mathematicos (Sextus Empricus), 29, 42, 126, 154

aether: and celestial powers and rays, 179, 181–83; and circular vs. rectilinear movement, 28, 49–50, 188; composition of, and spherical shape of the heavens, 125; location of, 181; and the soul, 148–50; theory of aethereal physics, 188–95, 203. *See also* element theory

aethereal bodies: as most complete and rational of all entities, 84; and movement of other aethereal bodies, 189–90; physical characteristics of, 187–95; as theological objects, 18, 81. *See also* celestial bodies

Against the Fifth Substance (Xenarchus), 181

air, 49–50, 148–50, 181. *See also* element theory

Albert the Great, 205

Albinus, 69, 176n22

Alcinous, 29, 59; and the contemplative and practical lives, 65–66, 68, 73; and contribution of astronomy to theology, 47; and criterion of truth, 154; definitions of the theoretical sciences, 31–33; and distinction between knowledge and opinion, 33–35; and the soul, 58, 150, 163, 164; and *telos* of becoming godlike, 69. See also *Didaskalikos*

Alexander of Aphrodisias, 20–21, 42, 70, 150, 151, 199, 199n80

Almagest (Ptolemy), 1–2, 1n1, 6, 10, 15–27, 45–51, 125–39; Alcinous and, 29–31; Aristotle and, 10, 15–17, 25; and arithmetic and geometry as indisputable methods, 115–16; and chronology of works, 7–8, 161; and contribution of mathematics to other sciences, 47, 49–50, 200; and defining and ranking the theoretical sciences, 10, 15–19, 81; and *epimarturêsis*, 126–27; and epistemology, 4–5, 10, 26–27, 29–31, 35, 37–38, 50–51, 114, 119, 169; and ethics, 52, 62–63, 65, 68–69, 72, 77–79, 112, 207; Hipparchus and, 133–34; and limits of observation, 131–34; literary style of (philosophical introductory scheme), 29–31, 176n22, 201; and metaphysics, 10, 15; and models of planetary movement, 128–39, 141; and perception, 22, 24; and *phantasiai*, 62–64, 68; and physical and mathematical objects, 20–22, 24, 103; and physics as conjecture, 202; and Prime Mover, 144, 203; and principles (matter, movement, and form), 80; and Ptolemy's preference for simplicity, 194; and Ptolemy's preference for the theoretical sciences, 62; and relationship between theoretical and practical philosophy, 53, 57, 76–78; and spherical shape of heavens, 125–26, 128; Theon of Alexandria's commentary on, 207. *See also* astronomy

Ammonius Hermeiou, 94

Ammonius Saccas, 3–4

antimarturêsis, 126–27

A NOTE ON THE TYPE

This book has been composed in Arno, an Old-style serif typeface in the classic Venetian tradition, designed by Robert Slimbach at Adobe.